McGRAW-HILL PUBLICATIONS IN
AGRICULTURAL ENGINEERING
DANIELS SCOATES, A.E., CONSULTING EDITOR

FARM GAS ENGINES AND TRACTORS

FARM GAS ENGINES

AND

TRACTORS

BY

FRED R. JONES, M.S.

*Professor of Agricultural Engineering, Agricultural and
Mechanical College of Texas; Fellow, American
Society of Agricultural Engineers*

SECOND EDITION

McGRAW-HILL BOOK COMPANY, INC.

NEW YORK AND LONDON

1938

THE MAPLE PRESS COMPANY, YORK, PA.

PREFACE TO THE SECOND EDITION

Numerous changes that have taken place in the farm-power picture during the past six years have necessitated a revision of this book. The outstanding development has been the extensive adoption and use of row-crop tractors and equipment as a result of constant improvement in design, thus providing a greater range of adaptability and ease and convenience of operation. Likewise, this has resulted in a continued displacement of animal power by mechanical power in agriculture. Other significant changes are the design and introduction of successful Diesel-powered tractors, the adoption of pneumatic tires for wheel tractors, and the improvements made in carburetion, ignition, lubrication, and general engine and transmission design.

In making the revision, the author, as a result of suggestions submitted by many users of the book, has combined the two parts as found in the first edition and has rearranged the subject matter in such a manner as to eliminate considerable duplication. It is hoped that this rearrangement will prove more satisfactory and better adapt the text to farm-power instruction work. Such subjects as Diesel-engine construction and operation, fuels and their characteristics, and electric starting and lighting equipment are treated in special chapters and in greater detail than in the first edition.

Realizing the great value and importance of good illustrations in a book of this nature, the author has attempted, as far as possible, to replace obsolete illustrations with up-to-date ones and to add others where advisable. Appreciation and thanks are extended to the numerous tractor and accessory manufacturers who so generously supplied these illustrations.

Special acknowledgment and appreciation are extended to Earl L. Consoliver and Ben G. Elliott for the use in Chap. XX of a number of illustrations taken from their book "The Gasoline Automobile"; to Don Christy for his assistance in the preparation of the material on wheat production costs; to H. P. Smith for numerous suggestions as to the preparation of the revised material; and to the farm-power instructors in the various agricultural engineering departments throughout the United States for their many valuable suggestions.

FRED R. JONES.

COLLEGE STATION, TEXAS,
July, 1938.

v

PREFACE TO THE FIRST EDITION

With the internal-combustion engine now firmly established as a most important and almost indispensable source of power for agricultural production, a need has arisen, particularly in agricultural colleges, for an up-to-date textbook dealing with this subject in all of its various phases.

The material presented has been collected and prepared by the author as a result of fifteen years' experience in teaching the subject of farm power to both collegiate and non-collegiate students. It is likewise anticipated that this text will prove useful and helpful to instructors in vocational work dealing with farm-type engines and tractors, to mechanics and service men, and to owners and operators of stationary engines and tractors.

In the preparation and arrangement of the subject matter, three important considerations have been kept in mind: First, that the material should be presented in sufficient detail and in as elementary and non-technical a form as possible; second, that a thorough understanding of gas-engine fundamentals as exemplified by the simple, stationary farm engine should precede a study of the more complicated tractor mechanism; and third, that any publication of this nature should also treat of such related subjects as power transmission, materials of construction, utilization of power, and so on.

The book is divided into two distinct parts. The first deals with the fundamentals involved in the construction and operation of the simple internal-combustion engine with particular application to the small, stationary farm-type engine. An introductory chapter of a more or less general nature discusses the relation of farm power to agricultural production and enumerates the primary sources of power with their adaptations and disadvantages. Following a brief discussion of early gas-engine development, such subjects as construction and operating principles, carburetion, ignition, and lubrication are taken up in complete detail with respect to both past and recent developments.

The second part, in a similar manner, covers the detailed construction and operation of the various types of farm tractors. Heretofore, with the farm tractor constantly undergoing radical changes in design, the advisability of publishing a textbook dealing with this subject was questionable. Apparently, however, tractor design has become more or less standardized and stabilized so that the material presented should not require as frequent revision as it formerly would have.

Because of the recent introduction and almost immediate extensive adoption of the all-purpose or cultivating-type tractor, a special chapter has been prepared which takes up the fundamental requirements of such a machine and describes briefly the different makes now available together with their outstanding features.

The final chapter deals with the tractor from the standpoint of selection and efficient utilization under different conditions.

In order to make the text better suited to use by advanced students, research workers, and designers, a number of appropriate references are given at the end of the more important chapters.

The author will be glad to receive any suggestions and criticisms involving arrangement of subject matter, omissions, inaccuracies of statements or illustrations, and typographical errors.

Acknowledgment is accorded George W. Hobbs, Earl L. Consoliver, and Ben G. Elliott for the use of certain illustrations taken from their book "The Gasoline Automobile," to H. P. Smith for numerous suggestions in the preparation of the manuscript, to certain engine-accessory manufacturers who have most willingly given valuable advice and information, and to the many concerns who have supplied illustrative material.

FRED R. JONES.

AGRICULTURAL AND MECHANICAL COLLEGE OF TEXAS,
COLLEGE STATION, TEXAS,
January, 1932.

ACKNOWLEDGMENTS

Realizing that the value of a textbook of this nature is to a large extent dependent upon a rather profuse use of good, clear illustrations, the author has attempted to secure from various sources, such as trade literature, instruction books, and catalogues, the most appropriate and up-to-date illustrations possible. In this connection, sincere thanks and appreciation are extended to the following concerns and particularly to those who, at the author's request, promptly supplied original photographic prints, electrotypes, and drawings: Deere and Company; International Harvester Company of America; Caterpillar Tractor Company; J. I. Case Company; Oliver Farm Equipment Company; Massey-Harris Company; Minneapolis-Moline Power Implement Company; Advance-Rumely Corporation; Huber Manufacturing Company; Cleveland Tractor Company; Allis-Chalmers Manufacturing Company; Rock Island Plow Company; Avery Power Machinery Company; Ford Motor Company; Heebner and Sons; Challenge Company; Westinghouse Electric and Manufacturing Company; General Electric Company; Fitz Water Wheel Company, A. B. Farquhar Company, Ltd.; Sears, Roebuck and Company; Appleton Manufacturing Company, Vacuum Oil Company; Tide Water Oil Sales Corporation; Standard Oil Company of Indiana; Clipper Belt Lacer Company; Wright Aeronautical Corporation; De La Vergne Machine Company; Fulton Iron Works Company; San Antonio Machine and Supply Company; Fairbanks, Morse and Company; Willys Overland Corporation; Bendix Stromberg Carburetor Company; Zenith-Detroit Corporation; Associated Manufacturers Corporation of America; Bessemer Gas Engine Company; Electric Storage Battery Company; Willard Storage Battery Company; Edison Storage Battery Company; Delco Products Corporation; Hobart Brothers Company; K. W. Ignition Company; Webster Electric Company; A-C Spark Plug Company; Hercules Motors Corporation; Splitdorf Electrical Company; Ensign Carburetor Company; Tillotson Manufacturing Company; Wheeler-Schebler Carburetor Company; United American Bosch Corporation; The Lunkenheimer Company; Madison-Kipp Corporation; L. S. Starrett Company; John Chatillon and Sons; Szekely Aircraft and Engine Corporation; Dayton-Dowd Company; Link-Belt Company; Chicago Pulley and Shafting Company; Buda Company; Yuba Manufacturing Company; Timken Roller Bearing Company; Hyatt Roller Bearing Company; Fafnir Bearing Company; Otto Engine Works; Pioneer Manufacturing Company; Gravely Motor Plow and Cultivator Company; Centaur Tractor Corporation; Eisemann Magneto Company; Wico Electric Company; Lauson Corporation; McCord Radiator and Manufacturing Corporation; United Air Cleaner Corporation; Vortox Manufacturing Company; Waukesha Motor Company; Algoma Foundry and Machine Company; Borg and Beck Company; International Textbook Company; Chain Belt Company; Foote Brothers Gear and Machine Company; Simms Magneto Company; Hercules Products, Inc.; Marvel Carburetor Company; Witte Engine Works; Alamo Engine Company; Muncie Oil Engine Company; Gilson-Bolens Manufacturing Company; Handy Governor Corporation; McQuay-Norris Manufacturing Company; Mallory Electric Corporation; S. L. Allen and Company; Studebaker Corporation.

CONTENTS

FARM GAS ENGINES
AND TRACTORS

CHAPTER I

POWER ON THE FARM—SOURCES AND UTILIZATION

The influence and effects of power and machinery upon agriculture are well summarized by Kinsman in *U. S. Department of Agriculture, Bulletin* 1348. He states as follows:

The adoption of labor-saving machinery made possible by the extensive use of power has been universally acknowledged as the outstanding feature of American agriculture during the past three-fourths of a century. Seventy-five years ago

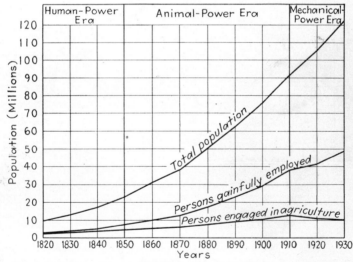

FIG. 1.—Total population, persons gainfully employed, and persons engaged in agriculture in the United States, 1820 to 1930.

the average agricultural worker would care for but 12 acres of crops; now, considering the United States as a whole, he can attend to at least 34 acres and in some States where large power units are common, the average is more than 100 acres, while on many individual farms it will run as high as 300 acres or more. At the same time the workers' hours have been considerably shortened, and much of the drudgery and monotony of farm work has been eliminated.

1

TABLE I.—TOTAL POPULATION, NUMBER OF FARMS, AGRICULTURAL WORKERS, ACRES OF HARVESTED CROPS AND VALUE OF POWER UNITS, IMPLEMENTS, AND MACHINERY FOR CENSUS YEARS, 1870–1930

Year	Total population	Farms	Agricultural workers[1]	Harvested crops			Value of power units, implements, and machinery				
				Total acres	Acres, per farm	Acres, per worker	Horses and mules	Implements	Total	Per farm	Per worker
1870	38,558,371	2,659,985	5,922,471	....	....	....	$ 583,150,000	$ 270,914,000	$ 854,064,000	$321	$144
1880	50,155,783	4,008,907	7,713,875	178,500,000	44.5	23.1	673,665,000	406,520,000	1,080,185,000	269	140
1890	62,622,250	4,564,641	9,148,448	234,000,000	51.3	25.6	1,230,358,000	494,247,000	1,724,605,000	378	189
1900	75,994,575	5,737,372	10,381,765	295,350,000	51.5	28.4	965,762,000	749,776,000	1,715,538,000	299	165
1910	91,972,266	6,361,502	12,659,082	324,500,000	51.0	25.6	2,648,573,000	1,265,150,000	3,913,728,000	615	309
1920	105,710,620	6,448,343	10,953,158	363,100,000	56.3	33.2	2,728,481,000	3,594,773,000	6,323,254,000	981	577
1930	122,775,046	6,288,648	10,482,323	371,719,000	59.1	35.5	1,348,647,000	3,301,655,000	4,650,302,000	739	444

[1] Persons 10 years old and over.

The increased efficiency in accomplishing farm work has greatly enhanced returns from farming and has released large numbers of workers from agriculture to other industries. This has resulted in greater production and a lower cost of comforts and luxuries, the enjoyment of which determines to a large extent the standard of living of a people. Undoubtedly these factors have played an important part in making possible the present standard of living of the people of the United States.

According to this bulletin and also a later study by the U. S. Department of Agriculture,[1] the great expansion and development in industry during the past century have resulted in a marked increase in the number of industrial workers while the number of persons engaged in agriculture has remained more or less the same. Figure 1 shows graphically the trends in this respect from 1820 to 1930 inclusive. In 1820 approximately 83 per cent of all persons in the United States ten years old and over, reported as being gainfully employed, were engaged in agriculture. By 1930 only about 21 per cent of such persons employed in all occupations in the United States were engaged in agriculture. The ratio of agricultural workers to total population also decreased from 21.5 per cent in 1820 to 8.5 per cent in 1930.

Table I, as well as Fig. 1, also shows that there was a continued increase in the number of persons engaged in agriculture until 1910 but

[1] U. S. Dept. Agr. Misc. Pub. 157.

that this increase was not in proportion to the increase in total population. Between 1910 and 1930, however, there was an actual decrease in the number of agricultural workers. Despite this decrease during this period, there was an increase in both acreage of harvested crops and in total crop production. The increase in production per worker was due, in a large measure, to the increased use of power and machinery on farms.

1. Value of Mechanical Farm Equipment.—Table I shows the value of implements, machinery, and power units on farms in the United States

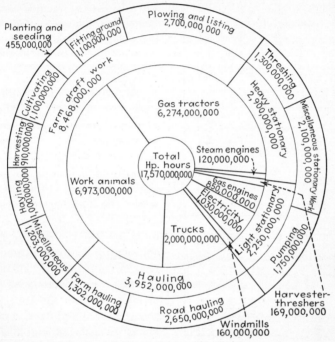

Fig. 2.—Estimated horsepower-hours of power developed annually by different kinds of farm power and amounts required for the principal farm operations in the United States, 1936.

during the period 1870 to 1930. Prior to the introduction of mechanical-power units, these figures, with some exceptions, represented the value of what is generally considered as farm implements. For later years the figures include also the mechanical-power units such as tractors, trucks, and automobiles.

From Table I it is observed that the value of power and machinery per agricultural worker has increased from approximately $144 in 1870 to $444 in 1930. Automobiles probably comprised at least 20 per cent of the value of power units and machinery on farms in 1930. Although the automobile is not considered as a source of power for farm work, the expenses involved must be borne by the farm.

2. Utilization of Power.—Table II summarizes the amount of power available, by types, on farms in the United States in 1936 and gives an estimate of the horsepower-hours developed annually. Of the total of nearly 18,000,000,000 hp.-hr. developed annually, animals furnish approximately 40 per cent and tractors approximately 35 per cent. Probably the most significant change that has occurred during the preceding 10 years has been the decrease in the number and use of work animals, and the increase in the utilization of tractors. This change has taken place largely as a result of the introduction of the light-weight all-purpose tractor, which is adaptable to performing a variety of operations on even the smaller sizes of farms.

TABLE II.—ESTIMATED POWER AVAILABLE AND POWER DEVELOPED ON FARMS IN THE UNITED STATES, 1936[1]

Kind of power	Power units				Power developed annually		
	Thou-sands	Aver-age size, hp.	Total rating		Avg. per. rated hp., hp-hr.[2]	Total	
			1,000 hp.	Per cent		1,000 hp.-hr.	Per cent
Work animals:							
Horses......................	10,374	1.00	10,374	11.0	500	5,187,000	29.5
Mules.......................	4,465	0.80	3,572	3.8	500	1,786,000	10.1
Windmills....................	1,000	0.40	400	0.4	400	160,000	0.9
Steam engines.................	10	40.00	400	0.4	300	120,000	0.7
Gas engines:							
Small......................	1,100	2.50	2,750	2.9	200	550,000	3.1
Large......................	15	20.00	300	0.3	1,000	300,000	1.7
Individual electric plants.........	250	3.00	750	0.8	160	120,000	0.7
Central station:							
Small motors.................	400	2.00	800	0.9	200	160,000	0.9
Large motors.................	50	15.00	750	0.8	1,000	750,000	4.3
Gas tractors:							
Drawbar....................	1,382	12.00	16,584[3]	17.6[3]	250	4,146,000	23.6
Belt.......................	1,382	22.00	30,404	32.2	70	2,128,000	12.1
Trucks......................	1,000	25.00	25,000	26.5	80	2,000,000	11.4
Harvester-threshers.............	75	30.00	2,250	2.4	75	169,000	1.0
Total......................	......	.....	94,334	100.0	.....	17,576,000	100.0

[1] Revised to conform to 1936 conditions according to author's estimates based upon data obtained from *U. S. Dept. Agr., Misc. Pub.* 157, February, 1937, issue of *Crops and Markets*, and Apr. 8, 1937, issue of *Farm Implement News*.
[2] Estimated equivalent hours operated at rated capacity.
[3] Duplicated in belt-power rating.

With the widespread use of mechanical-power units on farms there has also been a change in the type of work performed by horses. With a tractor available for heavy-drawbar work and a truck for hauling, the use of work animals on some farms is largely confined to light-drawbar work.

In some cases it has been possible to increase the number of hours of annual use per work animal by decreasing the number of such units per farm, and by the proper coordination of animal and mechanical-power units.

The various forms of power and their relative importance, application, and utilization are shown by Fig. 2.

The relative amount of power utilized varies greatly in different states owing to a number of factors such as type of agriculture and kind of crops raised, topography, climate, soil conditions, and kind and available supply of labor.   As indicated by Fig. 3, the type of agriculture seems to

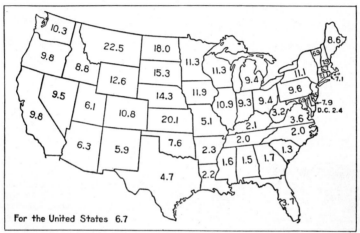

Fig. 3.—Average horsepower available per agricultural worker by states, 1930.

have a greater effect on the horsepower per agricultural worker than does any other one factor.   For example, it is observed that the wheat-producing or Great Plains region shows the greatest power available per agricultural worker, while the cotton-producing states in the southeastern and Gulf Coast areas show the lowest power available per agricultural worker.

Figure 4 shows the general relationship between the power available, the investment in mechanical equipment, and the average gross income derived from crops, livestock, and livestock products, per agricultural worker.   The states having the most available power usually lead in value of agricultural products per worker, but a number of exceptions will be noted.

According to Kinsman[1] the most serious difficulties encountered in the efficient use of power and labor in farm work are the extreme seasonal

[1] *U. S. Dept. Agr., Bull.* 1348.

demands of many of the crops, the diversity of the operations, the small
size of the usual power units, and the low load factor or small percentage

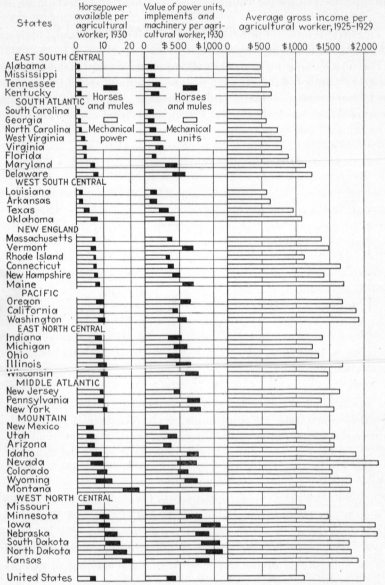

FIG. 4.—Power available and value of power units and machinery, 1930, and average gross
income, 1925 to 1929, per agricultural worker.

of time the power unit is used. The result is a relatively high cost per
unit of power produced.

The cost of using power equipment is also considerable and its adoption becomes profitable only if the net earnings of the owner or operator are increased through its use.

SOURCES OF FARM POWER—ADAPTABILITY AND COST

Power is required on the farm for doing two kinds of work, namely, tractive work requiring pulling or drawing effort; and stationary work, usually accomplished by means of a belt, gears, or some other form of power transmission. Tractive jobs include (1) plowing and land preparation, (2) planting and seeding, (3) crop cultivation, (4) harvesting, and (5) hauling. Stationary jobs include (1) water pumping, (2) threshing, (3) feed grinding, (4) hay baling, (5) ensilage cutting, and (6) wood sawing, in addition to numerous other jobs of a like nature.

3. Sources of Power.—There are five possible sources of power for doing the various kinds of work. In other words, it can be said that there are five prime movers available for the farmer. These are as follows: (1) domestic animals, (2) wind, (3) flowing water, (4) electricity, and (5) heat engines. Some of these are necessarily limited in use as will be mentioned later. In fact, up to this time only two of the five mentioned, namely, domestic animals and heat engines, have proved practical for supplying tractive power. Thus far wind, water, and electric power are confined entirely to stationary work.

4. Animal Power.—At one time, when farming was really not an industry or a business but was carried on by each individual or family merely as a means of providing food, clothing, and shelter for self-preservation, the power required for doing all work incident to planting and harvesting the crops and their manufacture into a finished product was supplied only by human hands. As the population increased, the demand for such essentials as food and clothing became greater and the necessity arose of growing crops on a larger scale, that is, producing more per man or per family. This required the adoption of a more suitable form of power, a form that would enable one individual to cultivate more land, hence produce a greater quantity of raw material from which to provide food and clothing for other individuals who might be engaged in some so-called nonproductive industry such as mining or milling or printing. Hence, the ox and later the horse and the mule were brought into use to pull the plows, harrows, planters, and harvesters and to haul the crop to market.

Horses and mules still supply a large portion of the tractive power required on the farms of the world, but animal power has proved satisfactory for, or adaptable to, stationary work to only a limited extent.

5. Power of Horses and Mules.—The power and pulling ability of horses and mules are matters that are frequently debated but seldom understood.   According to King[1] a horse working continuously for several hours and walking at the rate of 2½ m.p.h. should not be expected to pull more than one-tenth to one-eighth of its weight.   On this basis, a 1,000-lb. horse can develop 0.67 to 0.83 hp., a 1,200-lb. horse, 0.80 to 1.00 hp., and a 1,600-lb. horse, 1.07 to 1.33 hp.

Studies and tests made at the Iowa State College[2] demonstrated that:

1. It is possible for horses to exert a tractive effort of one-tenth to one-eighth of their own weight and travel a total of 20 miles per day without undue fatigue.

2. It is possible for horses weighing 1,500 to 1,900 lb. or over to pull continuously loads of 1 hp. or more for periods of a day or longer.

3. A well-trained horse can exert an overload of over 1,000 per cent for a short time.

Fig. 5.—A portable tread power.

4. For a period of a few seconds and over a limited distance of perhaps 30 ft. or less a horse can exert a maximum pull of from 60 to 100 per cent of its actual weight. Under such conditions one horse may develop as much as 10 hp. or more depending upon its size and pulling ability.

5a. Animals for Stationary Power.—Horses and mules and, to a lesser degree, cattle and sheep and other common farm animals can be utilized under certain conditions for supplying stationary power.   Certain stationary power-driven machines have been and still are being operated on farms by domestic animals.   The power for such machines is generated by the animal by the use of one or the other of two devices, namely, the treadmill and the sweep power or a modification of the latter.   The treadmill (Fig. 5) consists of a crate or pen with an inclined, slatted, endless apron as a floor as found in some manure spreaders.   This apron or tread passes over a drum or shaft at each end and is also supported underneath by two or more rows of rollers to eliminate friction.   A pulley, from which the power is taken off, is placed on one of the end shafts. The animal, such as a horse, bull, or sheep, is placed in the crate with the

---

[1] KING, "Physics of Agriculture."
[2] *Iowa Agr. Exp. Sta., Bull.* 240.

head at the high end, and, as it attempts to walk forward, it causes the tread to slide backward owing to the force of gravity. The animal will continue to tread forward and thus keep the apron in motion and rotate the pulley.

The use of the treadmill is limited, owing, primarily, to the comparatively small amount of power that can be generated. Experiments at the Montana Experiment Station[1] showed that a 2,060-lb. bull, walking at a speed of 1 m.p.h., developed from 0.75 hp. with a 20 per cent tread slope to 1.02 hp. with a 25.1 per cent slope. A 1,250-lb. bull, under the same conditions, developed from 0.42 to 0.62 hp. The power developed with a treadmill depends upon the weight of the animal, the rate of travel, and the slope of the tread. A treadmill will operate a cream separator, a milking machine, a water pump, or any similar light machine. On dairy farms it often serves a twofold purpose, namely, to provide exercise for the herd bull and to supply power at milking time.

Before the development of the modern steam and gas engine, the sweep power (Fig. 6) was widely used about the farm for operating larger stationary machines, such as threshers and wood saws. In this machine, the horses or mules, hitched to sweeps, traveled in a circle, and by means of an arrangement of gears and a rod called a tumbling rod, power was gener-

FIG. 6.—A two-horse sweep power.

ated and transmitted to the machine to be operated. The sweep power, in a modified form, is still being used to operate hay balers, cane mills, feed grinders and stump pullers.

6. Water Power.—In certain sections of the country are found many small streams, which can be harnessed up in such a manner as to develop useful power to be employed on near-by or conveniently located farms. Such sections of the country, however, are of course limited to comparatively rolling and hilly ground. Furthermore, a stream may furnish abundant power at one period of the year, but become so low in the driest season as to fail to supply sufficient power. The power developed by flowing water depends upon two factors, namely, the volume of water flowing per minute, and the head or vertical distance the water drops at the point where the power installation is located. The former can be measured either by the float method or by a weir (Fig. 7). The head

[1] *Mont. Exp. Sta., Circ.* 93.

is determined by measuring the difference in surface level before the water falls and after.

Fɪɢ. 7.—Measuring the stream flow with a rectangular weir.

For example, suppose that the following stream measurements have been made:

	Ft.
Average width	12
Average depth	2
Velocity per minute	15
Head	4

Knowing that water weighs 62.4 lb. per cubic foot and that 33,000 ft.-lb. per minute is equal to 1 hp., the theoretical power available from the stream is

$$\frac{12 \times 2 \times 62.4 \times 15 \times 4}{33,000} = 2.7 \text{ hp.}$$

Owing to frictional losses in the water wheel or other means used, the actual available horsepower would probably be somewhat less than 2.7. Devices used for converting water power into useful form are generally classed as either water wheels (Fig. 8) or turbines (Fig. 9).

7. Wind Power.—The energy of the wind for farm use, like that of flowing water, is more or less limited, chiefly because it cannot be controlled and is seldom available when needed. Consequently, the use of wind power on the farm is confined largely to water pumping, because whenever the wind blows, even if but once or twice a week, enough water can be pumped and conveniently stored to last several days, or until the wind blows again. The power of the wind is made available by means of the common windmill.

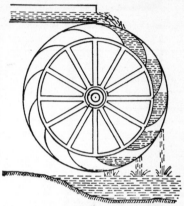

The power developed by this device depends primarily upon the size of wheel and the wind velocity. However, a number of other factors such as type of wheel, design of wheel and mill, and height of tower affect the performance of a windmill. The theoretical power of a stream of air passing through a circular area perpendicular to the direction of travel of the air is represented by the formula

$$Hp. = 0.00000525D^2W^3$$

Fig. 8.—An overshot water wheel.

where D = maximum diameter of wind wheel or circle.

W = wind velocity in miles per hour.

Fig. 9.—A water-turbine installation for generating electricity.

Owing to certain reactions between the wind and the revolving wheel and to mechanical imperfections, it is usually considered that the actual efficiency of the common multisail type of farm windmill based upon this formula will be approximately 30 per cent for wind velocities up to 10 m.p.h., 20 per cent for wind velocities of 15 and 20 m.p.h., and 15 per cent for wind velocities of 25 m.p.h. or more.   Table III gives the approximate horsepower developed by windmills of various sizes and at different wind velocities, based upon the above assumptions.

TABLE III.—POWER OF WINDMILLS FOR WHEELS OF DIFFERENT SIZES AND FOR DIFFERENT WIND VELOCITIES

Wind velocity, m.p.h.	Diameter of wind wheel, ft.					
	6	8	10	12	14	16
6	0.01	0.02	0.03	0.05	0.07	0.09
10	0.06	0.10	0.16	0.23	0.31	0.40
15	0.13	0.23	0.35	0.51	0.70	0.91
20	0.30	0.54	0.84	1.21	1.65	2.15
25	0.44	0.79	1.23	1.77	2.42	3.15
30	0.77	1.36	2.12	3.06	4.16	5.45

The airplane type wind wheel, commonly used where high speed is desirable, as in the case of wind-driven electric plants, likely has a higher efficiency than the ordinary farm-windmill type of wheel.

FIG. 10.—A portable electric motor for farm use.

8. Electricity.—Farms that are located on the outskirts of villages and cities or along cross-country electric transmission lines, can, in many cases, utilize this so-called central-station current for stationary-power purposes as well as for lighting.   Machines for converting electrical energy into useful mechanical energy are called electric motors.   Such motors (Fig. 10) have numerous advantages.   They are small, compact, and light in weight for the power developed, are made in any size desired, produce smooth, uniform power, make little noise, seldom require attention or give trouble, start readily, and are well adapted to varying as well as uniform loads. However, even though the farm is favorably located, the expense involved in transmitting the current or tapping the transmission line is often found to be excessive, and, therefore, the use of central-station electricity for power would in such cases be impractical. Much thought and attention are now being given to the problem of a wider use of electricity for farm power as well as for lighting, and the number of so-called electrified farms is rapidly increasing.

FIG. 11.—A complete stationary steam-engine layout.

9. Heat Engines.—Fortunately, about the time that the demand arose for larger power-producing units for operating such stationary farm machines as the thresher, the wood saw, the corn shredder, and the ensilage cutter, the steam engine and, later, the gas engine were invented. These engines are known by the engineer as heat engines. In either case, whether it is a steam engine or a gas engine, some kind of combustible material known as a fuel—particularly wood, coal, oil, or natural gas—is ignited, combustion takes place either slowly or rapidly, heat is produced and utilized in such a way as to create pressure, and the latter, when applied to certain movable parts of the apparatus, produces motion, and therefore energy and power.

The steam engine (Fig. 11) and the ordinary gas engine (Fig. 12) are the two common types of heat engines. In the former, the heat of the burning fuel is applied to water in a closed receptacle called a boiler. As the water becomes heated, it is converted into steam. As the heating continues, more steam is formed and high pressure results. This steam, under pressure, when conducted by a pipe into a cylinder behind the piston, places the latter in motion and thus generates power. Since the fuel is ignited and burned outside the cylinder and its heat energy applied indirectly to the piston by an intermediate medium, namely, water vapor, the steam engine is called an external-combustion engine.

The gas engine resembles the steam engine in that the pressure is applied to a piston sliding back and forth within a cylinder, but differs greatly in the combustion of the fuel and the application of the pressure resulting from the heat produced. In the case of the common gas engine and engines of a similar type the combustible fuel mixture is first placed

Fig. 12.—A stationary farm gas engine with flywheel weighted opposite the crank for balancing.

inside the cylinder in a gaseous condition and compressed before it is ignited. The ignition of this compressed mixture causes very rapid combustion and an instantaneous application of pressure to the piston, more commonly known as an explosion. The piston is consequently set in motion and power is generated. Since the fuel is ignited and burned inside the cylinder, the gas engine is known as an internal-combustion engine, and all engines that operate in a similar manner are likewise known as such.

ADVANTAGES AND DISADVANTAGES OF DIFFERENT FORMS OF POWER

10. Advantages and Disadvantages of Animal Power.—Kinsman[1] gives the following advantages and disadvantages of animal power for farm work:

Advantages:
1. Great reserve power for emergencies and temporary overloads.
2. Uses feed that is produced largely on the farm.
3. Great flexibility of size of power unit.
4. Adapted to practically all draft work.
5. Fairly good traction in wet or loose ground.

[1] *U. S. Dept. Agr., Bull.* 1348.

6. Lay up of one animal does not lay up entire power plant.

7. Can be reproduced on the farm.

8. Does not require constant attention in guiding.

9. Relatively cheap type of power in areas where a surplus of both grain and roughage is produced.

Disadvantages:

1. Requires feed and care when not working.

2. Work at heavy loads limited to short periods.

3. Requires frequent resting periods.

4. Cannot work efficiently in hot or sultry weather.

5. Working speed is limited.

6. Not efficient for stationary work.

7. Relatively large amount of time required to feed, harness, and care for.

8. Requires a relatively large space for shelter and feed storage.

9. Unwieldy when used in large units.

10. Requires the products from one-fourth of all crop land to feed them.

11. Advantages and Disadvantages of Tractors.—Mechanical power in the form of a gas tractor has the following advantages and disadvantages in doing farm work:

Advantages:

1. Can work continuously at heavy loads.

2. Not affected by hot weather.

3. Adapted to draft, belt and power-take-off work.

4. Has considerable range of working speeds.

5. Little attention required when not in use.

6. Requires no fuel when not in use.

7. Quickly available when needed in an emergency.

8. Requires small storage space.

Disadvantages:

1. Limited overload capacity.

2. Requires cash expenditure for fuel and lubricants.

3. Requires some mechanical skill for successful operation.

4. Inflexibility of size of power unit for economical power production under some conditions.

12. Gas vs. Steam Engines—Advantages and Disadvantages.—The gas engine is now used about the farm as a stationary power unit in preference to the steam engine for the following reasons:

1. More efficient, that is, a greater percentage of the heat and energy value of the fuel is converted into useful power. The efficiency of the internal-combustion engine varies from 10 to 20 per cent whereas that of the external-combustion engine is often as low as 3 and seldom exceeds 10 per cent.

2. Weighs less per horsepower.

3. More compact.

4. Original cost less per horsepower.

5. Less time and work necessary preliminary to starting.

6. Less time and attention required while in operation.

7. Can be made in a greater variety of sizes and types and adapted to many special uses, that is, has a greater range of adaptability.

However, the steam engine likewise has certain distinct advantages, which make it more adaptable to certain kinds of work.   Some of these are as follows:

1. Uses a greater variety of fuels.
2. Speed and corresponding power more flexible.   Internal-combustion engines must attain a certain speed before they can produce any power whereas steam engines can be operated under load at very low as well as at high speeds.   In fact, a gas engine must be relieved of its load before it can be started, but the steam engine will start even against the resistance of a heavy load.
3. Simpler mechanically.
4. Lubrication not so difficult.
5. Readily reversed and generates the same power when operated in either direction.
6. Readily started regardless of size, type, and use.

References

BOND: "Farm Implements and Machinery."
DAVIDSON: "Agricultural Machinery."
ELLIS and RUMELY: "Power and the Plow."
Iowa Agr. Exp. Sta., Bull. 297.
KING: "Physics of Agriculture."
Mont. Agr. Exp. Sta., Circ. 93.
N. D. Agr. Exp. Sta., Bull. 105.
Okla. Eng. Exp. Sta., Pub. 10.
POTTER: "Farm Motors."
RAMSOWER: "Equipment for the Farm and Farmstead."
U. S. Dept. Agr., Bull. 1348.
U. S. Dept. Agr., Farmers' Bull. 1430.
U. S. Dept. Agr., Farmers' Bull. 1658.
U. S. Dept. Agr., Misc. Pub. 157.
Wash. State Coll., Agr. Ext. Service, Bull. 124.
Water Power on the Farm, *Trans. Amer. Soc. Agr. Eng.*, Vol. 12.
Wis. Agr. Exp. Sta., Bull. 68.

CHAPTER II

HISTORY OF THE INTERNAL-COMBUSTION ENGINE—HISTORY AND DEVELOPMENT OF THE FARM TRACTOR

The first ideas concerning the operation and construction of an internal-combustion engine were based upon the action of the ordinary rifle or cannon, that is, the barrel served as a cylinder and the bullet or cannon ball acted as a piston. The difficulty encountered, however, was in getting the piston to return to its original position, thus producing a continuous back-and-forth movement to insure a continuous generation of power.

13. Early Ideas and Inventions.—Nothing of consequence was accomplished before the seventeenth century. In 1678, Hautefeuille, a Frenchman, proposed the use of an explosive powder to obtain power. He is said to be the first man to design an engine using heat as a motive force and capable of producing a definite quantity of continuous work. Huygens, a Dutchman, is credited with being the first man actually to construct an engine having a cylinder and piston. This device used explosive powder as fuel and was exhibited to the French minister of finance in 1680.

None of these early attempts was successful, however, and further efforts in the construction of an internal-combustion engine were abandoned for about a hundred years, owing to the fact that during the eighteenth century the possibilities of utilizing steam for power were recognized and developed, and the energies of the engineers of that time were turned almost entirely toward applications of the steam engine. About 1800 the thoughts of these investigators were again directed toward the possible design of a gas engine. During the period from 1800 to 1860, a number of engines were constructed, none of which were really successful. Some notable steps were made, however, among which were the use of compression and an improved system of flame ignition by Barnett in 1838 and the actual construction and manufacture of an internal-combustion engine on a commercial scale by Lenoir in 1860. The Lenoir engine later proved to be impractical.

14. Beau de Rochas.—Perhaps the individual making the first really important contribution toward the development of the present-day types of internal-combustion engines is Beau de Rochas, a French engineer. In 1862 this man advanced the actual theory of operation of all modern types

17

of internal-combustion engines. He first stated that there were four conditions which were essential for efficient operation. These were as follows:

1. The greatest possible cylinder volume with the least possible cooling surface.
2. The greatest possible piston speed.
3. The highest possible compression at the beginning of expansion.
4. The greatest possible expansion.

These principles are still considered as fundamental and extremely important in gas-engine design. Beau de Rochas proposed further that a successful engine embodying these principles must consist of a single cylinder and a piston that made a stroke for each of the four distinct events constituting a cycle as follows:

1. Drawing in of the combustible fuel mixture on an outward stroke.
2. Compression of the mixture on an inward stroke.
3. Ignition of the mixture at maximum compression producing an outward power or expansion stroke.
4. Discharge of the products of combustion on a fourth or inward stroke.

15. Otto and Clerk.—Beau de Rochas never succeeded in constructing an engine based upon his theories, but they were promptly accepted as being essential. Although considerable effort was expended in the design of an engine during the next few years, it was not until 1876 that Dr. N. A. Otto, a German, patented the first really successful engine operating on this four-stroke-cycle principle. The engine was first exhibited in 1878. This cycle, although originally proposed by Beau de Rochas, is commonly known as the Otto cycle.

The invention of the four-stroke-cycle engine by Otto was soon followed by the issue of a patent, in 1878, to Dugald Clerk, an Englishman, on the first two-stroke-cycle engine, that is, an engine producing one power impulse for every revolution instead of for every two revolutions. This particular engine was not marketed at once, however, and was not really perfected until 1881.

16. Diesel.—Another notable contribution to the development and varied application of the internal-combustion engine was the work of Dr. Diesel, a German engineer, who conceived the idea of utilizing the heat produced by high compression for igniting the fuel charge in the cylinder. In 1892 he secured a patent on an engine designed to operate in this manner. This first machine proved unsatisfactory, however, and it was not until about 1898 that the first successful Diesel-type engines were produced. During the past 15 years, rapid strides have been made in the development and utilization of the Diesel principle in internal-combustion engines for both stationary and tractor applications.

We thus observe that the invention of the internal-combustion engine is a comparatively recent one and that the many finely constructed modern types of single- and multiple-cylinder engines have been designed and developed almost overnight. But when it is considered that petroleum, now the almost universal source of fuels and lubricants for these engines, was first discovered about 1858 and that very little was known concerning the application of electricity in the operation of internal-combustion engines previous to the latter part of the past century, we readily perceive the explanation of the retardation of this invention—now considered an indispensable device to modern life throughout the world.

HISTORY AND DEVELOPMENT OF THE FARM TRACTOR

A tractor is defined specifically as a self-propelled machine that can be used for (1) drawing other machines, objects, or devices in the field

FIG. 13.—Steam tractor operating a grain thresher.

and along roads, and (2) for operating stationary machines. The two common types are the steam tractor, in which an external combustion or steam engine supplies the power, and the gas tractor, in which an internal-combustion engine serves as the source of power.

17. Steam Tractor.—The invention and development of the steam engine preceded the internal-combustion engine by 100 years or more. Consequently, the earliest known tractors were of the steam type. They first came into general use for operating threshers (Fig. 13) in the wheat and grain-growing sections of the country during the last two or three decades of the nineteenth century. Their self-propelling feature was utilized primarily for moving about from one threshing job to another. Later on, with the opening up of the large wheat farms of the West and

Northwest, steam tractors displaced animal power to a certain extent for preparing the land and sowing and harvesting the crop.

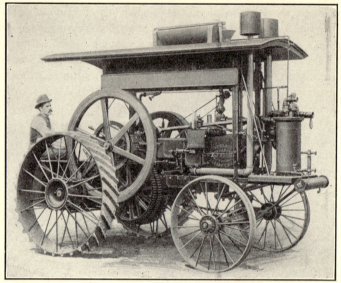

Fig. 14.—A gas tractor built in 1892.

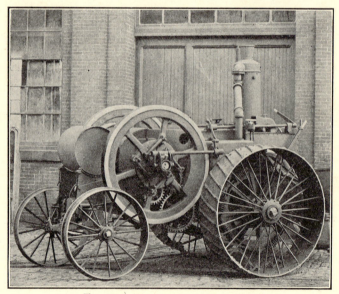

Fig. 15.—A gas tractor built in 1897.

The steam tractor for field work had its limitations. It was very heavy and slow moving, the fuel was bulky and difficult to handle, and the matter of boiler-water supply and fueling meant constant attention on the part of one man, with a second man to handle and guide the machine.

18. Early Gas Tractors.—Certain manufacturers, foreseeing a greater future demand for suitable mechanical power for field work, particularly in the wheat-growing sections of the country, started the construction of gas tractors even before the close of the nineteenth century. For example, Fig. 14 shows a machine that is said to have been built in 1892. Another tractor (Fig. 15) is reported to have been put into use in North Dakota in 1897, and Fig. 16 shows the first Hart Parr tractor, which was sold in 1902. These heavy, cumbersome-appearing machines were the forerunners of the present-day tractor industry,

Fig. 16.—A gas tractor sold in 1902.

which started soon after the opening of the present century and began to gain momentum about 1905.

According to Ellis and Rumely:[1]

By the spring of 1908 the builders of the first successful tractor had about 300 machines in the field and the sales of that year equaled those of the 5 years preceding. The following year the number in the field was again doubled and by the close of the year 1910 over 2,000 of these tractors were said to be in active service. Another company began to produce a small tractor in 1907 and by the close of the decade was selling several thousand yearly. Dozens of gas tractor factories sprang up and practically every manufacturer of steam traction engines either went out of business or added an internal-combustion engine to his line.

These early tractors consisted usually of a large one-cylinder gas engine mounted on a heavy frame placed on four wheels. The two rear

[1] Ellis and Rumely, "Power and the Plow."

wheels were connected by a train of heavy, exposed cast-iron gears to the crankshaft of the engine, thus making the machine self-propelling. Like the steam tractors, they were heavy, cumbersome, and powerful, and seemed to be designed as a mere substitute for the former. They possessed certain advantages, however. The fuel was easier to handle, there was less water to haul, and less time and attention were required for starting and during operation. One man was usually able to handle the largest outfit.

19. The Light-weight Tractor.—About 1910 the designers turned their attention toward the possibilities of a lighter weight gas tractor to meet the approaching demand of the smaller grain and livestock farmer for mechanical power. Consequently, about 1913, there began to appear on the market a number of machines, comparatively light in weight and differing greatly in construction and appearance. In most cases, they were equipped with two- and four-cylinder engines.

By 1915, the farmers were presented with an amazing array of types, models, and sizes, ranging from the giant one- or two-cylinder four-wheelers to a tractor attachment for a small automobile. No two machines resembled each other. Some had two wheels, some three, and some four. Some were driven from in front and others from the rear. Some had the plows attached under the frame while others pulled them in the usual manner. Competition became very marked and large sums of money were spent for public demonstrations. Many of these freak machines were sold. Some proved more or less successful while others gave unsatisfactory results and tended to destroy the faith of their owners in the future value of the tractor. True it was that a few of these first light-weight tractors were unreasonable in design and weak in construction. On the other hand, many of them were not. Their failure in the hands of the farmer could be attributed largely to the lack of knowledge of their operation and care.

The period of the freak tractor was short and, by 1917, many of them had disappeared from the market, and the more far-sighted designer observed that there were certain fundamentals of tractor design that had to be adhered to.

20. The World War and the Tractor.—The World War had a very pronounced effect upon farm-tractor development and use. Maximum agricultural production was urged. There was a shortage of labor and prices became abnormally high. All these factors meant an increased demand for labor and time-saving machinery, especially small farm tractors. As a consequence, a large number of small tractor manufacturers sprang up overnight, so to speak, and placed upon the market between 200 and 300 different makes and models of machines. These ranged from the small two-wheel garden tractor to the larger four-wheel types. The average and most popular size seemed to be a machine rated at about 10 to 15 drawbar hp. and 20 to 30 belt hp. The tendency in design was toward a four-wheel rear-drive tractor with a four-cylinder engine. Most of these tractors seemed to give much better satisfaction and service than those of earlier manufacture.

Thus the war proved to be a great stimulus to the adoption of mechanical power by the farmer. Farm tractors were being used successfully in every nook and corner of the country.

21. Effect of Agricultural Depression of 1920.—The tractor industry, like many others that were dependent upon the prosperity of the farmer, received a severe setback as the result of the unexpected agricultural depression in 1920. At this time more than 100 different companies were offering about 250 sizes, models, and types of machines. Many of these concerns were small and lacked the capital and the organization necessary to enable them to compete with the older, larger, and better-established companies. They attempted to struggle along; but with little or no surplus to draw upon, soon fell by the wayside. Within 2 years' time practically every tractor company that had been organized during the 5 or 10 years preceding for the primary purpose of manufacturing and selling farm gas tractors was compelled to quit. Even some of the older and better-established farm-equipment manufacturers, who had entered the tractor business, found little sale for their machines.

Those companies that were able to remain in the business realized that it was no longer a problem of convincing the small or average farmer that mechanical power, in the form of the gas tractor, was practical and economical. They saw that he was willing to pay the price for a well-built, sensibly designed machine that would actually do the work for which it was recommended. Consequently, in spite of the generally depressed and unprosperous condition of agriculture throughout the United States during the years 1921 to 1926, the demand for farm tractors continued to grow as shown by the fact that the number in use on farms increased from about 250,000 in 1920 to over 500,000 in 1925. An analysis of the figures by states (Table V) shows further that the greatest increase during this period was in the middle western, eastern, and southern states, indicating beyond a doubt, that the small or average farmer, as well as the large grain raiser, was now a tractor convert.

The agricultural depression of 1930–1935 was even more pronounced than that of 1920. The prices of agricultural products dropped to the lowest point reached in many years. Consequently, there was little demand for farm tractors and equipment and sales were relatively small, particularly in 1932 and 1933. However, the small, low-priced, all-purpose tractor was introduced at this time, and, as commodity prices improved, an immediate, widespread demand developed for such a tractor.   As a result, tractor sales increased steadily and manufacturers were operating at capacity by 1935.

22. Design Standardized.—It is true that very few of the machines sold during this period, 1920 to 1926, were adapted to doing everything about the farm, that is, they were not of the all-purpose type; and a really successful machine of this type was yet to appear on the market. Most of the machines were light-weight, two- or three-plow tractors that were used largely for plowing and land preparation and for belt work.   The owners were satisfied to have something that would prepare the land quicker and better and likewise relieve the horse of this heavy work.

Another outstanding fact was that, during this period, the design of these tractors, as a whole, seemed more stable and uniform.   In other words, there was less variation in design and construction than formerly, indicating that the manufacturers had now, after years of costly experience, settled upon many of the essentials of a successful farm tractor.

23. All-purpose Tractors.—The next and most logical step seemed to be the design of a light-weight, low-priced, all-purpose tractor that would do any kind of field or stationary work on the average farm, including plowing, harrowing, planting, cultivating, harvesting, threshing, or anything requiring similar power.   Several machines of this type were designed and sold during the period 1917 to 1921, but few of them ever proved what might be called successful.   In other words, it appeared to be a difficult problem to construct a tractor that would be heavy and rugged enough to plow and harrow the heaviest soil, and still be practical for such lighter jobs as planting and cultivating of row crops, mowing hay, and so on.

Efforts were continued, however, by certain manufacturers to build a really successful and practical all-purpose farm tractor, and in 1925 one of these concerns introduced a machine that came nearer to meeting the requirements of such a tractor than any that had been previously built.   Other companies have since designed and placed in the field one or more models of machines of this type.

24. Tractor Census.—A rather graphic picture of the development of the tractor is presented by Tables IV and V.   The figures making up Table IV are largely estimates based upon census and sales reports issued by the U. S. Department of Commerce and the *Farm Implement News*.   It will be noted that the first important upturn in the sale of farm tractors occurred about 1918 and continued until the reaction in 1921.   Heavy sales during this period were due in part to the introduction of a low-priced machine by the Ford Motor Company, and partly to the scarcity of labor and the war-period prosperity.   The demand remained below normal for a year or two and then began to increase, reaching the peak in 1927, when 195,000 tractors were manufactured and 156,000 were sold in the United States.   Since 1930, the sales of all-purpose tractors have far exceeded the sales of all other types. For example, a report of the Bureau of the Census, U. S. Department of

TABLE IV.—APPROXIMATE NUMBER OF TRACTORS IN USE ON FARMS IN THE UNITED STATES

January, 1915, to January, 1937

Year	Number of Tractors	Year	Number of Tractors
1915	25,000	1924	455,000
1916	35,000	1925	506,000
1917	50,000	1926	580,000
1918	85,000	1927	650,000
1919	160,000	1928	768,000
1920	246,000	1929	853,000
1921	350,000	1930	920,000
1922	360,000	1936	1,248,000
1923	400,000	1937	1,383,000

TABLE V.—NUMBER OF TRACTORS ON FARMS IN THE UNITED STATES BY STATES, 1920, 1930, AND 1936[1]

State	1920	1930	1936	State	1920	1930	1936
Illinois	23,102	69,628	124,882	New Jersey	946	8,088	11,012
Iowa	20,270	66,258	113,108	Maryland	1,525	7,208	10,882
Texas	9,048	37,348	88,306	Washington	2,635	8,388	10,740
Kansas	17,177	66,275	83,881	Tennessee	1,872	6,865	10,444
Ohio	10,469	52,974	73,751	Georgia	2,252	5,870	10,122
Indiana	9,230	41,979	73,643	Florida	680	5,618	9,559
Minnesota	15,503	48,457	72,509	Kentucky	2,029	7,322	9,747
Wisconsin	9,407	50,173	70,429	Louisiana	2,812	5,016	9,507
Nebraska	11,100	40,729	58,807	Alabama	811	4,664	7,779
California	13,852	44,437	58,766	Idaho	1,587	4,691	6,412
New York	7,497	40,369	55,197	Massachusetts	592	3,921	5,469
Michigan	5,884	34,579	49,896	South Carolina	1,304	3,462	5,229
North Dakota	13,006	37,605	47,289	Wyoming	1,075	4,110	4,715
Pennsylvania	5,697	33,513	46,740	Arizona	930	2,558	4,465
Missouri	7,889	24,999	42,634	Maine	635	3,410	4,314
South Dakota	12,939	33,837	41,643	New Mexico	491	2,497	4,078
Oklahoma	6,210	25,962	36,894	Connecticut	440	2,667	4,013
Montana	7,647	19,031	21,255	West Virginia	572	2,792	3,489
North Carolina	2,277	11,426	18,070	Vermont	444	2,426	3,216
Colorado	4,990	13,334	15,514	Utah	583	1,426	2,123
Mississippi	667	5,542	12,379	Delaware	239	1,600	2,763
Virginia	2,544	9,757	11,803	New Hampshire	207	1,096	1,468
Arkansas	1,822	5,684	11,358	Rhode Island	79	589	853
Oregon	3,070	9,838	11,312	Nevada	210	360	407

[1] *Farm Implement News*, Vol. 52, No. 45 and Vol. 58, No. 7.

Commerce, shows that, in 1936, this type made up 68 per cent of all tractors produced or 79.9 per cent of all wheel tractors manufactured in that year. As a result, the number of tractors on farms in the United States has increased markedly in the past few years as shown by Table V. Furthermore, the introduction and adoption of a successful all-purpose tractor have resulted in a partial or, in many instances, a complete elimination of animal power on many farms throughout the country. As a consequence, there has been a pronounced decrease in the horse and mule population.

The steam tractor is rapidly disappearing from the scene. A few are still being used in the grain-growing sections for operating threshers, but the demand has dropped off to practically nothing because of the greater utility of the gas tractor and the increased popularity of the small thresher and the combined harvester-thresher.

References

Development of Tractor Design, *Trans. Amer. Soc. of Agr. Eng.*, Vol. 10, No. 2.
Factors Influencing Tractor Development, *Trans. Amer. Soc. of Agr. Eng.*, Vol. 15.
Tendency of Farm Tractor Design, *Trans. Amer. Soc. of Agr. Eng.*, Vol. 9, No. 1.

CHAPTER III

MAKE-UP OF A SIMPLE GAS ENGINE—NOMENCLATURE AND DEFINITIONS

The ordinary single-cylinder stationary farm gas engine is a good example of the simplest type of internal-combustion engine, and yet its construction is essentially similar to that of the more complicated and powerful types. Therefore the following discussion will apply particularly to such an engine, and will likewise be applicable, to a large extent, to practically all other types. In other words, the various parts and systems which are discussed, are common to nearly all engines and necessary for their proper and efficient operation.

25. Principal Engine Parts.—The parts and systems making up a simple stationary farm-type gas engine are as follows:

1. Base or frame.
2. Cylinder.
3. Cylinder head.
4. Piston.
5. Piston rings.
6. Piston pin.
7. Connecting rod.
8. Crankshaft.
9. Flywheel.
10. Valve system.
    - a. Valves—intake and exhaust.
    - b. Cam gear and cam or camshaft.
    - c. Push rod.
    - d. Rocker arm.
11. Fuel-supply and carburetion system.
    - a. Fuel tank.
    - b. Fuel line.
    - c. Fuel pump.
    - d. Carburetor.
    - e. Manifold.
12. Ignition system.
    - a. Battery or magneto.
    - b. Coil.
    - c. Sparking device.
    - d. Timing mechanism.
    - e. Switch and wire connections.
13. Cooling system.
    - a. Tank, radiator, or water hopper.

   b. Pump (in forced circulation system only).
   c. Water jacket.
   d. Fan.
   e. Pipes and connections.
14. Lubrication system.
   a. Oil and grease cups.
   b. Oil pump.
   c. Oil lines.
   d. Oil gage.
15. Governing system.
   a. Weights and springs.
   b. Detent arm or finger (in hit-and-miss system only).
   c. Throttle butterfly (in throttle system only).

Fig. 17.—Base, cylinder, water reservoir, and cylinder head cast separately.

26. Base or Frame.—The base or frame of an engine (Fig. 17) is that part, usually made of cast iron, to which all other parts of the engine are directly or indirectly attached. It is equipped with holes, usually four in number, by which the engine may be firmly anchored to its foundation.

27. Cylinder.—Cylinders are ordinarily made of high-grade gray cast iron and may be cast with the engine base (Fig. 18) or cast separately (Fig. 17). The former construction is confined largely to the smaller sizes of engines, whereas the detachable cylinder is preferred in the medium and larger sizes. Attaching the cylinder and base together produces rigid construction but necessitates a greater outlay if the cylinder must be replaced when worn or damaged. This objection is some-times overcome by using a removable cylinder wall or lining as shown in Fig. 18. If the cylinder is not a part of the base but is bolted to it, care should be taken to keep the bolts tight and retain the proper cylinder alignment at all times, thus reducing breakage and wear.

28. Cylinder Manufacture.—In manufacturing cylinders and cylinder blocks, a number of distinct steps are involved, as follows:

1. Melting and casting.
2. Cleaning the rough casting.
3. Machining: planing and milling.
4. Boring, drilling, and tapping.
5. Grinding the wall.
6. Honing and polishing the wall.

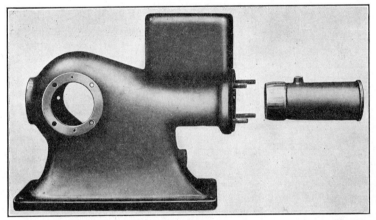

Fig. 18.—Base and cylinder in one piece with removable cylinder sleeve.

The grinding process consists of removing a thin layer of metal and producing an absolutely true bore by means of a somewhat coarse abrasive grinding wheel revolved at high speed. Honing is a finishing and polish-ing operation performed by a very fine-grained revolving stone. This operation produces a bore of exact dimensions and a mirror-smooth surface, which will provide the most efficient lubrication and the least possible wear.

29. Removable Cylinder Liners.—Some stationary engines, most tractor engines, and even some truck engines are now equipped with removable cylinder liners or sleeves (Figs. 18 and 58). The possible advantages are (1) overhauling and rebuilding the engine is rendered relatively simple, easy, reliable, and satisfactory at a reasonable cost, that is, by inserting new liners with new factory-fitted pistons, rings, and piston pins, the engine is practically restored to its original new condition so far as the cylinders and pistons are concerned; (2) in case one cylinder is scored or damaged, it can be easily repaired and the original bore and balance retained; and (3) a higher grade of cast iron or an alloy with better wear-resisting properties can be used for these liners, whereas to make the entire block of such material would be too expensive.

FIG. 19.—Cylinder and head cast together.

30. Cylinder Head.—Cylinder heads (Fig. 17) are made of cast iron. In most cases they are detachable from the cylinder but are sometimes cast with it (Fig. 19). When cast separately, a gasket of some nonburning material must be used to prevent loss of compression and leakage of the cooling liquid. The detachable type of cylinder head is preferable for the reason that it allows better accessibility; that is, it is more convenient to get to the valves and clean the carbon from the cylinder and piston.

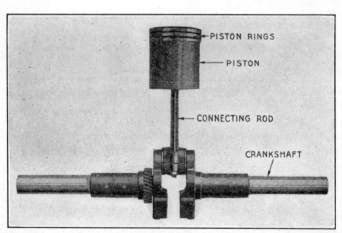

FIG. 20.—Piston and crankshaft assembly showing counterweighted crankshaft.

31. Piston.—Pistons (Fig. 20) are ordinarily made of cast iron, but aluminum-alloy pistons are frequently used in multiple-cylinder engines. Aluminum alloy offers the advantage of being lighter in weight but has a greater coefficient of expansion and, therefore, must be given more clearance than the cast-iron piston of the same size.

The approximate composition of a common piston alloy is:

Element	Per Cent
Aluminum	78.00 to 86.00
Silicon	11.25 to 15.00
Nickel	1.00 to 3.00
Iron, max	1.30
Copper	0.50 to 1.50
Magnesium	0.70 to 1.30

The principal characteristics of this alloy are (1) low coefficient of expansion as compared with other aluminum alloys, (2) hardness and resistance to wear, and (3) good mechanical properties at high temperatures.

Another piston alloy especially adapted to high-temperature conditions consists of:

Element	Per Cent
Aluminum	90.00 to 92.00
Copper,	3.75 to 4.25
Nickel	1.80 to 2.30
Magnesium	1.20 to 1.70
Iron, max	1.00
Silicon, max	0.70
Other impurities, max	0.20

The piston of the engine is made to move back and forth in the cylinder owing to the explosion of the fuel mixture. This back-and-forth movement of the piston is transmitted through the connecting rod and crankshaft to the belt pulley, and thus power is generated. The piston must fit in the cylinder in such a way as to move freely, but excessive looseness is undesirable.

The term *piston clearance* is applied to the space between the piston wall and cylinder wall. For cast-iron pistons, this clearance, at the closed end, when the engine is cold, should be about 0.001 in. for each inch of cylinder diameter. Most pistons are slightly tapered and larger in diameter at the open end and may have only 0.0005-in. clearance per inch of diameter at this end, owing to the fact that this portion of the piston is not exposed to so much heat as the closed end and therefore expands less. Aluminum-alloy pistons require about twice as much clearance as cast-iron pistons. Too much piston clearance will cause loss of compression, oil pumping, and piston slap. Too little clearance will cause the piston to stick or seize in the cylinder as the engine gets hot. However a piston may seize in a cylinder even though it has the proper clearance, owing to the improper action of the cooling system or to lack of cylinder lubrication.

32. Piston Rings.—Piston rings are made of cast iron. The number of rings per piston varies from three to seven, depending upon the type of engine and the compression desired. The ordinary types of engines

seldom have more than three or four rings, although five may be used in tractors and in similar heavy-duty engines. Diesel-type engines sometimes have as many as seven rings. The function of the piston rings is to retain com-

FIG. 21.—Oil-control piston ring with butt joint.

pression, and, at the same time, reduce the cylinder-wall and piston-wall contact area to a minimum, thus preventing friction losses and excessive wear.

Piston rings are classed as compression rings and oil rings (Figs. 21 and 22), depending upon their specific function and location on the piston. Compression rings are usually plain one-piece rings and are always

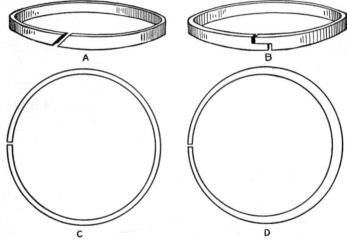

FIG. 22.—Types of piston rings and joints. *A.* Bevel joint. *B.* Step-cut joint. *C.* Concentric ring. *D.* Eccentric ring.

placed in the grooves nearest the piston head. Oil rings are grooved or slotted and are located either in the lowest groove above the piston pin or in a groove near the piston skirt. Their function, obviously, is to control the distribution of the lubricating oil to the cylinder and piston surfaces and to prevent unnecessary or excessive oil consumption.

When removed from the cylinder, piston rings are always slightly larger in diameter than the cylinder itself and must be compressed when inserted. Three kinds of joints are found—the bevel joint (Fig. 22*A*), the lap joint (Fig. 22*B*), and the plain butt joint (Fig. 21). Neither has any particular advantage over the other, although it is claimed that the lap joint is preferable as it offers less opportunity for any loss of compression.

Piston-ring clearance is the distance or space at the joint of the ring when it is in the cylinder. It is usually about 0.001 to 0.002 in. per inch of cylinder diameter. This clearance is necessary to allow for the expansion of the ring as it gets hot. Otherwise, without such clearance the ring would buckle and break and, consequently, injure the cylinder and piston of the engine. Too much piston-ring clearance is apt to produce leakage of compression and possibly waste of lubricating oil.

Piston rings may be of either the concentric type (Fig. 22C) or the eccentric type (Fig. 22D). The latter is thicker opposite the joint while

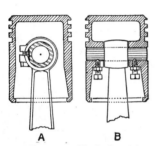

Fig. 23.—Methods of fastening piston pins. *A*, pin clamped to connecting rod. *B*, pin fastened to piston.

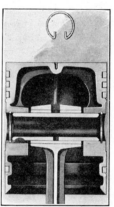

Fig. 24.—A floating piston pin.

the former is of uniform thickness. The advantage of the eccentric type is that it is less flexible and therefore hugs the cylinder wall better. However, the groove in the piston must be deep enough to receive the thickest part of the ring, and therefore is too deep for the thin part at the joint, where it may produce compression loss. The concentric type of ring seems to be more popular.

In fitting and replacing piston rings care should be taken to see that the ring is free in its groove but not too loose, that it has the proper clearance, and that the joints of the ring are not in line when the piston is inserted in the cylinder. Some pistons are equipped with anchor pins in the piston-ring groove, which prevent the ring from working around in it.

33. Piston Pin.—Piston pins are made of casehardened steel. The function of the piston pin is to connect the connecting rod to the piston and, at the same time, provide a flexible or hingelike connection between

the two.   Three different methods are used to anchor piston pins so
that they cannot work sideways and score the cylinder.   The first is to
clamp the pin to the connecting rod by means of a clamp screw or setscrew (Fig. 23*A*), so that the bearing will be at each end of the pin where
it fits in the piston.   In the second case, the pin is anchored to the piston
by means of setscrews and the bearing is in the connecting rod (Fig. 23*B*).
In the third method (Fig. 24), the piston pin is allowed to float, so to

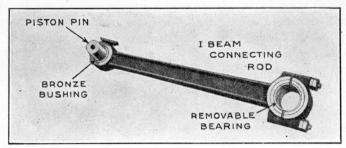

FIG. 25.—The most common type of connecting rod.

speak, or move in both the piston and connecting rod, but is held in place
by means of retaining rings at each end of the pin.

Some kind of a removable bushing, usually bronze (Fig. 25), is placed
either in the connecting rod or in the piston to receive the wear.   When
the piston-pin bearing becomes badly worn, this bushing can be removed
readily and replaced.   Frequently, however, it is necessary to replace
both the piston pin and the bushing.

FIG. 26.—Marine-type connecting rod.

34. Connecting Rod.—Connecting rods are made of what is known as
drop-forged steel.   They must be of some material that is neither
brittle nor ductile, that is, they must stand a twisting strain but should
not break or bend when subjected to such a strain.   The I-beam type
(Fig. 25) is the prevailing connecting-rod shape.   It gives strength with
less weight and material.

Figure 26 illustrates what is known as a *marine-type connecting rod.*
As shown, it is so made that both halves of the bearing are removable
from the connecting rod proper.   That end of the rod fastened to the
piston pin is known as the small end, and the other end, which is attached
to the crankshaft, is spoken of as the large end of the connecting rod.
The large end ordinarily has what is known as a *split bearing.*   The

bearing metal, usually babbitt, may be cast directly in the connecting rod or in a removable bronze or steel backing or shell.   When the bearing becomes worn and loose, it is adjusted by means of shims or thin metal strips, but replacement with a new bearing is often necessary.   The removable bearing-shell type is usually cheaper to replace than that having the babbitt cast in the connecting rod and bearing cap.   The two

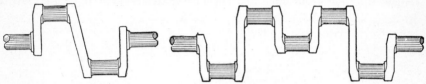

FIG. 27.—Two-cylinder crank-shaft with opposed cranks.

FIG. 28.—Typical four-cylinder crankshaft.

halves of the bearing are held together by either two or four bolts, depending upon the width of the bearing.

35. Crankshaft.—Crankshafts are made of drop-forged steel, and carefully machined, ground, and polished at the bearing points or journals.   The journals that support the crankshaft and hold it in position are called main journals, whereas the part to which the connect-

FIG. 29.—Flywheel hollowed out adjacent to crank for balancing.

ing rod is attached is spoken of as a crank journal.   Figures 20, 27 and 28 illustrate various types of crankshafts.

36. Flywheel.—Flywheels are made of cast iron and serve two purposes.   They assist in maintaining a uniform engine speed, especially in one-cylinder engines, and, second, provide a means of balancing the engine properly so that vibration and wear are reduced to a minimum. Uniformity of speed is maintained by the inertia of the heavy flywheel.   Correct balancing is obtained by making a certain portion of the flywheel just opposite the crank journal of the crankshaft heavier, as in Fig. 12, or by making that portion of the flywheel rim adjacent to the crank journal lighter, as in Fig. 29. Correct mechanical balance may be obtained also by placing what are known as counterweights on the crankshaft of the engine, as illustrated in Fig. 20.

From the outline given (Par. 25) it is noted that the remainder of the engine consists of six distinct systems. These various systems will be taken up in detail in succeeding chapters.

References

Cornell Univ., N. Y. State Coll. Agr., Ext. Bull. 85.
Cornell Univ., N. Y. State Coll. Agr., Junior Ext. Bull. 40.
DUELL: "The Motor Vehicle Manual."
"Dyke's Automobile and Gasoline Engine Encyclopedia."
ELLIOTT and CONSOLIVER: "The Gasoline Automobile."
HELDT: "Automotive Engines."
HELDT: "The Gasoline Automobile," Vol. I.

CHAPTER IV

PRINCIPLES OF OPERATION

37. Cycle of Operations.—Any internal-combustion engine, regardless of size, number of cylinders, use, and so on, is either of one or the other of two general types. These are known as the four-stroke-cycle type and the two-stroke-cycle type, or, as more commonly stated, four-cycle and two-cycle. A cycle consists of the events taking place in each and every cylinder of an engine between two successive explosions in that cylinder. These events or operations in the order in which they occur, are as follows: (1) the taking in of a combustible mixture, (2) the compression of this

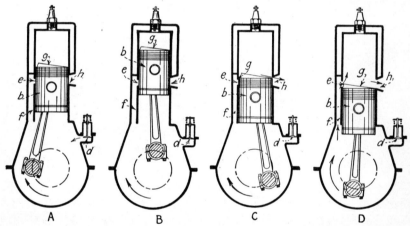

Fig. 30.—Two-port, two-stroke-cycle engine operation. (*Courtesy of Vacuum Oil Company.*)

mixture, (3) the ignition of the compressed mixture, (4) the expansion of the burnt gases producing the power and (5) the exhaust of the products of combustion.

In the two-stroke-cycle engine, two strokes of the piston or one revolution of the crankshaft are required to complete this cycle. In the four-stroke-cycle type, four strokes of the piston or two complete revolutions of the crankshaft are needed, and the four strokes are frequently referred to as suction, compression, power, and exhaust, respectively. It must be kept in mind that in engines of more than one cylinder, this cycle of events must be carried out in each of the cylinders, that is, one cylinder cannot perform one event and another perform the second, and so on. Also the events can take place only in the order

37

stated.   A better understanding of these principles of operation may be obtained by referring to the following explanation and illustrations.

38. Two-stroke-cycle Operation.—Two important characteristics of two-stroke-cycle construction must be noted and kept in mind: first, that ports or openings in the cylinder walls at some distance below the head serve as intake and exhaust valves; and second, that the open or crank end of the engine cylinder is completely enclosed.   Referring to Fig. 30*A*, the piston in its upward travel has closed the ports *e* and *h* and is compressing the charge.   At the same time the crankcase volume is being increased and the fuel mixture is drawn into the crankcase through an opening *d* to be compressed on the next downward stroke of the piston

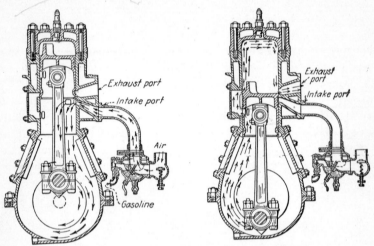

Fig. 31.—Three-port, two-stroke-cycle engine operation.

and forced through a connecting passage *f* into the combustion space when the inlet port *e* is uncovered.   Near the end of the upward or compression stroke of the piston the spark is produced and the compressed charge is fired.   The explosion and resulting expansion send the piston downward on its power stroke (Fig. 30*B*), near the end of which the two ports are uncovered as shown by Fig. 30*C*.   Even though the piston is nearly at the end of its stroke, considerable pressure remains in the cylinder thus forcing the products of combustion out through the exhaust port *h*.   At the same time, the fresh mixture in the crankcase has been compressed on the downward stroke and passes upward through the intake port to the combustion chamber as shown.   The piston has now completed two strokes, and the crankshaft has made one revolution, thus completing the cycle.

Figure 31 illustrates the three-port type of two-stroke-cycle engine in which the fuel mixture enters the crankcase through a third port, which is uncovered by the piston on its upward travel. Figure 149 is a cross-sectional illustration of a heavy-duty two-stroke-cycle oil engine.

Further explanation of questions that may arise in the mind of the reader concerning the operation of this type of engine will be made later.

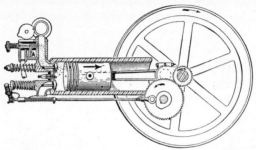

Fig. 32.—Four-stroke-cycle operation—beginning of the intake stroke, with the intake valve open and air and fuel being sucked in through the intake passage.

39. Four-stroke-cycle Operation.—Figures 32, 33, 34, and 35 show the usual four-stroke-cycle construction and operation. It will be observed, first of all, that an enclosed crankcase is unnecessary and that valves located in the cylinder head are used instead of ports. Starting with the piston at the head end of the cylinder (Fig. 32), it moves toward the crank end, drawing in a fuel mixture through the open inlet valve.

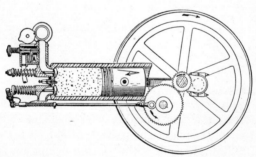

Fig. 33.—Four-stroke-cycle operation—starting the second or compression stroke in which the piston compresses the fuel mixture, both valves being closed so that the mixture cannot escape.

Soon after the end of this stroke, the inlet or suction valve closes and the mixture is compressed as the piston returns to the head end of the cylinder (Fig. 33). Near the completion of this stroke a spark is produced that ignites the charge causing an explosion, which, in turn, sends the piston on the third or power stroke (Fig. 34). Near the end of this stroke the exhaust valve is opened, and the burned residue is completely removed from the cylinder as the piston travels backward toward the cylinder head on what is known as the exhaust stroke (Fig. 35). The piston has now passed through four complete strokes, the crankshaft has made two revolutions, and a cycle has been completed. It is thus observed that the two-stroke-cycle single-cylinder engine has a power impulse for each revolution of the crankshaft; and that the four-stroke-cycle single-cylinder engine gives one power impulse in two revolutions.

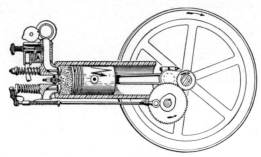

Fig. 34.—Four-stroke-cycle operation—with both valves still closed, a spark ignites the compressed charge which burns rapidly, creating heat and gases under high pressure which push against the piston and produce the third or power stroke.

This explains why heavy flywheels are necessary on one-cylinder engines. Were it not for these flywheels, such an engine when under load would have a tendency to lose speed between explosions. These flywheels, owing to their inertia, carry the piston through these so-called idle strokes in spite of the resistance offered by the load and thus maintain

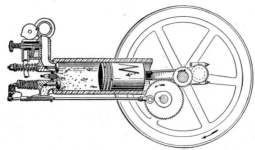

Fig. 35.—Four-stroke-cycle operation—the fourth or exhaust stroke. The exhaust valve is opened and the burned material is driven out by the piston as it moves backward into the cylinder.

uniformity of speed. A one-cylinder two-stroke-cycle engine will not require so heavy a flywheel as the four-stroke, owing to the fact that it fires twice as frequently.

40. Multiple-cylinder Operation.—As previously stated, these principles of operation likewise apply to multiple-cylinder engines. That is, a four-cylinder four-stroke-cycle engine is nothing more than four single-cylinder four-stroke-cycle engines built into a single unit; or a two-cylinder two-stroke-cycle engine is nothing more than two single-cylinder two-stroke-cycle engines built into a single unit.

A better understanding of the application of these principles to multiple-cylinder engines may be obtained by reference to the following illustrations and explanations: Fig. 37*A* illustrates the usual construction

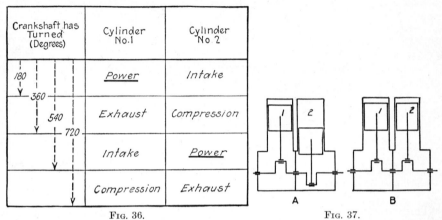

Crankshaft has Turned (Degrees)	Cylinder No.1	Cylinder No 2
180 360 540 720	Power	Intake
	Exhaust	Compression
	Intake	Power
	Compression	Exhaust

FIG. 36.   A   B   FIG. 37.

FIG. 36.—Chart showing occurrence of events in a two-cylinder four-stroke-cycle engine with crank arrangement shown in Fig. 37*B*.

FIG. 37.—Crank arrangements for two-cylinder, twin engines. *A*. Opposed cranks. *B*. Twin cranks.

of a two-stroke-cycle two-cylinder engine. Note that the crankshaft consists of two opposite cranks; therefore, the pistons will move in opposite directions at all times. A cycle will be completed in each cylinder at every revolution of the crankshaft, but a power impulse or explosion will occur at every half turn, or there will be two explosions per revolution. Figure 37*B* illustrates a typical two-cylinder four-stroke-cycle engine layout. Note that the cranks on the crankshaft are arranged side by side, and that the two pistons move together. By reference to the accompanying chart (Fig. 36), it will be noted

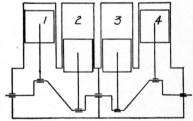

FIG. 38.—Crankshaft and cylinder arrangement for a four-cylinder four-stroke-cycle engine.

that there will be two power impulses during two revolutions of the crankshaft or one power impulse for each revolution; that is, when piston 1 is on the power stroke, piston 2 is on intake, and so on. It is possible so to construct this engine that both pistons would be doing the same thing at the same time and, consequently, both explosions would come at the same instant, but such construction is never used in practice.

Figure 38 illustrates the usual four-cylinder four-stroke-cycle construction. Four-cylinder crankshafts are always arranged as shown, that is, cranks 1 and 4 are together and opposite cranks 2 and 3. Referring to the chart (Fig. 39) and assuming that piston 1 has just received

Crankshaft has Turned (Degrees)	Cylinder No. 1	Cylinder No 2	Cylinder No.3	Cylinder No.4
180 / 360	*Power*	Compression	Exhaust	Intake
540 / 720	Exhaust	*Power*	Intake	Compression
	Intake	Exhaust	Compression	*Power*
	Compression	Intake	*Power*	Exhaust

Fig. 39.—Chart showing occurrence of events in a four-cylinder four-stroke-cycle engine having a 1-2-4-3 firing order.

a power impulse so that it is on the power stroke, piston 4 must necessarily be on the intake stroke. If piston 2 is now placed on the compression stroke, piston 3 must necessarily be on the exhaust stroke. Completing the chart by filling in the strokes for the four pistons in the correct order, the *firing order* becomes 1–2–4–3.

Crankshaft has Turned (Degrees)	Cylinder No.1	Cylinder No 2	Cylinder No.3	Cylinder No. 4
180 / 360	*Power*	Exhaust	Compression	Intake
540 / 720	Exhaust	Intake	*Power*	Compression
	Intake	Compression	Exhaust	*Power*
	Compression	*Power*	Intake	Exhaust

Fig. 40.—Chart showing occurrence of events in a four-cylinder four-stroke-cycle engine having a 1-3-4-2 firing order.

A second firing order is likewise possible as shown by the chart (Fig. 40): that is, again assuming that piston 1 is on the power stroke and piston 4 on the intake stroke, piston 2 could be on exhaust and piston 3 on compression. Thus the firing order becomes 1–3–4–2. These are the only firing orders possible, however, with such a crankshaft and crank arrangement. Both firing orders are used in four-cylinder four-stroke-cycle engines, and neither has any particular advantage over the other.

The particular firing order used is determined by the order in which the intake or exhaust valves operate and by the order in which the sparks are delivered to the cylinders. Given two four-cylinder four-stroke-cycle engines, one having a firing order of 1–2–4–3, and the other of 1–3–4–2, the principal mechanical difference will be in the arrangement of the cams on the camshaft (Fig. 73), which operate the valves. These principles of operation likewise apply to engines having six, eight, or any number of cylinders, and the proper distribution of the explosions and the firing order are obtained by the correct crank arrangement and valve operation.

41. Firing Interval.—It should now be clear that the advantage of an engine having more than one cylinder is that it fires more frequently; that is, there are more explosions in the same number of revolutions of the crankshaft, and consequently the power is said to be more uniform and steady. The greater the number of cylinders, the shorter the distance moved by the crankshaft between two successive explosions in the engine. The interval between successive explosions in an engine is known as the *firing interval* and is determined in a four-stroke-cycle engine by dividing 720, the number of degrees in two revolutions, by the number of cylinders; or, in a two-stroke-cycle engine, by dividing 360 deg. by the number of cylinders; that is, we know that a circle consists of 360 deg.; therefore, in making one revolution, the crankshaft would turn that number of degrees; or in making two revolutions it would turn two times 360 or 720 deg. The firing interval for a one-cylinder two-stroke-cycle engine would be 360 deg., and for a single-cylinder four-stroke-cycle engine it would be 720 deg.

42. Two-stroke-cycle-engine Characteristics.—Two-stroke-cycle engines have the following distinguishing mechanical characteristics: (1) the crankcase is enclosed and must be as airtight as possible; (2) ports or openings in the side of the cylinder, opened and closed by the piston sliding over them, take the place of valves; (3) no valve-operating mechanism of any kind is necessary; (4) the fuel mixture usually enters and passes through the crankcase on its way to the cylinder; and (5) the cylinder is usually vertical.

Some of the important advantages of two-stroke-cycle engines are as follows: (1) lighter in weight per horsepower, (2) simpler in construction, (3) greater frequency of working strokes or power impulses, and (4) usually operate in either direction. Some disadvantages are: (1) fuel mixture controlled with difficulty, (2) inefficient in fuel consumption, (3) do not operate satisfactorily under fluctuating loads, (4) speed and corresponding power not readily controlled, and (5) cooling and lubrication more difficult.   Two-stroke-cycle engines are usually inefficient in fuel consumption and frequently give considerable trouble in starting and during operation, owing largely to the fact that complete exhaust of the burned fuel residue is extremely difficult.   Likewise, the problem of producing the correct fuel mixture and placing it in the cylinder is a difficult one.   Unless such an engine is in perfect condition and correctly adjusted at all times, trouble will be apt to develop.

It might be assumed that a two-stroke-cycle engine of a certain bore, stroke, and speed will develop twice as much power as a four-stroke engine of the same size and speed, because it would have twice as many power strokes in a given time.   Such is not the case, however, because it is impossible to get as effective a charge of fuel mixture into the cylinder.

The removal of the burned residue from the cylinder is dependent entirely upon the somewhat greater pressure remaining in the cylinder than that existing on the outside when the exhaust port is opened.   There is not a distinct exhaust stroke as in the four-stroke-cycle engine to push this material out; therefore its complete expulsion is unlikely and the incoming fuel charge may be thus contaminated and its effectiveness reduced.   To prevent the possible escape of a portion of the incoming fuel mixture through the exhaust port before the piston has closed the latter, a projection known as a deflector, as shown by g in Fig. 30, is placed on the closed end of the piston at the point where it passes the intake port.   As the name implies, this device deflects the charge in an upward direction toward the cylinder head.

References

Cornell Univ., N. Y. State Coll. Agr., Ext. Bull. 85.
Cornell Univ., N. Y. State Coll. Agr., Junior Ext. Bull. 40.
DUELL: "The Motor Vehicle Manual."
"Dyke's Automobile and Gasoline Engine Encyclopedia."
ELLIOTT and CONSOLIVER: "The Gasoline Automobile."
HELDT: "The Gasoline Automobile," Vol. I.

CHAPTER V

STATIONARY-ENGINE TYPES—ENGINE SPEEDS—PISTON DISPLACEMENT—COMPRESSION

Internal-combustion engines are made in various types and sizes according to the purpose for which they are to be used. They may vary according to the number of cylinders from one to twelve or even more. Engines for stationary farm use are usually of the single- or two-cylinder type. Farm tractor engines have two, four, and six cylinders. For automobile engines four, six, eight, and twelve cylinders are used.

43. Cylinder Arrangement.—According to cylinder arrangement, engines may be designated as horizontal, vertical, V-type, and so on.

FIG. 41.—A vertical four-cylinder engine for stationary work.

Most of the single-cylinder farm engines are horizontal (Fig. 12). The vertical type is used to a limited extent. The horizontal engine has the advantage of offering greater accessibility and possibly simpler construction. The vertical-cylinder engine wears more uniformly about the piston and cylinder walls and is less affected by vibration, owing to the fact that the vibratory action is vertical rather than horizontal.

Multiple-cylinder engines are usually of the vertical type, as shown by Fig. 41. For eight- and twelve-cylinder engines, the V-construction is often used as illustrated by Fig. 42, although the eight-in-line type is now quite commonly found in automobiles. Figure 43 illustrates the

45

radial type of engine, now used extensively in airplanes. Another method of arranging the cylinders, now seldom used, is illustrated in Fig. 44. This is known as the single- or double-opposed arrangement.

44. Four-stroke-cycle Types and Uses.—Engines of the four-stroke-cycle type have a wide range of utility. Some of the more or less distinct types are as follows:

FIG. 42.—A V-type eight-cylinder engine.

I. *Stationary Engines:*

    A. Farm Types:

        1. Light-weight, single-cylinder, medium- to high-speed engines.
        2. Medium-weight, single-cylinder, low-speed engines (Fig. 12).
        3. Multiple-cylinder, medium- to high-speed engines (Fig. 41), used for heavier work about the farm such as wood sawing, feed grinding, and so on.

    B. Heavy-duty, low-speed oil engines for operating electric power plants, pumping plants, cotton gins, and the like.

        1. Single cylinder (Fig. 45).
        2. Multiple cylinder (Fig. 46).

II. *Automotive Engines:*

    Multiple-cylinder, medium- and high-speed, light-weight engines burning gasoline and kerosene and used in automobiles, trucks, tractors, motorcycles, airplanes, motorboats, and the like (see Figs. 41, 42, and 43).

45. Two-stroke-cycle Types and Uses.—The utility of the two-stroke-cycle engine is limited as compared to the four-stroke. Many of the small, light-weight, high-speed engines built for the propulsion of motorboats and launches are of this type. Two-stroke-cycle engines are especially adapted to marine purposes for the reason that the load is uniform and cooling is not a difficult problem.

Heavy-duty, two-stroke-cycle, single- and multiple-cylinder oil-burning engines, as illustrated by Figs. 47 and 48, are used extensively to supply power for operating electric generators, large pumping units, cotton gins, and the like. These engines, however, differ considerably in construction from the lighter types, burn heavier fuel, and give more efficient results. Owing to the limited use of the two-stroke-cycle engine, no further reference of any consequence will be made to it, and the

Fig. 43.—A radial-type, air-cooled, four-stroke-cycle, airplane engine.

balance of this text will be devoted largely to the construction and operation of the four-stroke-cycle type of engine.

46. Engine Speeds.—Ordinarily, the fewer the number of cylinders and the heavier the engine, the slower its crankshaft speed, and such engines are therefore known as low-speed engines. Examples are the common farm-type, stationary, single-cylinder engines, which have a speed varying from 300 to 600 r.p.m., and the large, heavy-duty, stationary oil engines made in both single- and multiple-cylinder types that operate at from 200 to 400 r.p.m. Some small, light-weight, single-cylinder engines have speeds running as high as 1,000 to 2,000 r.p.m. Four-stroke-cycle engines, operating at so-called high speeds, are usually multiple-cylinder engines, such as those used in automobiles, tractors, trucks, and airplanes. These engines may operate at a very low speed when idling but when under load usually attain speeds varying from 750 to 4,000 r.p.m. or more.

47. Compression.—Some time before the internal-combustion engine was invented, and before its possibilities were given even serious consideration, the discovery was made that if the fuel mixture were placed under pressure in the cylinder it would generate more power and exert a greater force when burned; that is, a combustible mixture of vaporized fuel and air, when placed under pressure, burns more rapidly and generates more power than when ignited at atmospheric or very low pressure.

For this reason the fuel charge in all types of internal-combustion engines is compressed to a certain degree before being ignited.

Fig. 44.—Horizontal opposed engines.   *A.* Single opposed.   *B.* Double opposed.

Fig. 45.—A 50-hp. single-cylinder four-stroke-cycle oil engine.

When the fuel mixture enters the cylinder, the piston travels toward the open end, and the charge completely fills the space between the piston and the cylinder head as shown in Fig. 33.   The existing pressure is now about the same as that of the outside atmosphere. Before the mixture is fired, the valves or ports are closed and the piston moves toward the

FIG. 46.—A multiple-cylinder four-stroke-cycle oil engine.

cylinder head pushing the charge into what is known as the combustion space or clearance space (Fig. 34) and placing it under considerable pressure. This combustion or clearance space is that space remaining

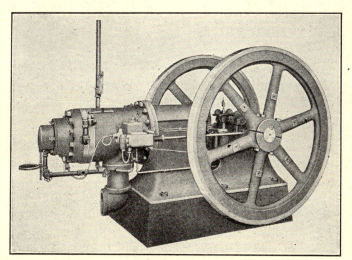

FIG. 47.—A single-cylinder two-stroke-cycle oil engine.

behind the piston when it has reached the end of the compression stroke. The total cylinder volume is the total space existing between the piston and cylinder head when the piston is on crank dead center.

48. Piston Displacement.—Piston displacement is the space swept through by the piston in moving from one end of its stroke to the other and is expressed in cubic inches. For a single-cylinder engine, it is the cross-sectional area of the piston or cylinder multiplied by the length of stroke of the piston. For a multiple-cylinder engine, the total piston displacement is equal to the piston displacement of one cylinder, multiplied by the number of cylinders. For example, the size or dimensions of an engine are usually expressed in terms of the bore of the cylinder and the stroke of the piston. A $4\frac{1}{2}$- by 6-in. cylinder is one having a $4\frac{1}{2}$-in.

Fig. 48.—A multiple-cylinder two-stroke-cycle oil engine.

bore or diameter and a 6-in. piston stroke. If it had but one cylinder, the piston displacement would be calculated as follows:

$$\text{Area of cylinder} = (\tfrac{1}{2}\text{ bore})^2 \times \pi$$
$$= (2.25)^2 \times 3.1416$$
$$= 15.9 \text{ sq. in.}$$
$$\text{Piston displacement} = \text{area of cylinder} \times \text{stroke}$$
$$= 15.9 \times 6$$
$$= 95.4 \text{ cu. in.}$$

A four-cylinder engine would have 4×95.6 cu. in. or 382.4 cu. in. displacement.

The compression pressure measured in pounds per square inch depends upon the ratio of the piston displacement in cubic inches to the total cylinder volume in cubic inches; that is, the nearer the piston displacement of a cylinder approaches the total cylinder volume of that cylinder, the smaller the clearance, or combustion space and the greater the compression. Or, expressed in another way, the clearance or combustion space is equal to the difference between the total cylinder volume and the piston displacement. Therefore, the smaller the difference becomes, the greater the compression pressure.

49. Compression Ratio.—The compression ratio of an engine is the total cylinder volume divided by the clearance volume. The greater this ratio, the higher the compression pressure in the cylinder. Also the clearance space may be expressed as a certain percentage of the piston displacement, and, the greater this percentage, the lower the compression pressure. Table VI gives the compression pressures in pounds per square inch for different compression ratios and percentages of clearance space.

TABLE VI.—COMPRESSION RATIOS AND PRESSURES

Compression ratio	Cylinder clearance of piston displacement, %	Compression pressure (gage pressure), lb. per square inch	Compression ratio	Cylinder clearance of piston displacement, %	Compression pressure (gage pressure), lb. per square inch
3.6	38.5	57.9	6.4	18.5	137.3
3.8	35.7	63.1	6.6	17.9	143.4
4.0	33.3	68.4	6.8	17.2	149.7
4.2	31.2	73.7	7.0	16.6	155.9
4.4	29.4	79.1	8.0	14.3	188.0
4.6	27.8	84.7	9.0	12.5	221.4
4.8	26.3	90.2	10.0	11.1	255.8
5.0	25.0	95.9	11.0	10.0	291.2
5.2	23.8	101.6	12.0	9.1	328.0
5.4	22.7	107.4	13.0	8.3	365.2
5.6	21.7	113.3	14.0	7.7	406.5
5.8	20.8	119.2	15.0	7.1	441.8
6.0	20.0	125.2	16.0	6.7	482.6
6.2	19.3	131.2	17.0	6.3	523.3

50. Compression Pressure.—The compression pressure used in different engines depends largely upon the fuel that the engine burns. For engines using gasoline, kerosene, distillate, or similar heavy fuels, it usually varies from 60 to 120 lb. per square inch. For certain types of engines, much higher pressures are required, as will be described under the subject of ignition. Theoretically, considerably higher pressures than those mentioned will give more efficient results with gasoline or kerosene, but if these particular fuel mixtures are subjected to excessively high pressures a detonation or knocking results, and its control is difficult. Water is frequently used with kerosene fuel mixtures to overcome this trouble. The use of water with kerosene is explained later.

References

DUELL: "The Motor Vehicle Manual."

"Dyke's Automobile and Gasoline Engine Encyclopedia."

ELLIOT: "Automobile Power Plants."

ELLIOT and CONSOLIVER: "The Gasoline Automobile."

FAVARY: "Motor Vehicle Engineering—Engines."

FRASER and JONES: "Motor Vehicles and Their Engines."

HELDT: "Automotive Engines."

HELDT: "The Gasoline Automobile," Vol. I.

KERSHAW: "Elementary Internal Combustion Engines."

STREETER and LICHTY: "Internal-combustion Engines."

CHAPTER VI

TRACTOR-ENGINE TYPES AND CONSTRUCTION

The first gas tractors built were equipped with large slow-speed horizontal engines having only one or two cylinders. Obviously, these engines required a strong frame, large wheels, and other supporting parts. Consequently the tractors themselves were very heavy, difficult to start and handle, and had considerable vibration. Later on, when the possibilities of lighter weight tractors were observed, designers turned their attention toward the use of a higher speed, lighter weight power plant, having at least two and, possibly, four cylinders, such as those used in trucks and automobiles. The result is that the trend in the tractor

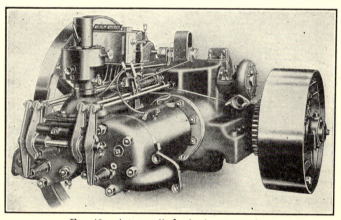

Fig. 49.—A two-cylinder horizontal engine.

industry, as in automotive design, has been toward engines having two, four, and even six cylinders, with lighter weight and higher operating speeds. At the present time, of the ten principal tractor manufacturers, nine use four- and six-cylinder engines, and one uses a two-cylinder engine.

50a. Tractor-engine Characteristics.—Tractor power plants may be classified as two-cylinder twin horizontal engines and as four- and six-cylinder vertical engines. All tractor engines are either of the four-stroke-cycle heavy-duty carbureting type or of the Diesel type. Gasoline, kerosene, distillate, and certain so-called tractor fuels are the predominating fuels.

51. Horizontal Two-cylinder Engines.—The horizontal, two-cylinder-type engine (Fig. 49) has been used by a number of leading manu-

FIG. 50.—Two-cylinder opposed crankshaft.

facturers in the past and is still found in some makes.   The advantages claimed for it are (1) fewer and heavier parts with slower speed and,

Crankshaft has Turned (Degrees)	Cylinder No. 1	Cylinder No. 2
180 360	Power	Compression
540 720	Exhaust	Power
	Intake	Exhaust
	Compression	Intake

FIG. 51.—Chart showing occurrence of events in a two-cylinder four-stroke-cycle engine with opposed cranks.

therefore, less wear and longer life, (2) greater accessibility, and (3) belt pulley is driven direct and not through gears, because crankshaft is placed crosswise on the frame.

FIG. 52.—Two-cylinder twin crankshaft.

Figure 50 shows the usual two-cylinder crankshaft with opposed cranks.   Referring to Fig. 51, such an arrangement produces an unequal firing interval, namely, 180 and 540 deg.   It would appear that this uneven firing would be undesirable and produce an uneven power flow.

However, with such an engine running at 600 to 900 r.p.m., there would not be any noticeable effect.

The crank arrangement shown in Fig. 52 gives an equal firing interval —360 deg. according to the diagram (Fig. 53)—but is less preferable because of difficulty in securing mechanical balance and, therefore, the least possible vibration.

Vibration in any engine is caused largely by the reciprocating movement of certain parts such as the pistons. The sudden stopping and reversing of the direction of travel of a piston when it reaches dead center cause a jerk, especially if it is very large and heavy. However, if two pistons attached to the same crankshaft are moving in opposite directions and reach opposite dead centers at exactly the same instant, then

Crankshaft has Turned (Degrees)	Cylinder No. 1	Cylinder No. 2
180 360	*Power*	*Intake*
540 720	*Exhaust*	*Compression*
	Intake	*Power*
	Compression	*Exhaust*

Fig. 53.—Occurrence of events in a two-cylinder twin engine with twin crankshaft.

the action of one opposes the action of the other and vibration is almost entirely eliminated. The crankshaft is likewise practically balanced without the need of counterweights required for the twin crankshaft (Fig. 52). It may be said, therefore, that the uneven firing produced by the opposed crank arrangement in a two-cylinder engine is more than offset by the almost complete elimination of vibration.

The two-cylinder opposed type of engine with opposed cranks (Fig. 44) was used in some early tractors. This construction permits an equal firing interval and provides good mechanical balance but offers other serious objections such as (1) difficult manifolding and fuel-mixture distribution owing to the scattered arrangement of the cylinders; (2) inaccessibility of bearings and other working parts; and (3) lack of compactness of mechanism, which is desirable from the standpoint of lubrication and adjustment. The horizontal double-opposed engine (Figs. 44 and 54) has been used by only one tractor manufacturer. It provides a well-balanced mechanism with little vibration and likewise gives good weight distribution, which is very important in good tractor design. However it has the same undesirable characteristics as the two-cylinder opposed engine.

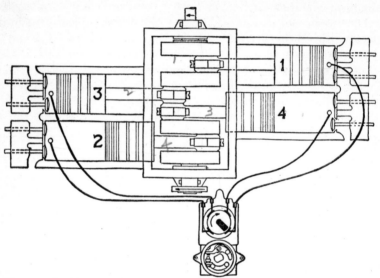

Fig. 54.—Double-opposed engine showing crank arrangement and firing order.

52. Vertical Three-cylinder Engines.—One tractor manufacturer makes one model having a three-cylinder vertical engine. Figure 55 shows the crankshaft for this engine. The three cranks make a 120-deg. angle with each other, and the firing interval is 240 deg.

Fig. 55.—A crankshaft for a three-cylinder engine.

53. Four-cylinder Engines.—The vertical four-cylinder type of engine, placed lengthwise of the frame, predominates in the tractor field at the present time. Some of the reasons for this are: (1) Good weight distribution is secured. (2) Manifolding and fuel-mixture distribution are simplified. (3) Uniform and positive cylinder and bearing lubrication is facilitated. (4) Vibratory effects caused either by moving parts or by the explosions act in a vertical plane and are thus less noticeable and nullified by the engine and tractor weight itself. (5) Valve mechanisms, ignition devices, and other parts are made more accessible. (6) The clutch and transmission parts can be assembled in such a way as to give the entire machine a better balanced construction and more symmetrical appearance. _Vert._

In view of the fact that the same crank arrangement is used in all four-cylinder engines, only two firing orders are possible: namely, 1–2–4–3 and 1–3–4–2. Both are used and neither offers any advantage over the other. The firing order is determined primarily by the order in which the

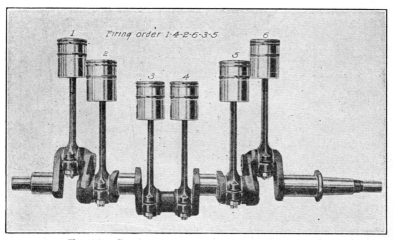

Fig. 56.—Crank arrangement for a six-cylinder engine.

valves are operated. Consequently, the principal difference between two four-cylinder engines having different firing orders would be in the arrangement of the cams on the camshaft.

54. Six-cylinder Engines.—The six-cylinder engine is being used in a few makes and models of tractors. Some manufacturers who make several sizes of tractors, for example, standardize on one size of cylinder (bore and stroke). Therefore, in their larger machines, the necessary additional power is obtained by using six cylinders instead of four but with the same bore and stroke per cylinder in each case. A six-cylinder engine is said to produce somewhat smoother power and less vibration than an engine with fewer cylinders.

Six-cylinder crankshafts are constructed as shown in Fig. 56; that is, crank 1 is paired with No. 6; No. 2 with No. 5; and No. 3 with No. 4. The three pairs are then arranged so as to make 120 deg. with each other, as shown by Fig. 57. If, when facing the front end, cranks 1 and 6 are up and 3 and 4 extend to the right, it is said to be a right-hand, six-cylinder crankshaft. The firing order used in this case is 1–5–3–6–2–4. If with cranks 1 and 6 up, Nos. 3 and 4 extend to the left, it is called a left-hand crankshaft, and the firing order used is 1–4–2–6–3–5.

55. Features of Tractor-engine Design.—There is now a much greater similarity in tractor-engine construction and design than in

Fig. 57.—Left- and right-hand six-cylinder crankshaft.

former years. Such features as valve-in-head construction, removable cylinder sleeves, and cylinders cast en bloc are used in a majority of cases. Tractor engines are classed as heavy-duty engines; that is, castings, pistons, crankshafts, bearings, valves, and practically all important parts are made heavier and larger than would be necessary in an automobile

Fig. 58.—Cylinder block for four-cylinder engine showing removable sleeve.

engine, in order to withstand the unusually heavy strains and continuous heavy loads to which they are subjected.

Cylinders are cast en bloc, that is, all in one casting (Fig. 58), for the small and medium sizes of tractors. Larger engines usually have the cylinders cast in pairs.

Tractor-engine cylinders are subjected to rather rapid wear, which eventually causes excessive power losses and oil pumping. This trouble can be overcome in one or the other of two ways, namely, (1) by replacement of the cylinders, or (2) by reboring of the old cylinders together with the use of oversize pistons. Either procedure is more or less

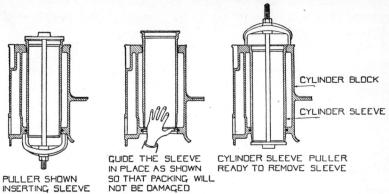

PULLER SHOWN
INSERTING SLEEVE

GUIDE THE SLEEVE
IN PLACE AS SHOWN
SO THAT PACKING WILL
NOT BE DAMAGED

CYLINDER SLEEVE PULLER
READY TO REMOVE SLEEVE

CYLINDER BLOCK

CYLINDER SLEEVE

Fig. 59.—Method of removing and replacing cylinder sleeve.

expensive and time consuming. Therefore, a number of tractor engines are now equipped with removable cylinder walls or sleeves (Fig. 58). These can be readily removed as shown in Fig. 59, without removing the engine from the tractor or disassembling it to any extent. Furthermore, the complete set of sleeves will cost considerably less than a new block or a good reboring job. It is always advisable to install a new set of pistons along with the sleeves.

Cylinder heads for tractor engines are cast separately from the block. This permits ready access to the combustion chamber for cleaning out the carbon deposits. With the valves located in the cylinder head, valve adjustment and grinding are easier to handle. Figure 60 shows one type of cylinder head with a specially shaped combustion chamber. This is known as the Ricardo head. It is claimed

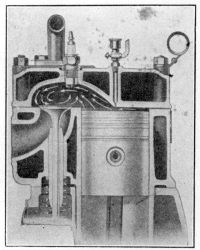

Fig. 60.—Ricardo-type combustion chamber.

that this particular construction produces a better turbulence of the fuel mixture and, therefore, causes it to ignite and explode in such a manner as to generate the greatest possible power.

Tractor pistons are in most cases made of cast iron with three to seven rings.

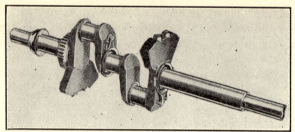

Fig. 61.—Two-cylinder, three-bearing, counterbalanced crankshaft.

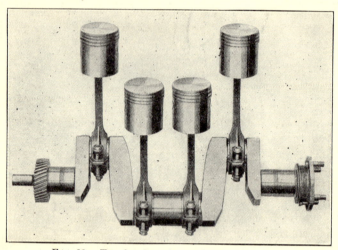

Fig. 62.—Two-bearing, four-cylinder crankshaft.

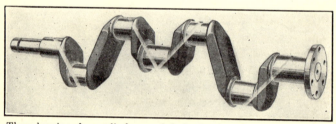

Fig. 63.—Three-bearing, four-cylinder crankshaft showing oil passages to crank bearings.

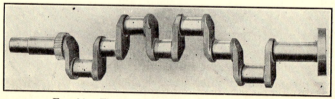

Fig. 64.—Five-bearing, four-cylinder crankshaft.

Piston pins are made hollow with the bearing either in the piston bosses (Fig. 23*A*), or in the connecting rod (Fig. 23*B*). Figure 24 shows the floating piston pin used in some tractors. Side movement is prevented by the two retainer springs, which fit in small grooves in the piston. Piston-pin wear is usually overcome in tractors by replacement of the bronze bushings.

Fig. 65.—Two-bearing, four-cylinder crankshaft equipped with ball bearings.

56. Crankshafts.—All tractor crankshafts are one-piece drop forgings, which have been carefully machined and heat treated. Two-cylinder crankshafts may have two main bearings (Fig. 50), or three main bearings (Fig. 61).

Four-cylinder crankshafts may have two, three, or five main bearings as shown by Figs. 62, 63, and 64. The fewer the bearings, the larger the shaft diameter. The three-bearing crankshaft is most common. Both the main and the crank journals are very accurately machined and ground to within 0.001 or 0.002 in. of the specified dimension.

Fig. 66.—Precision-type bearing shell.

57. Bearings.—All crankshaft bearings—both main and crank—on most tractor engines consist of a pair of removable babbitt-lined shells having either a bronze or a steel backing. A high-grade babbitt is used and is poured and finished in such a manner that the bearing surface has a mirrorlike finish, Fig. 66. These are sometimes known as precision-type bearings for the reason that they wear very slowly and require no scraping or special fitting procedure when replaced. Thin shims may or may not be used to obtain the correct bearing fit.

Figure 65 shows a two-bearing tractor crankshaft with ball bearings as used with success by one manufacturer who guarantees both the crankshaft and the bearings against breakage and defect during the life of the machine. This bearing possesses certain advantages such as

(1) very little friction loss, (2) adjustment is unnecessary and (3) bearing is easily lubricated.

58. Accessory Shaft.—The magneto, water-circulating pump, and governor are placed alongside the engine block and are driven by a special gear and shaft. In some cases this gear and shaft are on the camshaft side. The gear then meshes with the cam gear but has only one-half as many teeth and, therefore, runs at crankshaft speed. In other engines, the accessory shaft is placed opposite the camshaft and driven directly off the crankshaft.

59. Tractor-engine Speeds.—The operating speed of a tractor engine depends primarily upon the size of the engine and the number of cylinders. Two-cylinder engines have a larger bore and a longer stroke and operate at speeds varying from 450 to 900 r.p.m. The speed of four- and six-cylinder engines varies from 650 r.p.m. in the largest sizes, to 1,500 r.p.m. in the light-weight machines. All tractors are equipped with governors; therefore, the operating speeds are somewhat lower than automobile-engine speeds, and the range of speed variation is considerably less.

60. Garden-tractor Engines.—Garden tractors use two general types of engines, depending on whether the tractor is of the small or light-weight type (Fig. 430), or of the heavier type (Figs. 431 and 432). The former uses a vertical single-cylinder air-cooled high-speed engine. It may be a two-stroke-cycle or a four-stroke-cycle engine. The heavier type garden tractor is equipped usually with a vertical two-cylinder water-cooled engine.

VALVES AND VALVE OPERATION

The one-piece medium-carbon or nickel-steel poppet valve is used in all tractor engines. Since these valves must operate under rather abnormal conditions, such as high temperatures and continuous heavy loads, they must be correctly designed and well made of the best materials. A tractor engine should be tested frequently for compression leakage around the valves. The latter should be carefully ground if necessary.

61. Camshafts and Camshaft Drives.—Engines of the usual type equipped with two valves per cylinder require a single camshaft with two cams per cylinder (Fig. 73). Since these engines operate on the four-stroke-cycle principle, all cylinders complete a cycle in two crankshaft revolutions. Therefore, the camshaft must make one complete revolution per cycle or travel at one-half crankshaft speed.

Camshafts, in most cases, are driven by plain or spiral spur gears (Fig. 291). To insure the valves being properly timed, certain teeth are marked. Figure 471 shows a silent chain camshaft drive that is used extensively in automobile engines.

62. Valve Location.—Both the valve-in-head arrangement (Fig. 74) and the L-head arrangement (Fig. 73) are found in tractor engines. The former is used in all the horizontal-type engines and in a majority of

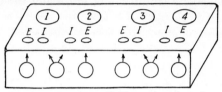

FIG. 67.—Valve arrangement and manifold openings for a four-cylinder engine.

the vertical engines. With the valves located in the cylinder head, the mechanism is more accessible both from the standpoint of valve removal for grinding and in adjusting for wear. Another advantage, especially

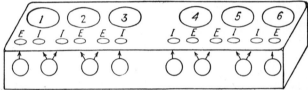

FIG. 68.—Valve arrangement and manifold openings for a six-cylinder engine.

important in tractors, is that uniform and satisfactory cooling of the valves is greatly facilitated.

The L-head arrangement requires less operating mechanism; but, with the valves and valve openings in the cylinder block, should a valve seat become chipped or damaged, an entirely new block might be necessary.

Figure 67 shows the arrangement of the intake and exhaust valves with respect to each other and to the different cylinders for a four-cylinder engine. Figure 68 is for a six-cylinder engine. It will be noted that the end valves in each case are exhausts. It is desirable to keep the intake valves as near to the center of the block as possible in order to permit the use of shorter intake manifolds.

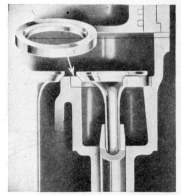

FIG. 69.—Valve-seat insert.

63. Valve-seat Inserts.—The exhaust valve openings of some tractor engines are equipped with seat inserts (Fig. 69). The inserts are made of a steel alloy that is harder than the cylinder block or head material. Therefore, there is less danger of damage or excessive and rapid valve-seat wear owing to the high temperatures to which the exhaust valve and its seat are ordinarily subjected.   Furthermore, in case of such wear and damage, the renewal of the seat is simple and inexpensive.   The inserts are pressed in place but are easily removed.   Since they are made of hardened steel, a special grinding tool is usually needed to properly fit and seat the valve.

64. Valve Operation and Adjustment.—For detailed information relative to tractor-valve operation, timing, and adjustment, the reader is referred to Chap. VII.

References

Duell: "The Motor Vehicle Manual."
"Dyke's Automobile and Gasoline Engine Encyclopedia."
Elliott and Consoliver: "The Gasoline Automobile."
Favary: "Motor Vehicle Engineering—Engines."
Heldt: "Automotive Engines."
Heldt: "The Gasoline Automobile," Vol. I.
Streeter and Lichty: "Internal-combustion Engines."

CHAPTER VII

VALVES AND VALVE OPERATION

As stated in the preceding discussion, a single-cylinder engine has at least two ports or two valves, depending upon whether it is of the two- or the four-stroke-cycle type, respectively. The intake port or valve allows the fuel mixture to enter the combustion chamber on the intake stroke. The exhaust port or valve allows the products of combustion to escape following the explosion and expansion. As previously mentioned, some two-stroke-cycle engines have three ports, the extra port serving as a passage for the fuel from the fuel tank and carburetor to the crankcase.

Since the ports in a two-stroke-cycle engine are nothing more than openings in the cylinder wall that are opened and closed by the piston in its movement, no special valve-operating mechanism is required, and the problem of valve timing is not present. The following information, therefore, applies to the four-stroke type of engine entirely.

65. Types of Valves.—Most valves are of the poppet or mushroom type (Fig. 71); that is, the valve itself consists of a flat head or disk with a beveled edge, called the face, and the stem. The cheaper types of valves are made in two pieces, whereas the better grades are made in one piece. In the two-piece valve, the head or disk is made of cast iron and has a steel stem welded to it. The one-piece type usually has a head of nickel steel, tungsten steel, or other steel alloy, is lighter and more heat resistant, and less brittle. The exhaust and intake valves are usually alike in size and are interchangeable. A spring serves to hold the valve firmly to its seat during the compression and expansion strokes of the piston so that there will be no possible loss of compression around the valve. Continued use causes the valve to wear, however, and grinding is necessary to restore this leakproof condition between the valve face and its seat.

Another type of valve used on some engines is known as the sleeve valve. This type offers two advantages over the poppet valve. First, very little leakage develops so that grinding or replacement is unnecessary; second, it is practically noiseless. The mechanism, however, is somewhat more complicated and the cost of manufacture is greater. Figure 70 illustrates the usual sleeve-valve construction and operation. As noted, the mechanism consists of two sleeves placed between the piston and the cylinder wall. The sleeves have an up-and-down movement produced by the eccentric shaft and the short connecting rods attached to them. At the desired time, openings in the sleeves register with the intake or exhaust manifolds, as the case may be, and thus allow the fuel mixture to enter and the burned charge to escape. In order to reduce the wear and friction losses to a minimum, thorough lubrication is provided by grooves and small holes in the sleeves. Sleeve-valve engines consume more lubricating oil than other types.

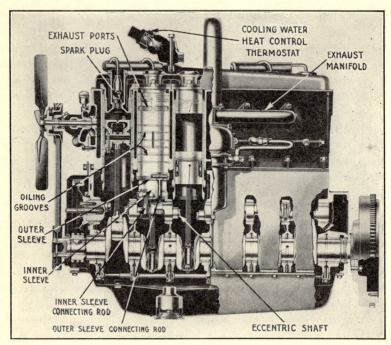

Fig. 70.—Sleeve-valve construction and operation.

66. Valve Operation—Automatic Intake Valves.—Both the intake and exhaust valves on most types of engines are operated mechanically, but on many single-cylinder farm engines the intake valve is said to operate automatically. When thus operated, it is equipped with a very light spring. On the suction stroke of the piston, the valve is drawn open by the suction or decreased pressure in the cylinder, and the fuel charge is drawn in through the open valve. Near the end of the stroke, as the suction decreases, the light spring causes the valve to return to its seat. This method of intake-valve operation is very simple, but unless the valve is in perfect condition it may not seat properly, creating loss of compression. It is also noisy and not well adapted to high-speed engines, for the reason that there is no control of the time of opening and closing of the valve. Therefore the engine may not receive a good, full charge of fuel mixture.

67. Mechanically Operated Valves.—In many single-cylinder engines and in all multiple-cylinder types the intake as well as the exhaust valves are operated mechanically. The mechanism (Fig. 71) consists usually

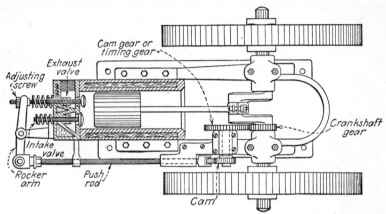

Fig. 71.—Valve-operating mechanism for a single-cylinder farm engine.

of the following parts: (1) a small gear on the crankshaft, (2) a half-time or cam gear, (3) a cam, or a camshaft bearing a number of cams in the case of multiple-cylinder engines, (4) a push rod, and (5) a rocker arm. As the cam gear rotates the cam, the high point on the latter strikes the push rod, and it, in turn, operates the rocker arm that opens the valve. The usual shape of the cam is illustrated in Fig. 72*A*. On certain farm types of engines using what is known as the make-and-break system of ignition, the cam will have the shape as illustrated in Fig. 72*B*. The high point *C* moves the push rod enough to trip the sparking mechanism; then the second high point *D* moves the push rod still farther and opens the exhaust valve.

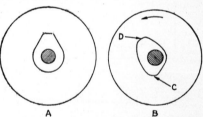

Fig. 72.—Valve cam shapes. *A*. Plain cam. *B*. Combined exhaust valve and ignitor-trip cam.

Single-cylinder engines having an automatic-type intake valve require but one cam, which is usually cast with the cam gear so that a camshaft is unnecessary. Engines having two or more valves, all operated mechanically, have the cams placed on a camshaft (Fig. 73) and distributed about it in such a manner as to open the valves in the correct order.

68. Location of Valves.—There are three types of valve arrangement: (1) valve-in-head (Fig. 74), (2) L-head (Fig. 75), and (3) T-head (Fig. 76).

The valve-in-head arrangement is found in many types of engines and is used almost entirely in single-cylinder, farm-type as well as in heavy-

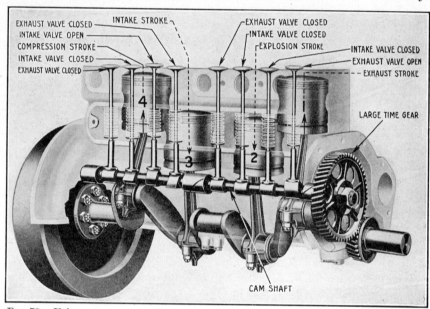

Fig. 73.—Valve arrangement and operation for a typical multiple-cylinder, L-head engine.
(*Courtesy of Ford Motor Company.*)

duty multiple-cylinder engines. Some advantages given are: (1) the valves are removable with the cylinder head and, therefore, are accessible

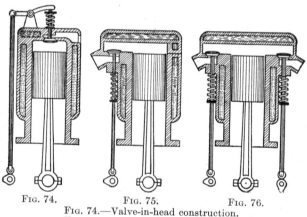

Fig. 74.   Fig. 75.   Fig. 76.
Fig. 74.—Valve-in-head construction.
Fig. 75.—L-head construction.
Fig. 76.—T-head construction.

and easily ground and replaced; (2) in case of a damaged valve seat, it is necessary to purchase only the cylinder head. In other types it might be necessary to replace the complete cylinder or cylinder block; and (3) the entire fuel charge is directly over the piston and the latter receives the direct effect of the explosion.

The principal disadvantages of the valve-in-head arrangement are: (1) a more complicated valve-operating mechanism, (2) more places for wear, and (3) more frequent adjustment and more noise.

The L-head arrangement is found in many multiple-cylinder engines, particularly in automobiles. The valve-operating mechanism is simple, as a rocker arm is unnecessary; that is, the cam and push rod act directly against the valve stem.

In the T-head arrangement, the intake valves are placed on one side of the cylinder block and the exhaust valves on the other, and two cam-shafts are necessary. This arrangement is practically out of use.

69. Speed of Cam Gear and Camshaft.—It has already been stated that in any four-stroke-cycle engine, there are four strokes of the piston per cycle, the first stroke being known as the *intake stroke,* and the last stroke of the cycle as the *exhaust stroke.* Therefore, the intake valve must open and close on the intake stroke, and the exhaust valve must open and close on the exhaust stroke; that is, each valve must operate one time per cycle. Consequently, the cam gear and cams that open and close these valves must make one revolution per cycle. The cam gear, however, is driven by a gear on the crankshaft that makes two revolutions per cycle. Therefore, the speed of the cam gear will be one-half that of the crankshaft, and it will have twice the number of teeth found on the crankshaft gear. For this reason, the cam gear is often referred to as a half-time gear. Likewise, in multiple-cylinder engines having more than two valves, the speed of the cam gear will be one-half the speed of the crankshaft, because, as previously explained, all cylinders complete a cycle during the same two revolutions of the crankshaft. Therefore, all valves must open and close once during any two revolutions, and the camshaft, in order to open and close all these valves one time, must make one revolution to two of the crankshaft.

70. Valve Timing.—It would be assumed, ordinarily, that a valve would start to open at the beginning of its stroke and be completely closed at the end. Such, however, is not necessarily true. It has been found that most engines operate efficiently only when the valves open and close at certain points in the cycle and remain open a certain length of time. Therefore, the timing of the valves in any engine is very important.

Figure 77, commonly known as a valve-timing spiral or diagram, illustrates the timing of the average engine. The horizontal line might be called the dead-center line with crank dead center (C.D.C.) at the right and head dead center (H.D.C.) at the left. The crankshaft and piston may be said to be on dead center, when the centers of the piston pin, crankshaft, and crankpin all fall in this line.   The spiral represents the travel of the crankpin or connecting-rod journal.   Starting at H.D.C., the crank rotates in the direction of the arrow toward C.D.C. moving the piston through the intake or suction stroke, followed by compression from C.D.C. to H.D.C., expansion from H.D.C. to C.D.C., and exhaust from C.D.C. to H.D.C., thus completing the cycle.

Referring to the figure, it is noted that the intake valve should begin to open when the crankshaft has rotated about 10 deg. past H.D.C., and that it remains open throughout the stroke and does not close until the

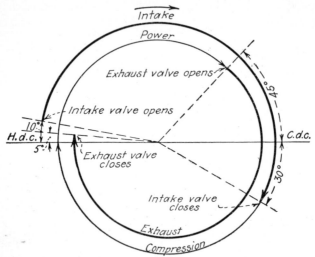

Fig. 77.—Diagram showing usual timing of valves.

crankshaft has reached a point about 30 deg. after C.D.C.   The exhaust valve begins to open before the piston reaches the end of the power stroke, that is, when the crankshaft is making an angle of about 45 deg. before C.D.C. with the piston on the power stroke.   It remains open through the exhaust stroke and closes when the crankshaft has rotated about 5 deg. after the piston has reached H.D.C.   The figures given are not correct for all engines but are very near the average.

71. Range of Operation.—The number of degrees through which the crank rotates from the point where the valve starts to open, to the point where it is just completely closed, is known as the *range of operation* of that valve.   The average range of operation for an intake valve, according to the figure, is 200 deg.   For the exhaust valve it is 230 deg.

An adjusting screw, in one end of the rocker arm (Fig. 71), permits adjusting the range of operation whenever necessary.   This screw when turned toward the valve stem permits the rocker arm to strike it sooner and therefore open the valve earlier. The valve also closes later. If this screw is turned away from the valve, the latter will open later and close sooner, thus decreasing the range of operation.

Even though the range of operation is correct, the valve may open too early, and therefore close too early by the same amount; or it may open too late, and therefore close too late by the same amount. This is due to the fact that the cam gear is out of time; that is, the cam is not coming around at the correct time to open and close the valve at the proper time with respect to the travel of the piston. To time the valve, the cam gear must be removed and rotated the correct amount in the right direction and remeshed with the gear on the crankshaft. This shifting of the gear will not change the range of operation of the valve.

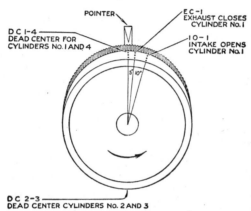

FIG. 78.—Flywheel timing marks.

The usual procedure, then, in checking up the valve timing of the ordinary, single-cylinder, farm gas engine is: First, check the range of operation and adjust it correctly by means of the screw on the rocker arm. Second, check the opening and closing of the valve and adjust this if necessary by shifting the cam gear and cam.

As previously stated, the intake valve on this type of engine is ordinarily operated automatically, so that the problem of timing is confined to the exhaust valve only.

Most manufacturers place marks on their engines (Fig. 78), by which the correct valve and ignition timing of the engine may be determined and checked. However, the system of marking used is not standard; that is, every manufacturer has a somewhat different system. Consequently, the instruction book supplied with an engine should be studied carefully so that the method used will be well understood.

72. Effects of Incorrect Intake-valve Timing.—As previously discussed, the intake valve opens at the beginning of the intake stroke, and remains open until the crankshaft has rotated considerably past crank dead center. At first thought, it would seem more reasonable to close the intake valve at the end of the intake stroke, but experience has proved that allowing this valve to remain open, after the piston has started back on the compression stroke, enables the engine to generate more power and operate more efficiently. The explanation is that the charge rushing into the cylinder has not completely filled the space when the piston reaches the end of the intake stroke; but if the valve is allowed to remain open somewhat longer, the fuel mixture, due to its inertia, will continue to pour in, and thus a more complete charge will enter the cylinder and greater power is generated. There is some variation in the closing of this valve in different types of engines, but, in general, the point is determined by the speed of the engine; that is, a high-speed engine would have the intake closing possibly as much as 45 to 50 deg. after C.D.C.

73. Effects of Incorrect Exhaust-valve Timing.—In practically all types of engines, regardless of size, number of cylinders, speed, or other factors, the exhaust valve opens before the end of the power stroke of the piston; that is, when the crankshaft lacks about 45 deg. of having completed the stroke. Experience has proved that this early opening gives more satisfactory results in that the engine generates more power, runs cooler, and uses less fuel than when the valve is opened later, or at a point near the end of the stroke. It would seem that, in opening the exhaust valve before the completion of the expansion stroke, this power would be lost or wasted. But what actually happens is that this unspent pressure is utilized to better advantage to force the exhaust gases out, thus producing better cleaning of the cylinder.

Overheating of the engine is one of the most pronounced effects of too late exhaust-valve opening. This overheating develops owing to the fact that the hot gases are allowed to remain in the cylinder longer and to follow the piston through the entire stroke, thus coming in contact with a greater cylinder area. The closing of the exhaust valve occurs shortly after the completion of the exhaust stroke. If the valve were closed before or even at the end of the stroke, a small amount of burned residue might remain in the cylinder.

74. Importance of Correct Valve Timing.—The correct timing of the valves in any internal-combustion engine is very important and essential to the proper operation of the engine, if efficient fuel consumption, maximum power, and smooth running are desired. In the design of an engine, the manufacturer determines accurately the correct timing for his engine and sees that the valves open and close so as to give the best possible results. In many engines, certain teeth on the cam gear and crankshaft gear are marked or center-punched so that, when removed and replaced (if the marks are placed together), the valves will be in time. Marking of the timing gears in multiple-cylinder engines is important and almost universally practiced. However, they may not be marked in farm types of engines. Therefore, before dismantling such an engine, some marks should be placed on these gears if they are not already there.

75. Tram Method of Valve Timing.—The following method of checking and timing the exhaust valve of farm engines is often used:

First, determine the two dead centers and mark the points on the flywheel using a tram (Fig. 79) of any convenient length, say 15 to 30 in., depending upon the size of the engine. To do this, turn the engine over until the edge of the piston is exactly flush with the edge of the cylinder wall (at open end), with the crank just above C.D.C. With center punch or chalk establish a base mark on some stationary part of the engine, such as top of water hopper or cylinder. Place one point of the tram on this mark and mark the point where the other end of tram touches the flywheel rim. Turn the engine over until the piston is again flush with the edge of the cylinder and

Fig. 79.—A tram for timing farm-engine valves.

crank is just below C.D.C. Again place one point of the tram on base mark, and mark point where other end touches the rim of the flywheel. Measure distance between the two marks on the flywheel and then mark a point on the flywheel exactly halfway between them. With one tram point on the base mark, turn the engine until the other end of tram touches this halfway mark. The engine is now on C.D.C. To determine H.D.C. measure exactly halfway around the flywheel from the C.D.C. point, and mark on the flywheel. Then, with one end of the tram on the base mark, place the other end on this mark on the flywheel and the engine will be on H.D.C.

Second, measure the circumference of the flywheel in inches and divide 360 by this number of inches to obtain the number of degrees per inch of flywheel circumference.

Third, determine the range of operation of valve as follows: Rotate engine in the correct direction until valve just starts to open; that is, when the rocker arm is just ready to push the valve open, place the tram on base mark and mark flywheel where the other point touches it. Rotate engine in direction of operation until valve just closes. Again place tram on base mark and mark point on flywheel. Measure the distance between these two points in inches and convert to degrees. Care should be taken in measuring between the two points to measure around the flywheel in the correct direction. If it is found that this range of operation is 225 to 230 deg., it is correct and no adjustment is necessary. On the other hand, if the range of operation is incorrect, it must be adjusted by means of the screw on the rocker arm, as previously described. In making this adjustment, care must be taken to turn the screw in the correct direction, depending upon whether it is desired to increase or to decrease the range of operation. Also after each adjustment of the screw, the exact opening and closing of the valve and, therefore, its range of operation must be remeasured as previously described until the correct range in degrees has been obtained.

Fourth, if the range of operation is now correct, the valves may still be out of time; that is, they may not open and close at the proper time in the crankshaft rotation and travel of the piston. Ordinarily, it will be found that if the valve opens a certain number of degrees too early or too late, it will likewise close too early or too late by the same number of degrees. Therefore, to place it in time, the cam gear is removed and rotated or shifted the correct amount, and remeshed with the gear on the crankshaft. It may be necessary to repeat this several times before the correct timing is obtained.

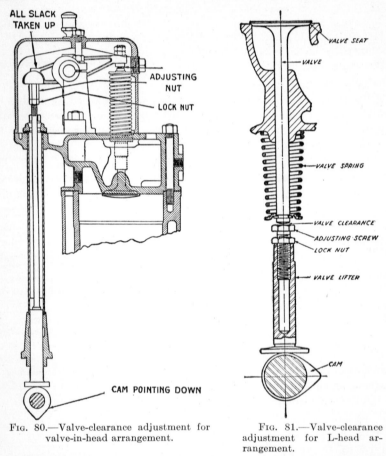

FIG. 80.—Valve-clearance adjustment for valve-in-head arrangement.

FIG. 81.—Valve-clearance adjustment for L-head arrangement.

Fifth, the ignition timing may now be checked and the timing of the spark properly adjusted if such is necessary. The subject of ignition timing is discussed in detail in Chap. XVIII. In most cases the spark timing cannot be checked and adjusted until the valves are properly timed, due to the fact that the cam gear not only operates the valve, but also operates the spark-producing mechanism. If the valves are first correctly timed and adjusted, the spark will be found to be nearly in time. If not quite correct, and if any adjustment is necessary, it can be made by some simple means on the sparking mechanism.

76. Tractor-engine Valve Timing.—The exact time of opening and closing of the intake and exhaust valves of most tractor engines, with respect to the crankshaft rotation and piston travel, conforms very closely to the following figures:

> Intake opens 0 to 10 deg. after H.D.C.
> Intake closes 30 to 50 deg. after C.D.C.
> Exhaust opens 35 to 50 deg. before C.D.C.
> Exhaust closes 5 to 10 deg. after H.D.C.

The size and type of engine, number of cylinders, speed, and other characteristics have no particular effect on this timing.

77. Multiple-cylinder Valve Timing—Valve Clearance.—The timing of the valves in multiple-cylinder engines is a comparatively simple matter, consisting first, of the meshing of the timing or cam gears so that the marked teeth are together, and, second, providing the correct valve clearance. Valve clearance is defined as the space allowed between the end of the valve stem and the rocker arm (Fig. 80), or in the case of the L-head type, between the valve stem and the push rod (Fig. 81). This clearance varies from 0.005 to 0.025 in., depending upon the size and type of engine. The valve-in-head construction requires more mechanism than the L-head. Therefore, slightly more wear is possible, and more frequent adjustment is usually necessary. Figure 80 shows the usual mechanism involved and the method of adjustment. Figure 81 shows the L-head arrangement and the usual provisions for adjusting valve clearance.

A certain amount of valve clearance is necessary to allow for the expansion of the valve stem and engine parts when they get hot, and thus permit the valves to seat properly. At the same time, too much clearance affects the timing of the valves and the smooth operation of the engine and also produces noise. In some cases the adjustment is made when the engine is cold while in others the engine should be warmed up or at its normal running temperature.

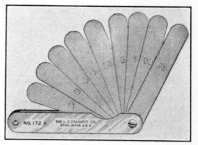

Fig. 82.—Thickness gage.

The importance of correct valve-clearance adjustment cannot be overstressed because, if neglected, more or less faulty operation may result. The following rules for adjusting valve clearance apply to either the L-head or to the valve-in-head type and should be observed carefully:

1. Read the instruction book supplied with the tractor and see whether the manufacturer advises making the adjustment when the engine is hot or when cold.

2. When making the adjustment always be sure that the push rod and cam follower are resting on the low part of the cam (see Fig. 80).

3. Some manufacturers mark the flywheel as shown by Fig. 78. If this is done, check the valve adjustment and timing by referring to these flywheel marks. The instruction book will explain just how to do this.

4. The correct clearance depends largely upon the size of the engine. Small and medium-size engines require about 0.010 to 0.015 in., which is about $\frac{1}{64}$ in. Larger engines may require as much as 0.020 to 0.030 in. In some cases, it is recommended that the exhaust valves be given slightly more clearance than the intakes because they get hotter.

5. Valve clearance should always be adjusted whenever the valves are ground. In valve-in-head engines always readjust the clearance whenever the cylinder head bolts are tightened.

6. For accurate adjustment use a standard thickness gage (Fig. 82).

7. Always see that lock nuts are tightened well.

References

Cornell Univ., N. Y. State Coll. Agr., Ext. Bull. 85.
Cornell Univ., N. Y. State Coll. Agr., Junior Ext. Bull. 40.
DUELL: "The Motor Vehicle Manual."
"Dyke's Automobile and Gasoline Engine Encyclopedia."
ELLIOTT and CONSOLIVER: "The Gasoline Automobile."
FAVARY: "Motor Vehicle Engineering—Engines."
HELDT: "Automotive Engines."
HELDT: "The Gasoline Automobile," Vol. I.
STREETER and LICHTY: "Internal Combustion Engines."

CHAPTER VIII

FUELS AND COMBUSTION

One disadvantage of the internal-combustion engine is that the fuels required for its successful operation must be of a specific nature and, therefore, are somewhat limited with respect to source and supply. Under certain conditions these fuels may be difficult to secure, and their use becomes economically prohibitive. Some of the essential characteristics of such fuels are:

1. They must have a reasonably high energy value.
2. They must vaporize at least partially at comparatively low temperatures.
3. The fuel vapors must ignite and burn readily when mixed in the proper proportions with oxygen.
4. Such fuels and their combustion products should not be unduly harmful or dangerous to human health or life.
5. They must be of such a nature that they can be handled and transported with comparative ease and safety.

78. Classification of Fuels—Gaseous Fuels.—Fuels for internal-combustion engines may be classified, first, as either gaseous or liquid according to the physical state before going into the engine cylinder; and, second, as artificial or natural, according to whether the fuel is obtained from natural sources or is a manufactured product.

The only gaseous fuel that is obtained from nature is natural gas, obtained from the earth either directly from so-called gas wells, or indirectly as the result of the distillation of petroleum. Natural gas is a product having the same general chemical composition as petroleum and its many other products. Its use as an engine fuel is confined largely to those sections of the country in which natural gas is produced.

In certain industries, particularly in the manufacture of iron and steel, certain so-called by-product gases are produced. The most common of these are blast-furnace gas, coke-oven gas, and producer gas. These artificial fuels are likewise limited in use to the production of power in or near these plants and are not available for general use. Illuminating gas, manufactured by the destructive distillation of coal and sometimes called coal gas, may also be used as a fuel in internal-combustion engines but is considered expensive and impractical if other more common fuels are available.

79. Liquid Fuels.—Liquid fuels have the advantage over gaseous materials, of being more concentrated and readily portable. That is, if natural gas, for example, were used as a fuel in an automobile, it would be necessary to place it under high pressure in a strong airtight tank.

Liquid fuels are obtained almost entirely from petroleum, a product of nature. The only artificial liquid fuel of any consequence for internal-combustion engines is alcohol. It has never been used extensively, however, and will be discussed later. The more common liquid fuels are gasoline, kerosene, fuel oil, and crude oil. Gasoline and kerosene are lighter fuels obtained from petroleum, largely by distillation. Distillates are heavier products resulting from the distillation and refining of crude oil. Some crude oils may likewise be used without being subjected to any special treatment, and engines burning distillates or crude oil are known as oil engines.

Crude oil and the products obtained from it are known as hydro-carbons, that is, they are products made up almost entirely of carbon and hydrogen, having such general chemical formulas as C_nH_{2n}, C_nH_{2n+2}, etc. A simple hydrocarbon, therefore, would be CH_2; another, CH_4; a third, C_2H_6, and so on. A careful analysis of crude oils and the common fuels obtained from them indicates that they contain about 84 per cent carbon and 15 per cent hydrogen by weight, with the remaining 1 per cent made up of nitrogen, oxygen, and sulphur.

Table VII gives some of the characteristics of the common liquid fuels. It should be noted that there is very little difference in the heating value of crude oil and its products, even though there is a marked variation in the weight of a given volume of each. As shown, alcohol has about the same gravity as kerosene but a much lower heating value.

TABLE VII.—CHARACTERISTICS OF LIQUID FUELS

Fuel	A.P.I.[1] test, degrees	Specific gravity	Weight per gallon, lb.	Initial boiling point, °F.	End point, °F.	Heating value, B.t.u.[2] per pound	
						Low	High
Gasoline.................	65 to 56	0.720 to 0.755	6.00 to 6.30	95 to 105	300 to 435	19,000	21,000
Kerosene................	45 to 40	0.802 to 0.825	6.70 to 6.90	340 to 360	500 to 550	18,000	20,000
Distillate:							
No. 1.................	50 to 43	0.780 to 0.810	6.50 to 6.80	200 to 375	460 to 540	18,000	20,000
No. 2.................	42 to 35	0.816 to 0.850	6.80 to 7.10	340 to 380	475 to 550	18,000	20,000
Special tractor fuel.......	50 to 38	0.780 to 0.835	6.50 to 7.00	200 to 350	440 to 520	18,000	20,000
Diesel tractor fuel........	38 to 27	0.835 to 0.893	6.90 to 7.40	380 to 460	575 to 725	18,000	20,000
Crude oil...............	57 to 10	0.750 to 1.000	6.25 to 8.00	....	....	18,000	20,000
Denatured alcohol........	45 to 40	0.802 to 0.825	6.70 to 6.90	150	....	10,000	13,000

[1] A.P.I. (American Petroleum Institute, see Par. 90).
[2] B.t.u. (British Thermal Unit) see Par. 97.

80. Crude Oil.—Crude oil is known as a mineral product and is obtained from natural deposits existing usually at depths varying from several hundred to thousands of feet below the surface of the earth. These oil deposits are found in certain sections of the United States and in other parts of the world. There is considerable variation in the character of crude oil obtained in these different sections of the country, but they are generally placed in one or the other of three classes, namely: asphalt-base crudes, paraffin-base crudes, and mixed-base crudes. An asphalt-base crude is usually a heavy dark liquid and, when distilled, produces an end product of a black, tarry nature. Paraffin-base crudes are lighter in weight and color, and upon distillation produce a residue that is light in color and resembles paraffin. The ordinary paraffin wax is obtained from such crudes. A few crude oils have been found that contain both asphaltic and paraffin materials and are therefore known as mixed-base crudes.

The first crude oil of any consequence was discovered in Pennsylvania in 1859 and came from a well having a depth of about 70 ft. This state continued to be the only producer for a number of years, but eventually oil was discovered in Ohio, Indiana, and Illinois, and later, and in more recent years, in California, Texas, Oklahoma, and Kansas. Crude oils, as obtained from these various sections of the country, differ considerably from the standpoint of the final products of distillation and vary greatly in the amount of the lighter products of distillation that they yield, such as gasoline and kerosene. The so-called light crudes, that is, those which are comparatively thin and lighter in weight per unit volume, produce a higher percentage of gasoline and kerosene and a lower percentage of lubricating oil and heavy products, whereas the heavier crude oils contain less gasoline and kerosene and produce a greater quantity of lubricating oil and heavy products. It was thought originally that the gasoline from certain crude oils was superior to that obtained from others, but with the modern refining processes the fuel products produced from these different crudes do not vary greatly in character or quality.

81. Petroleum Refining.—The production of crude oil and its conversion into hundreds of commercial products by the various distillation and refining processes have become one of the greatest industries of the age. As a result of economic and other factors, changes and improvements in refining methods are being constantly introduced so that the manufacture of the numerous crude-oil products has developed into a highly complex industry.

Likewise, the processes involved are not greatly standardized, and it is difficult to convey to the average layman, by any simple or limited explanation, just how the various products are made. A study of a typical flow chart (Fig. 83) shows that the first stage in the refining

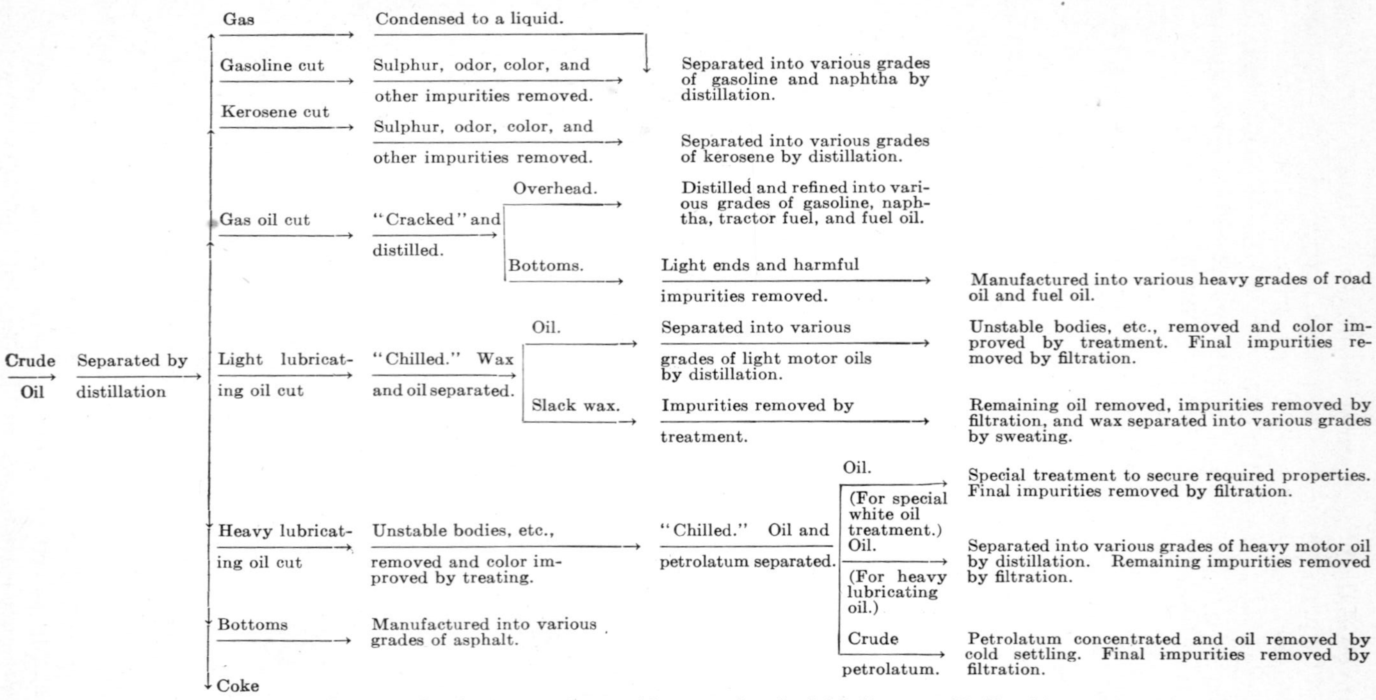

Crude Oil → Separated by distillation →

- **Gas** → Condensed to a liquid.
- **Gasoline cut** → Sulphur, odor, color, and other impurities removed. → Separated into various grades of gasoline and naphtha by distillation.
- **Kerosene cut** → Sulphur, odor, color, and other impurities removed. → Separated into various grades of kerosene by distillation.
- **Gas oil cut** → "Cracked" and distilled. →
  - Overhead. → Distilled and refined into various grades of gasoline, naphtha, tractor fuel, and fuel oil.
  - Bottoms. → Manufactured into various heavy grades of road oil and fuel oil.
- **Light lubricating oil cut** → "Chilled." Wax and oil separated. →
  - Oil. → Light ends and harmful impurities removed. → Separated into various grades of light motor oils by distillation. → Unstable bodies, etc., removed and color improved by treatment. Final impurities removed by filtration.
  - Slack wax. → Remaining oil removed, impurities removed by filtration, and wax separated into various grades by sweating.
- **Heavy lubricating oil cut** → Unstable bodies, etc., removed and color improved by treating. → Impurities removed by treatment. → Oil. → "Chilled." Oil and petrolatum separated. →
  - (For special white oil treatment.) Oil. → Special treatment to secure required properties. Final impurities removed by filtration.
  - (For heavy lubricating oil.) Crude. → Separated into various grades of heavy motor oil by distillation. Remaining impurities removed by filtration.
  - petrolatum. → Petrolatum concentrated and oil removed by cold settling. Final impurities removed by filtration.
- **Bottoms** → Manufactured into various grades of asphalt.
- **Coke**

Fig. 83.—Crude-oil flow chart. (*Courtesy Standard Oil Company [Indiana].*)

process involves the breakdown of the crude, by straight distillation, into about eight groups, the lightest of which is entirely gaseous in nature, the heaviest being a semisolid or solid such as asphalt and coke.

The second stage consists of converting the fractions into their respective end products by means of distillation, chemical action, and special

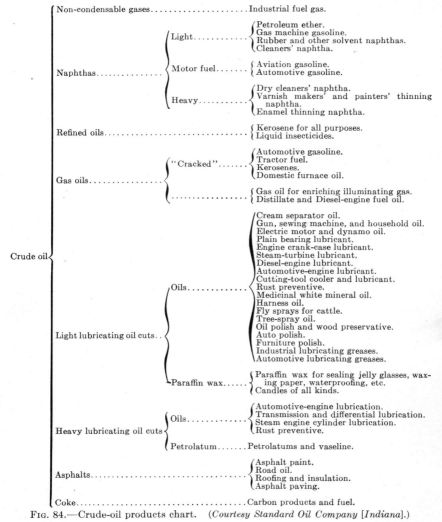

Fig. 84.—Crude-oil products chart. (*Courtesy Standard Oil Company* [*Indiana*].)

treatment. Referring to Fig. 83, it should be noted that, of the eight groups into which the crude is separated, three eventually produce gasoline. The first, or gaseous fraction, is condensed to form a light, highly volatile gasoline. The second, or "gasoline cut," after being treated to remove certain undesirable impurities and redistilling, becomes straight-run gasoline. The fourth, or "gas oil cut," is converted into gasoline by "cracking" as explained later.

The lighter lubricants for various purposes are secured from the "light lubricating oil cut," which follows the "gas oil cut." The heavier automotive lubricating oils for both engine and transmission lubrication are derived from another still heavier fraction known as *heavy lubricating oil cut.*

The products chart (Fig. 84) shows in more detail the large number and variety of products obtainable from crude petroleum and the particular "cut" yielding the different products.

GASOLINE

82. Gasoline Manufacture.—Gasoline originally was a product obtained only by the distillation of the crude oil. It was formerly somewhat lighter and more volatile than the average gasoline now found on the market. As indicated in Fig. 83, it is now produced in three distinct ways and is classified as straight-run, cracked, or casing-head, depending upon the manner of production. So-called straight-run gasoline is produced by the ordinary distillation of crude oil. The crude is heated to a given temperature for a certain length of time and the resulting product that distills over is gathered in a condenser. This distillation product is not ready for use, however, but is known by the refiner as crude benzene or crude naphtha. Before being placed on the market, it must be subjected to a refining process, which consists of heating it with chemicals and redistilling it to separate out any heavy materials that may have passed over during the first distillation and to remove any impurities.

Cracked gasoline is manufactured from heavier distillation fractions, particularly kerosene and gas oil, and, as the term implies, consists of breaking up the so-called heavy molecules into lighter ones that make up gasoline. The process consists essentially of placing these heavier distillation products in receptacles and heating them under pressure. This results in certain lighter products that pass off, are condensed and further treated and refined, and eventually resemble straight-run gasoline.

Casing-head gasoline is manufactured from the gas that issues from oil wells or is obtained in the distillation of the crude oil. The process consists of compressing and liquefying the gases and then distilling these liquids under pressure, thus producing a very light and volatile product. This gasoline makes up only a very small percentage of the total amount manufactured. Ordinarily it is not used in its original condition, owing to its great volatility, but it is mixed with heavier, lower quality gasoline to form what is known as a blended product. Much of the gasoline now found on the market is known as blended gasoline and consists in many cases of all three types: straight run, casing head, and cracked, as mixed together in the refining process.

83. Impurities in Gasoline.—Wilson[1] gives the following information concerning gasoline impurities:

The two chief impurities in gasoline that may cause trouble are sulphur and gum. While small traces of sulphur seldom cause noticeable damage, and even a large amount may not result in serious difficulties in warm weather or under conditions where there is no moisture present, where appreciable condensation occurs so that sulphuric acid is formed, a small amount of sulphur may raise havoc. In exceptional cases such as with the present mufflers in which the number of baffle plates has been increased in an effort to eliminate exhaust noises, gasoline of only 0.015 per cent sulphur has resulted in serious corrosion of the muffler in a single winter. The following effects of sulphur contained in gasoline have been determined by the U. S. Bureau of Mines:

1. Excessive sulphur in gasoline does cause serious corrosion of moving parts within the crankcase, especially in cold weather and when frequent stops are made.

2. There is no evidence that such corrosion occurs in warm weather or to any great extent on cars or parts that are kept continuously running during the day and in heated garages at night.

3. There is absolutely no evidence to indicate that anything other than sulphur in gasoline is to blame. (Sulphur combined in the lubricant is not to blame except in so far as it may actually be burned in the cylinder.)

4. The corrosion is undoubtedly caused in the first place by sulphuric acid, the formation of iron sulphide and oxide being secondary. As all forms of organic sulphur are burned to sulphur dioxide which is later changed to sulphuric acid, it is quite evident that the original form of sulphur in the fuel is of no importance.

5. The amount of corrosion is not proportionate to the sulphur content of the fuel.

6. Tests indicate that a sulphur content of 0.040 per cent does no harm, that 0.151 per cent sulphur does appreciable harm, and that 0.458 per cent does very great harm. Just where to draw the line is hard to say. It would seem from the data given that 0.10 per cent would be a fair dead line.

The gum-forming tendencies of certain gasolines have become of increasing interest in late years because of the increased popularity of the cracked gasolines and especially those which have been vaporphase cracked. The objection to gum in gasoline is that it tends to build up objectionable deposits. When the gasoline is sprayed from the carburetor, any dissolved gum present may be deposited on the heated portions of the intake manifold and on the inlet valves. Imperfect engine operation and loss of power may result either from a partial clogging of the manifold or delayed seating of the inlet valves. In severe cases, engine failure may be caused by complete sticking of an inlet valve. No other deleterious effects on the engine have been noted. Only a fraction of the dissolved gum in the gasoline is deposited in the intake system, the percentage of the total depending on the design and operating conditions of the engine. The undeposited gum is carried into the combustion chamber and burned with the charge. Carbon deposits seem to bear no relation to the quantity of gum in the fuel.

[1] *Univ. Wis., Eng. Exp. Sta., Bull.* No. 78.

Gum has been defined as a residue that is left by the evaporation of gasoline, while the residue which separates from the liquid has been called resin.   Resin is especially apt to form in the presence of copper.   Gum exists in gasoline in a dissolved state and may be formed during storage and remain in solution until the gasoline is used, at which time it causes trouble.   Gum formed during storage is referred to as "potential gum."   Hence, a cracked gasoline may be free from gum immediately after refining and yet contain an appreciable quantity of dissolved gum when it is used.   Certain tests are used which are intended to indicate "potential gum," the gum which may be expected to form later.   The results of such tests are intended to indicate the proportional amount of gum that may give trouble after a rather extended period of storage.   Some gum tests will show an exceedingly high gum value for straight-run gasoline, and yet it is ordinarily considered as nil for such gasoline.   The terms "actual gum," "preformed gum," and "residual gum" are generally considered identical in meaning, and they refer to the gum which is already present as dissolved gum in gasoline.

The exact nature of the compounds in gasoline that deposit as resin or gum is not known.   In some cases, it has been found that the same compounds cause the yellow color or strong varnishlike odor in some cracked gasolines.   Cracked gasolines from heavy crudes and fuel oils have a more pronounced color which increases with time.   However, insofar as the operation of a motor car is concerned, it has never been established that either colors or odors are important factors, except as possible indices of the qualities already discussed.   Contrary to popular belief, gasolines may be yellow without containing gum, or may be colorless and yet contain large quantities of gum.

The problem of predicting how long a gasoline may be stored without excessive gum formation is difficult, for not only is the gum stability a function of the particular gasoline under consideration, but it is also dependent upon conditions of storage.   The process of refining has the greatest single effect upon the stability as regards gum formation.   Cracked gasolines from the same gas oil, but made by different processes, vary widely in this respect.   The rate of gumming is increased by aeration, by rise in temperature, by the action of light, and by the presence of deleterious substances.   Cracked gasolines from the lighter crudes have very little gum, while those from heavy asphaltic crudes have a tendency to lose their color and form sediments.   Even in extreme cases, gum does not exceed 1 or 2 per cent.

HEAVY FUEL—SPECIAL FUELS

84. Kerosene.—Kerosene is a heavier product of the distillation of crude oil than gasoline.   Its initial boiling point is about 350°F. and the end point is about 550°F.   It has been used extensively in the past as a stationary-engine and tractor fuel, but, owing to its excessive detonation characteristic, it is being supplanted by distillates and special low-grade tractor fuels having a better antiknock quality and costing even less per gallon.

85. Distillates.—Distillates are crude oil products greatly resembling a low-grade kerosene but usually having a different color and odor. Owing to differences in crude oils, in the refining processes, and in trade demands, distillates produced by different refiners vary greatly in gravity, distillation range, and other qualities. However, the better grades of distillates are usually cracked and partly refined products made from selected crudes in order to secure high antiknock qualities and freedom from gum. Certain low-grade distillates are usually straight-run products with higher boiling and end points than the No. 1 grade.

86. Special Tractor Fuel.—To meet the demand for a satisfactory low-cost fuel of the nature of distillate, some refiners supply a special distillate-type fuel of controlled uniformity for tractor use. These fuels are prepared from special crudes and treated and refined in such a manner that they have a reasonably low initial boiling point and end point, a high antiknock characteristic, and a low gum and sulphur content.

87. Diesel Tractor Fuel.—Fuels for compression-ignition engines resemble a low-grade distillate, particularly with respect to gravity and distillation range. However they are largely straight-run rather than cracked products.

In general, large, slow-speed Diesel engines can utilize satisfactorily a rather wide variety of heavy fuels, whereas the smaller, multiple-cylinder, high-speed Diesel engines require fuels having certain specific qualities. The usual specifications for fuels for Diesel tractor engines and similar high-speed types are:

1. A.P.I. gravity.......................... 27 to 38°
2. Initial boiling point..................... 380 to 460°F.
3. End point............................. 575 to 725°F.
4. Viscosity at 100°F. (Say. Univ.)........... 35 to 42 sec.
5. Sulphur............................... 1.5 per cent (max.)
6. Carbon residue......................... 0.25 per cent (max.)
7. Moisture and sediment.................. 0.05 per cent (max.)
8. Flash point........................... 150°F. (min.)
9. Cetane number........................ 45 (min.)

Two important qualifications of Diesel fuels, particularly those used in high-speed engines are (*a*) freedom from solid matter, sediment and moisture; and (*b*) viscosity. An absolutely clean fuel is highly essential to the satisfactory operation of the injection mechanism, including the injection pumps and nozzles. Even slight traces of fine sediment or moisture may cause trouble.

A fuel having a certain viscosity range is likewise necessary to permit the injection mechanism to handle it properly and, at the same time, to provide some lubrication. A fuel having free-flowing characteristics at all temperatures at which the engine may be operated is also important.

88. Cetane Rating of Diesel Fuels.—The marked increase in the application of high-compression ignition to high-speed engines of the automotive type has disclosed the necessity of utilizing, in these engines, fuels having certain specific physical and chemical qualities that will provide the necessary combustion characteristics when injection occurs.

For the purpose of comparing and designating the ignition qualities of such fuels, the so-called cetane rating method has been developed. The cetane rating of a Diesel fuel is its designated ignition quality as determined by comparing it with a standard reference fuel consisting of a given blend of cetane ($C_{16}H_{34}$) and alpha-methylnaphthalene ($C_{11}H_{10}$). For example a 40 cetane fuel is one having the same ignition qualities as a blend containing 40 per cent cetane and 60 per cent alpha-methyl-naphthalene. The actual rating of a given fuel must be determined by means of a standard test engine and standard conditions.

89. Alcohol and Alcohol-Gasoline Blends.—Two kinds of alcohol might be used as gas-engine fuel, namely, methyl alcohol (CH_4O); and ethyl or grain alcohol (C_2H_6O). Neither of these is available, however, in its pure form, but a third, which is known as denatured alcohol, is manufactured in considerable quantities and would make a satisfactory fuel under certain conditions. Denatured alcohol consists largely of grain alcohol with some wood or methyl alcohol mixed with it, together with pyridine, which gives it a distinct odor and color. Perhaps the principal reason that alcohol is not used as a fuel at the present time is that it offers no advantages over hydrocarbon fuels, and the cost of manufacture is much greater. Other possible objections are: (1) It has only about one-half the heating value of other fuels, hence a greater quantity is required to generate the same amount of power for the same period of time. (2) It does not vaporize so readily as gasoline. (3) It requires a higher compression pressure for the best results.

The blending of alcohol with gasoline was brought to the attention of the American public by overproduction and the low prices prevailing for agricultural products. The use of alcohol as engine fuel and for other purposes would be an important factor in increasing the demand for agricultural products, thereby increasing the price.

Anhydrous or absolute ethyl alcohol will mix with gasoline in all proportions. Commercial alcohol (alcohol of 95 per cent purity) will not mix with gasoline unless it constitutes at least 50 per cent of the mixture. It can be made to blend with gasoline in any proportion by adding blending agents, such as propyl, butyl, and amyl alcohols, benzol, or acetone.

When added to gasoline, alcohol will raise the octane rating of the fuel. Its use with higher compression ratios shows a corresponding improvement in efficiency. A mixture containing 10 per cent absolute alcohol and 90 per cent gasoline has an octane value between that of ordinary gasoline and the standard ethyl gasoline.

Alcohol has a distinct advantage in that it burns with an absence of smoke and without disagreeable odors. It produces no carbon in the engine cylinders and may even be regarded as a carbon remover. It should be run with considerable excess of air to avoid the formation of corrosive products such as aldehyde or acetic acids.

Reported results of engine tests of alcohol-gasoline blends vary because of differences in test conditions. However, they do indicate that the variation in power over a straight gasoline will likely be less than 5 per cent, and it is generally conceded that any increase in power does not offset the increased cost of the fuel.

90. Gravity Method of Testing Fuels.—Heretofore it has been the practice to express the comparative quality of fuels, especially gasolines, in terms of gravity; that is, the owner or operator of an automotive vehicle such as an automobile or a tractor was led to believe that a so-called high-test gasoline was a higher quality product and, therefore, more desirable than a low-test gasoline. In the past few years, however, the demand for gasoline has become so great that it has become necessary for the oil refiner to meet the situation by converting more of the heavier fractions of crude oil into gasoline. Thus the commercial gasoline now being produced is a slightly heavier product than formerly and possibly somewhat less volatile. However, this apparent lowering of the quality of the fuel has had very little effect upon satisfactory engine operation, owing to the fact that the engine manufacturers have promptly taken steps to adapt their products to these fuels. In fact, the present-day automotive engines operate more smoothly and with less trouble on these lower gravity fuels than they did, a few years ago, on fuels that were considered as more desirable in quality.

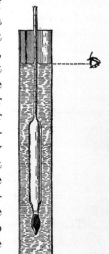

Fig. 85.—Use of hydrometer in determining gravity of fuels.

The gravity of a fuel or an oil may be expressed as *specific gravity* or as A.P.I. (American Petroleum Institute) gravity, formerly known as Baumé gravity. The latter scale is preferred in the United States in connection with petroleum products. The *specific gravity* of a liquid is the ratio of its weight to the weight of an equal volume of water at 60°F.

The relationship of the A.P.I. gravity scale to specific gravity is expressed by the following formula:

$$\text{Degrees A.P.I.} = \frac{141.5}{\text{sp. gr. } 60°/60°F.} - 131.5$$

The device used for testing the gravity of a liquid is known as a hydrometer. Usually two hydrometers are necessary, one for liquids lighter than water, and another for liquids heavier than water. Figure 85 illustrates how the hydrometer is placed in the liquid and read.

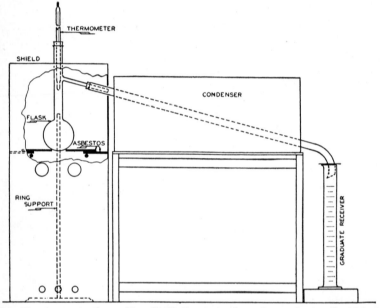

FIG. 86.—Apparatus for making distillation tests of gasoline and other liquid fuels.

Table VII gives the A.P.I. gravities of the more common liquid fuels. The values vary from about 70° for a high-test, very volatile gasoline to 30° for a low-grade distillate. It is thus observed that the lighter the fuel and the lower its specific gravity, the higher the A.P.I. gravity value in degrees. The A.P.I. gravity of pure water is 10°.

91. Distillation Test of Liquid Fuels.—The gravity method of determining the value of liquid fuels is rapidly being displaced by what is known as the distillation test. In making such a test a given quantity (100 cc.) of the fuel is placed in a flask (Fig. 86), heated, and its so-called initial boiling point noted from the thermometer inserted in the top of the flask. Then, as the heating is continued, a certain amount vaporizes, passes off, and is condensed. The temperatures at which certain percentages of the fuel pass off and, finally, the end point is observed; that is, the temperature at which the last drop evaporates. A better idea of such a test may be obtained from the curves (Fig. 87).

92. Detonation—Preignition—Knock Rating.—The tendency in recent years of designers of automotive and tractor-type engines to resort to higher compression pressures, in order to obtain more power and speed and increase efficiency, has resulted in pronounced fuel-knocking effects in many of these engines. This knocking, properly termed detonation, causes an unpleasant, sharp, clicking sound that is most noticeable when the motor is operating at low speed with wide-open throttle. Detonation

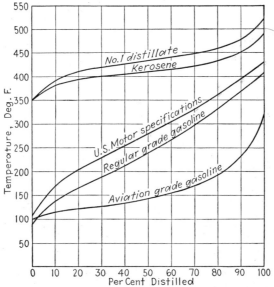

Fig. 87.—Distillation-test curves of common fuels.

is often confused with preignition, but there is a distinction. Preignition occurs when the charge is fired too far ahead of the compression-dead-center position of the piston, owing either to excessive spark advance or to premature, spontaneous ignition resulting from excessive heat in the cylinder. The resulting noise is that of the pistons and bearings as the piston completes its stroke against abnormally high pressure. Ordinarily, retarding the spark will eliminate this type of knock.

Detonation occurs during the process of combustion of the mixture within the cylinder after ignition has taken place. It is often referred to as "knock," "pink," or "ping." According to Wilson:[1]

While there is still much to be learned concerning just what takes place in a combustion chamber during the one-hundredth of a second period of combustion,

[1] *Univ. Wis., Eng. Exp. Sta., Bull.* No. 78.

the theory of detonation has been quite definitely established in the last few years. It is known that a combustible mixture of fuel and air, when ignited by a spark, will burn in a progressive manner, and an average value for the rate of spread of the flame has been determined as approximately 70 feet per second. It takes time for the flame to spread to the charge occupying the most distant portion of the combustion chamber. In the meantime things are happening to this unburned portion of combustible mixture. It is being compressed by the burned charge, which is rapidly increasing in volume as heat is liberated by the combustion process. This compression raises the temperature of the unburned charge. Moreover, this "last-to-burn" portion is being heated by radiation of heat from the flame itself and from any hot spots in its immediate vicinity. Once the detonating temperature (1400 to 1500°F.) is reached, the "last-to-burn" portion goes off with a bang before the orderly progression of the flame front ever reaches it. This sets up violent pressure waves within the combustion chamber. The force of these waves creates hammerlike blows against the interior walls and surfaces of the combustion chamber.

The problem of control of detonation is being handled in a number of ways as follows:

1. Use specially designed cylinder heads and pistons to so shape the combustion space that the last-to-burn portion of the charge will be spread out into a thin sheet and its temperature held down more effectively.
2. Provide more effective water circulation and cooling around the cylinder head and exhaust valves.
3. Use cylinder head and piston materials—particularly aluminum alloys—that provide more rapid heat dissipation.
4. Use a properly designed spark plug and locate it in the "hot region" and preferably near the exhaust valve.
5. Eliminate or reduce carbon deposits. Carbon is an excellent heat insulator and thereby induces detonation by creating ineffective cooling of the combustion space.
6. Maintain the correct mixture of fuel and air and the proper spark setting and valve adjustment.
7. Use specially treated or so-called "doped" or antiknock fuels.
8. Add water as in the case of tractors using heavy fuels having excessive detonating characteristics (see Par. 95).

93. Antiknock Fuels.—Wilson[1] states: The nature of the fuel has an important effect upon the tendency to detonate, and there are two factors which determine the extent of this tendency. They are the temperature at which the air-fuel mixture will autoignite, and the rate at which pressure builds up during combustion. The antiknock value of gasoline is dependent upon several factors. These include the source of the crude oil and the method of cracking and distilling. The gasoline produced from naphthenic base crudes such as those from California and Gulf Coast fields have higher natural antiknock qualities than those from

---

[1] *Univ. Wis. Eng. Exp. Sta., Bull.* No. 78.

the Pennsylvania fields. The cracking process can be controlled to produce gasoline of very high antiknock properties.

Certain chemicals may be added to gasoline to reduce its tendency to detonate. The effect of these so-called "dopes" upon combustion is to raise the autoignition temperature and also to slow down the combustion process. This affords time for the piston movement to provide some increase in volume for the hot gases. There are many substances which will successfully slow down combustion but are unsatisfactory for one reason or another. "Motor benzol" is very effective as far as the elimination of detonation troubles is concerned. It is a mixture of benzol and gasoline in a ratio of about 2 parts benzol to 3 or 4 parts gasoline. The benzol, a fuel in itself, is a coal-tar product recovered as a by-product in the manufacture of gas and coke. It usually contains an appreciable portion of toluene. This lowers the freezing point of the mixture, which is a desirable feature, since pure benzol has too high a freezing point for winter use. Unless carefully purified, benzol is likely to have too large a sulphur content.

Tetraethyl lead compound is the most popular gasoline knock suppressor in use today. Gasoline treated with this compound is marketed under the trade name "Ethyl." Ethylene dibromide is added to prevent the formation of lead oxide which would otherwise deposit on spark plugs, valve seats, and valve stems. The red aniline dye serves only for identification. Tetraethyl lead compound is added to gasoline in the ratio of about 1 part in 1,200 of gasoline. Although they are very effective in slowing the combustion process and preventing detonation, the addition of either ethyl or benzol increases the price of the fuel, which, at the low gasoline prices in the United States, is only partly offset by the increased fuel economy resulting from the use of the higher compression ratios.

94. Antiknock Rating.—The method of measuring and designating the antiknock quality of fuels is described by Wilson[1] as follows:

One of the big problems in the field of detonation studies has been the matter of establishing a means of rating motor fuels. It became necessary to establish, not only the equipment and procedure, but also the scale of measurement. *Octane number* has been established as the standard scale of measurement. Isooctane and normal heptane are two pure hydrocarbon liquids that can be reproduced. The octane has very high antiknock qualities, while the heptane is such a violent knocker that it cannot be used for fuel in engines with a compression ratio greater than 3.75 to 1. Numerically, the octane number is the percentage by volume of isooctane in a mixture of isooctane and normal heptane that matches the antiknock value of the fuel. This furnishes a scale from 0 to 100.

[1] *Univ. Wis. Eng. Exp. Sta., Bull.* No. 78.

Many attempts have been made to find some physical or chemical property of motor fuels that would predict the knocking tendency that the fuel would have when used in a motor. These attempts have been only partially successful, and hence, the standard method of rating consists of using the fuel to be rated in a standard test engine under prescribed

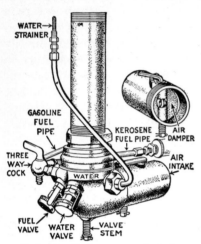

WATER→
STRAINER

GASOLINE
FUEL
PIPE.

KEROSENE   AIR
FUEL PIPE   DAMPER

THREE
WAY→   AIR
COCK   INTAKE
WATER

FUEL
VALVE   WATER   VALVE
         VALVE   STEM

FIG. 88.—Carburetor and fuel system for a single-cylinder, kerosene-burning engine.

procedure. In order to eliminate errors due to changes in atmospheric and engine conditions from day to day, the knock rating is accomplished by bracketing the unknown fuel between two standard blends of known octane rating. Because of the expensive nature of octane and heptane, secondary standard reference fuels are rated and used for routine test work. The octane number of a fuel is ascertained by comparing the knock intensity for the fuel with those for various blends of the reference fuels until two blends differing in knock rating by not more than two octane numbers are found, one of which gives a higher knock intensity than the unknown sample, and the other a lower knock intensity.

In recent years, the antiknock qualities of gasolines have shown a decided increase. Regular gasoline now shows an octane rating of 65 to 70 while premium grade fuels usually show 72 to 80. The octane rating of third-grade gasolines is usually about 50 to 60.

95. Use of Water with Heavy Fuels.—When kerosene, distillate, or similar heavy fuels are used in tractor engines, better vaporization, and, consequently, more power and greater efficiency often result if a comparatively high compression pressure—75 to 90 lb. per square inch—is main-

TABLE VIII.—AUTOIGNITION TEMPERATURES FOR LIQUID FUELS WHEN VAPORIZED
AND MIXED WITH AIR[1]

Fuel	Temperature, Degrees Fahrenheit
Coal-tar oil	1,076
Benzol	968
Ethyl alcohol	950
Gasoline	779
Kerosene	716
Gas oil	662

[1] *Purdue Univ., Eng. Exp. Sta., Bull. 27.*

tained. However, pronounced detonation occurs, particularly at heavy loads for the reason, principally, that such heavy-fuel mixtures ignite at a lower temperature than a gasoline-and-air mixture, as shown by Table VIII. It has been found that this knock can be eliminated and smooth action produced by permitting a certain amount of water vapor to enter the fuel mixture. This is done by equipping the carburetor with a water-jet nozzle and needle valve, similar to the fuel nozzle and valve, as illustrated in Fig. 88. This valve is opened and the water turned on only when the knock develops; that is, the engine may operate smoothly on kerosene and, therefore, require no water at certain loads. When the knock does develop, usually at heavy loads, the water valve should be opened just enough to eliminate the noise and no farther. The action of the water has never been clearly determined but the following theories seem most plausible:

1. It produces a slower burning mixture, that is, one that may be ignited at the usual time, but which burns slower than the same mixture without water vapor. Therefore, the explosion occurs later, or when the piston has reached the end of the stroke.

2. It reduces the temperature in the combustion space thereby retarding or preventing the preignition of the mixture.

3. It results in some sort of a chemical action whereby compounds are formed that do not ignite so readily at the existing temperature.

96. Vapor Lock.—Vapor lock, gas lock, or air lock is the partial or complete interruption of the fuel flow in the fuel-feed system as a result of vaporization of the fuel and the formation of vapor or gas bubbles at some point. It may be due to the use of a fuel having a too high percentage of light or volatile material or to the location of the fuel tank, lines, pump, and so on, with respect to the hotter parts of the engine. Vapor lock often occurs with engines operated at high altitudes as a result of the lower boiling point.

Difficulties from vapor lock show themselves most commonly in failure of the engine to idle after a fast, hot run or in traffic; sometimes in intermittent or uneven acceleration after idling; and sometimes by irregular operation during a sustained high-speed run. Complete stoppage of the engine rarely occurs, but when it does, the cause may be traced either to vapor lock in the line leading to the carburetor or to boiling in the filter bowl.

97. Heating Value of Fuels.—The heating value or the amount of heat energy contained in any fuel is measured by means of the heat unit known as the *British thermal unit* (B.t.u.), which is the quantity of heat required to raise the temperature of 1 lb. of water 1°F. The heating values of the common liquid fuels are given in Table VII.

98. Combustion of Hydrocarbon Fuels.—The term *combustion* is applied to the process by which a fuel unites chemically with oxygen producing what is known as an oxide and often generating heat of considerable intensity, and sometimes light. It may be a very slow or a very rapid action. For example, the rusting of a piece of iron is a comparatively slow process resulting in the union of the iron with oxygen, forming what is known as red iron oxide or rust. In a common wood or coal stove there is a more rapid union of oxygen with the carbon of the wood or coal, resulting in a high temperature and a heating effect.

In the internal-combustion engine, rapid combustion takes place; that is, the fuel, when mixed with the proper amount of oxygen and ignited, burns instantaneously resulting in the production of gaseous oxides—largely carbon monoxide and dioxide—and water. These gases, being confined in a very small space, produce high pressure and consequently exert great force on the piston of the engine and thus generate power.

Oxygen, therefore, is necessary for combustion in all cases, and the chemical action taking place during the combustion of the fuel mixture in a gas engine may be represented by the following chemical equation:

$$CH_4 + 2O_2 = CO_2 + 2H_2O$$

The oxygen, in all cases, is obtained from the atmosphere; that is, a certain amount of air is mixed with the fuel before ignition takes place. Although air is only about 22 to 23 per cent oxygen, the other 77 per cent is largely nitrogen, which does not have any effect upon combustion.

99. Correct Fuel Mixtures.—Knowing the chemical composition of a fuel, the atomic weight of the principal elements involved, namely, carbon, hydrogen, and oxygen, and the percentage of oxygen in the atmosphere, we can readily calculate the amount of air necessary to produce perfect combustion in the gas-engine cylinder. Such a calculation will show that a correct mixture of fuel and air is made up in the proportion of 1 part by weight of fuel to about 15 parts by weight of air; that is, 1 lb. of fuel requires 15 lb. of air for perfect combustion.

A fuel mixture containing less than the required amount of air is known as a rich mixture; that is, there is not enough oxygen present in the mixture to produce complete combustion, and the most noticeable effect is the production at the exhaust of the engine of black smoke, which is nothing more than free carbon. The usual indications of a too rich fuel mixture are: (1) black smoke at the exhaust; (2) lack of power; and (3) overheating of the engine.

A mixture containing more than the required amount of air is known as a lean mixture, and is best indicated by what is known as backfiring through the intake passage and carburetor. Such a mixture is very slow burning and produces: (1) uneven firing, (2) lack of power and (3) overheating.

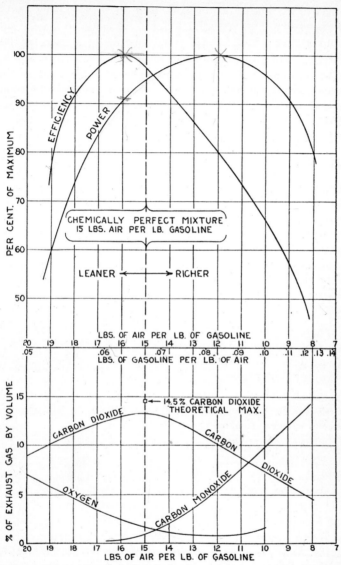

Fig. 89.—Curves showing the effect of air-fuel ratio upon power, economy, and exhaust-gas analysis.

100. Effects of Fuel Mixtures on Efficiency and Power.—The effects of fuel mixture on engine performance are discussed by Wilson[1] as follows:

[1] *Univ. Wis., Eng. Exp. Sta., Bull.* **78.**

The mixture ratio can vary over a wide range without causing the engine to stop. However, for good performance and economy of operation the ratio should not differ much from 13 to 15 lb. of air per pound of gasoline, and this ratio should be maintained for all speeds and all loads except full load. Figure 89 shows the effect of varying the air-fuel ratio by changing the carburetor adjustments or the size of the gasoline jets. The point of maximum efficiency (16 lb. of air per pound of gasoline) occurs for a carburetor adjustment leaner than that required for maximum power (12 lb. of air per pound of gasoline). Hence, the proper adjustment is somewhere in between these two limits. Maximum power may be obtained only at the expense of supplying an excess amount of gasoline per horsepower delivered. An adjustment of the carburetor for about 14¾ lb. of air per pound of gasoline would mean about 96 per cent maximum efficiency. Maximum efficiency may be anywhere between 17 and 24 per cent of the heat applied in the fuel, depending upon the design and condition of the engine.

As the throttle approaches the wide-open position, some increase in the richness of the mixture is desirable in order to permit the development of maximum horsepower. This can be accomplished only at the expense of economy, for the use of a rich mixture of 12 lb. of air to 1 lb. of gasoline results in a decrease in efficiency to 80 per cent of the maximum.

In addition to the lack of economy resulting from the use of overrich mixtures, there is the still more serious effect of the resulting deadly carbon monoxide, which is the poisonous element in the exhaust gas. It is colorless, almost odorless, and has no irritating effect upon the lungs. The presence of only 1 part in 5000 parts of air taken into the lungs for a period of 5 or 6 hr. will cause severe headache, while 1 part in 500 for a period of 1 hr. is sufficient to cause death. It acts largely by poisoning the brain cells through the displacement of oxygen in the blood. Figure 89 shows that the amount of carbon monoxide is practically negligible for air-fuel ratios of 16:1 or greater. For rich mixtures there is not enough oxygen present for complete combustion. Carbon dioxide measurements are being used as a good indication of the completeness of combustion, for the carbon of the fuel must be burned to carbon dioxide in order to liberate the maximum amount of energy.

Proper carburetor adjustment is usually the only means by which a correct fuel mixture may be obtained. Some of the factors affecting it are: (1) the load on the engine; (2) the speed of the engine; (3) the size and type of the engine; (4) the type of cooling system used; and (5) the kind of fuel used.

References

A Study of Explosions of Gaseous Mixtures, *Univ. Ill., Eng. Exp. Sta., Bull.* 133.
Alcohol-gasoline Blends as Engine Fuel, *Agr. Eng.*, Vol. 15, No. 3.
Alcohol-gasoline Engine Fuels, *Idaho Agr. Exp. Sta., Bull.* 204.
Car Carburetion Requirements, *Purdue Univ., Eng. Exp. Sta., Bull.* 17.
Carburetion of Kerosene, *Purdue Univ., Eng. Exp. Sta., Bull.* 27.
Certain Technical Aspects of Motor Fuels, *Agr. Eng.*, Vol. 14, No. 8.
Comparative Tests of Fuels in Low-compression Tractors, *Agr. Eng.*, Vol. 18, No. 7.
Cross: "Handbook of Petroleum, Asphalt, and Natural Gas."

Distillate as a Tractor Fuel, *N. D. Agr. Coll., Agr. Ext. Circ.* 94.

Fuels for Internal Combustion Engines, *Trans. Amer. Soc. Agr. Eng.*, Vol. 10, No. 2.

Fuels for Spark-ignition and Compression-ignition Engines, *Agr. Eng.*, Vol. 15. No. 4.

HELDT: "The Gasoline Automobile," Vol. IV.

Kerosene as a Fuel for Tractors, *Trans. Amer. Soc. Agr. Eng.*, Vol. 12.

Low Gravity Fuels for Internal Combustion Engines, *Trans. Amer. Soc. Agr. Eng.*, Vol. 5.

Lubricants and Fuels for Tractor and Motor Vehicle Engines, *Agr. Eng.*, Vol. 11, No. 7.

Motor Gasoline, Properties, Laboratory Methods of Testing and Practical Specifications, *U. S. Bur. Mines, Tech. Paper* 214.

Oil and Gasoline Information for Motorists, *Univ. Wis., Eng. Exp. Sta., Bull.* 78.

Proposed Standards for Tractor Fuels, *Agr. Eng.*, Vol. 18, No. 8.

STREETER and LICHTY: "Internal-combustion Engines."

THOMSEN: "Practice of Lubrication."

The Carburetion of Gasoline, *Purdue Univ., Eng. Exp. Sta., Bull.* 5.

The Effect of Speed on Mixture Requirements, *Purdue Univ., Eng. Exp. Sta., Bull.* 11.

The Relation of Motor Fuel Characteristics to Engine Performance, *Univ. Mich., Eng. Research Bull.* 7.

The Refiner's View of the Tractor Fuel Problem, *Agr. Eng.*, Vol. 18, No. 8.

The Significance of Tests of Petroleum Products, *Amer. Soc. Testing Materials Bull.*

The Tractor Fuel Situation in Kansas, *Agr. Eng.*, Vol. 17, No. 6.

The Use of Alcohol and Gasoline in Farm Engines, *U. S. Dept. Agr., Farmers' Bull.* 277.

Tractor Fuels in Relation to Tractor Operating Costs, *Agr. Eng.*, Vol. 18, No. 8.

CHAPTER IX

FUEL-SUPPLY AND CARBURETION SYSTEMS

The fuel-supply and carburetion system for an ordinary compression, carbureting-type engine consists essentially of (1) a fuel-supply container or tank, (2) a carburetor, (3) the necessary connecting lines, pump, filter, etc., and (4) the intake manifold to conduct the mixture from the carburetor to the cylinder. The usual systems are:

1. The suction system as used on some single-cylinder stationary farm engines.
2. The force-feed system as used on both single- and multiple-cylinder stationary and automotive-type engines. This system may use (*a*) an overflow-type carburetor with fuel-return line or (*b*) a float-type carburetor with a special fuel pump (Fig. 95).

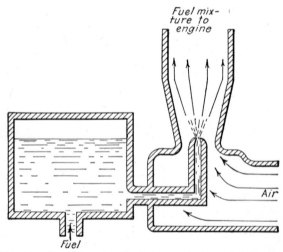

FIG. 90.—Operation of a simple jet-nozzle type of carburetor.

3. Gravity feed with float-type carburetor and elevated fuel tank, used on both single- and multiple-cylinder stationary and automotive-type engines (Fig. 93).

101. Principles of Carburetion.—The functions of a carburetor are: (1) to assist in properly vaporizing the fuel, (2) to mix the vaporized fuel in the correct proportions with air, and (3) to supply the engine with the proper quantity of this mixture depending upon the load, speed, temperature, and other conditions present.

The fundamental principle employed in practically all types of gasoline and kerosene carburetors is illustrated by Fig. 90. A short tubular device is fastened to the intake passage or manifold. When the intake valve opens, air flows at a high velocity past a fuel nozzle placed in the tubular passage, drawing the fuel from it in the form of a spray. The liquid fuel is thus broken up into fine particles and mixed with air to form an explosive mixture. The following conditions, therefore, are of fundamental consideration in the proper functioning of any carburetor operating on this spray-tube or jet-nozzle principle:

1. A constant fuel level must be maintained in the nozzle.
2. There must be at least partial if not complete vaporization of the liquid fuel, regardless of surrounding temperatures.
3. The correct mixture of vaporized fuel and air must be maintained at all times, regardless of engine load, speed, temperature, and other operating conditions.

102. The Suction System.—The fuel is placed in a tank that is located below the carburetor, with a fuel line connecting the two (Fig. 91).

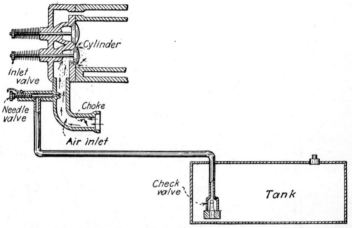

Fig. 91.—The suction system of fuel supply and carburetion.

On the suction stroke of the piston, owing to the vacuum produced in the cylinder, fuel is drawn through the fuel line to the carburetor. At the same time, air is drawn through the air passage. The fuel mixture thus formed passes into the combustion space of the cylinder and is ignited. The correct mixture is obtained by the proper adjustment of the fuel needle valve, which controls the amount of fuel passing through the jet nozzle. With this system, as well as with all others, it is necessary to maintain a certain fuel level in the nozzle of the carburetor. This is done by placing some kind of a check valve in the fuel line, which prevents the fuel from running back into the tank at the end of the suction stroke, so that, at the beginning of the next intake stroke, fuel is available at the nozzle and is immediately drawn from it.

103. Force-feed System.—The principle of operation and arrangement of the force-feed system with overflow is illustrated in Fig. 92. It consists of (1) a tank, located below the carburetor; (2) a pump to force the fuel from the tank up to the carburetor; and (3) an overflow line, to conduct the excess fuel back to the tank. A mechanically operated pump, usually of the plunger type, forces the fuel from the tank to the carburetor. and an overflow passage or arrangement of some kind, placed

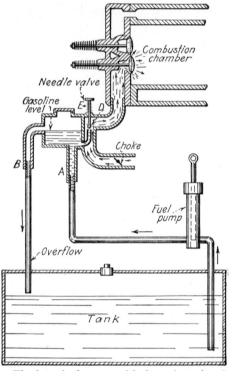

Fig. 92.—The force-feed system of fuel supply and carburetion.

at a certain level in the carburetor, maintains a constant level of fuel at the jet nozzle. The advantage of this system is that, regardless of the quantity of fuel in the tank, the level of fuel in the carburetor will always remain exactly the same. This system is, therefore, better adapted to the use of kerosene because the latter fuel is most sensitive to slight changes in fuel level, quality of mixture, and so on.

All automobile engines and some tractors are now equipped with a special type of fuel pump (Fig. 94), which permits locating the fuel tank below the float-type carburetor but requires no return line.

The operation of the pump is as follows: Movement of the rocker arm *D*, which is pivoted at *E*, by a cam on the camshaft, pulls the diaphragm *A* downward against the pressure of spring C and creates a vacuum in pump chamber M. Fuel from the tank enters the sediment bowl K through strainer L and passes through the check valve N

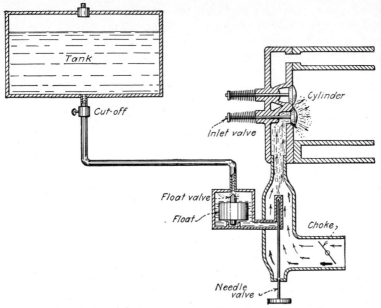

FIG. 93.—The gravity system of fuel supply and carburetion.

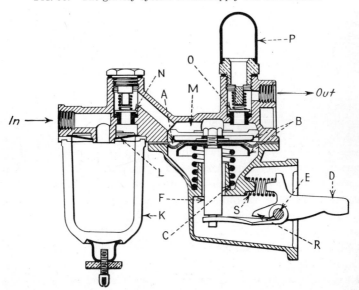

FIG. 94.—Sectional view of the automobile-type fuel pump.

into pump chamber M. On the return stroke, spring pressure pushes the diaphragm upward forcing fuel from the chamber M through outlet valve O into vapor dome P

and out to the carburetor.   When the carburetor bowl is filled, the float in the float chamber will close the float valve, thus creating a pressure in pump chamber M.   This pressure will hold diaphragm A downward against the spring pressure, and it will remain in this position until the carburetor requires further fuel and the float valve opens.   The rocker arm D is in two pieces and is split at R.   Therefore the movement of the rocker arm is absorbed by this "break" at R when fuel is not required.   Spring S is merely for the purpose of keeping the rocker arm in constant contact with the actuating cam.   Figure 95 illustrates the complete system using such a pump.

104. Gravity System.—The third, or gravity system, as illustrated in Fig. 93, consists of: the tank, placed above the carburetor; the fuel line; and the carburetor, fed by gravity.   The quantity of fuel that is allowed to flow into the latter is controlled by a float of either cork or hollow metal.

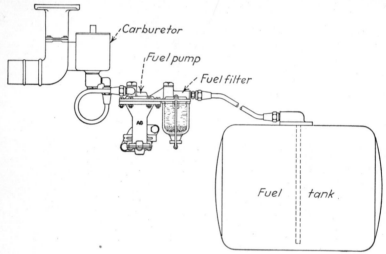

FIG. 95.—Float-type carburetor with pump feed.

This float is attached to a float valve in such a way as to allow the fuel to enter the carburetor at the same rate at which it is being consumed by the engine.   This system likewise maintains an absolutely uniform level of fuel in the carburetor, regardless of the quantity in the tank, and therefore, is adapted to both gasoline- and kerosene-burning engines. As will be seen later, carburetors used with the gravity system of fuel supply are, for certain reasons, more complicated than those used with the two other systems.

105. Carburetor Types and Construction.—The suction type of carburetor, as illustrated in Fig. 91, is the simplest possible device that can be used.   It consists essentially of a cast-iron body, the jet nozzle, a fuel needle valve, and the choke.   Its use is confined almost entirely to single-cylinder stationary farm engines burning gasoline.   Since these engines operate at one speed only, and usually at a more or less constant load, the correct mixture can be controlled by means of the needle valve alone.

The narrow part of the air passage around the nozzle of a carburetor (Fig. 90), usually called the *Venturi*, is for the purpose of creating a greater air velocity and, therefore, a stronger sucking effect at this point. The purpose of the choke is to shut off the air so that a greater amount of fuel may be drawn in, when the engine is being started, because the slow piston movement provides only limited suction. Ordinarily, the choke should be used for starting only and not for adjusting the mixture during operation.

The overflow type of force-feed carburetor is likewise used on single-cylinder one-speed farm engines. In addition to the parts found in

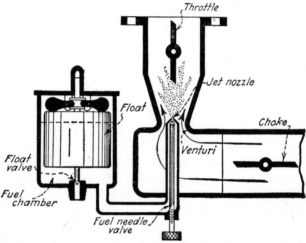

Fig. 96.—Simple type of float-feed carburetor showing essential parts. (*Courtesy of Vacuum Oil Company.*)

the suction type, this carburetor also has a fuel or overflow chamber and an overflow opening so that a constant fuel level may be maintained at all times at the jet nozzle. In other respects the carburetor works exactly like the suction type, and a correct fuel mixture is maintained in the same way.

The float-type carburetor (Fig. 96) is more complicated and is usually made up of the following essential parts: (1) jet nozzle; (2) fuel needle valve; (3) choke; (4) float; (5) float valve; (6) throttle butterfly; and (7) auxiliary air valve, or other compensating device. The float, together with the float valve, controls the fuel entering the float or fuel chamber of the carburetor from the tank and thus maintains a certain fuel level in this chamber and at the jet nozzle. Floats may be of either the cork type or hollow-metal type. The latter is used almost exclusively because of its durability.

The throttle butterfly is for the purpose of controlling the quantity of mixture allowed to enter the cylinders. It, therefore, controls the engine speed and power. With a low speed or a light load, less fuel mixture would be required, and the throttle would be only partly open. For a high speed or a heavy load, the throttle would be wide open in order to permit the cylinders to receive a full charge. The opening and closing of this device, according to the engine speed and power output required, are effected by either a hand or a foot lever as in automobiles, or mechanically by means of a governor on the engine as in farm tractors.

106. Auxiliary Air Valves and Other Compensating Devices.—The principal underlying reason why the float or gravity type of carburetor is more complicated than others is that it is used on engines operating at changing speeds and loads; that is, it has been found that in order to maintain a constant and correct fuel mixture under such conditions, a carburetor equipped with some kind of a so-called compensating device or arrangement is necessary. The following explanation will make this clear: Suppose that an engine is equipped with a simple carburetor, as shown in Fig. 96, and is operating at a comparatively low speed and light load, but with the proper fuel mixture. Obviously, the throttle butterfly will be nearly closed. Now suppose that the throttle is suddenly opened with the idea of increasing the engine speed or power output. The increased throttle opening itself permits a greater suction at the nozzle, causing more fuel to be taken up by the air stream. Likewise, the velocity of the air through the carburetor increases so that it would seem likely that the correct air-fuel proportion would be maintained. However, such is not the case. Experience as well as theory has proved that the quantity of air drawn in does not increase at a rate great enough to maintain the correct air-fuel mixture, and it becomes too rich. Now suppose, with the engine running at this increased speed, that the needle valve is readjusted so that the mixture is again correct; that is, the needle valve is closed somewhat. If the engine were again slowed down by closing the throttle, the mixture would be too lean, owing to the fact that the decreased suction does not permit the required amount of fuel in proportion to the air to be drawn from the nozzle. Therefore, if automobile, truck, and tractor engines, for example, were equipped with such a carburetor, considerable difficulty and inconvenience would be encountered in maintaining the proper fuel mixture and the much-desired smooth operation at any speed or load. Obviously, to be constantly readjusting the fuel needle valve would be impractical.

The float type of carburetor, therefore, is equipped with some sort of a compensating device or arrangement or is so constructed that the mixture adjusts itself instantly to any changes in engine speed or load and thus remains properly proportioned. Several different methods are employed by carburetor manufacturers to bring about this action. The more important of these are explained in the following descriptions of a number of well-known makes of carburetors.

FLOAT-TYPE CARBURETORS

107. Kingston Model L.—The Model L Kingston carburetor is being used on a number of tractors. As shown by Fig. 97, it is a compact device of the concentric type, that is, it has a doughnut-shaped cork float, with the spray nozzle and needle valve located practically in the center of the float chamber and float. With this construction, the tilting of the tractor and carburetor in any direction has little effect on

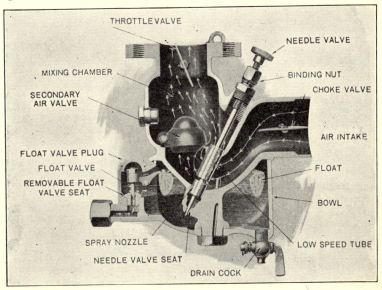

Fig. 97.—Sectional view of Kingston Model L carburetor.

the mixture, since the fuel level at the nozzle will remain at about the same point as it would if the carburetor were level.

The Kingston Model L carburetor operates as follows: Referring to Fig. 97, fuel enters around the float valve and passes into the bowl causing the float to rise and close the valve. At low or idling speeds, air enters at the air intake and picks up a small quantity of fuel from the spray nozzle. The resulting mixture passes upward through the low-speed tube as shown by the arrows and goes to the engine. Owing to the low engine speed and nearly closed throttle, there is insufficient suction to lift the secondary air valve, but enough fuel mixture can pass through the low-speed tube to keep the engine running.

As the throttle is opened and the vacuum increases, the secondary air valve, which is made of a heavy metal and hinged as shown, is lifted from its seat allowing more air to pass through. This air also passes across the spray nozzle with high velocity and becomes impregnated with atomized fuel. If the needle valve is correctly adjusted, a correct air-fuel mixture will be supplied to the cylinders at all speeds and loads, the so-called automatic air valve being so designed that it will open the correct amount at any engine speed and load.   In adjusting the needle valve, open it up about 1½ turns and start the engine, getting it well warmed up.   Then screw the needle valve down very slowly until the engine begins to lose speed and backfire through the carburetor. Then open the needle valve one-eighth to one-fourth turn, and the engine should run smoothly.   When the correct adjustment is secured, tighten binding nut to keep needle valve from turning due to vibration.

108. Marvel Model E Carburetor.—The Marvel Model E Carburetor, Fig. 98, is known as a *double-nozzle* type.   At idling speed, all air must

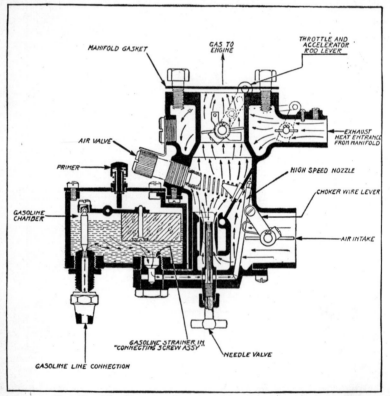

Fig. 98.—The Marvel carburetor with exhaust-heat jacket.

pass around the centrally located primary fuel nozzle where the fuel is picked up and the mixture formed.   As the engine speed is increased, the flat spring-controlled air valve is drawn open by the increased suction, and some fuel is drawn from the high-speed nozzle.   The two mixtures, thus produced at high speeds, combine to form the correct mixture that enters the cylinder.   In regulating this carburetor, the needle valve is adjusted for low speeds; then the throttle is opened and the tension on the air valve is adjusted by means of a screw.

One of the distinct features of the Marvel carburetor is the hot-air jacket surrounding the mixing chamber, which permits the exhaust gases to heat the carburetor to a certain degree, depending upon the speed, load, and outside temperature. A tube leading from the exhaust manifold conducts a portion of the exhaust gases to the carburetor. The heating effect is controlled by means of a damper, which is connected to the throttle butterfly control so that when the engine slows down, the

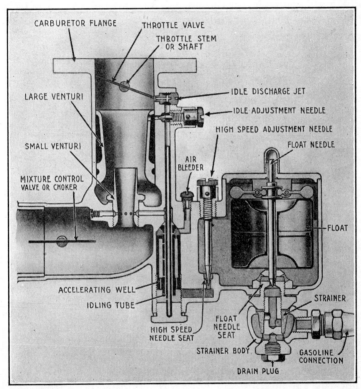

Fig. 99.—Sectional view of Stromberg Model M carburetor.

damper opens and releases more heat; but when the speed or load increases with a greater throttle opening, the damper closes and partly shuts off the heat. Another adjustment permits further regulation of the heating effect according to weather conditions and quality of fuel.

109. Stromberg Model M Carburetor.—The Stromberg carburetor (Fig. 99) is known as a *plain-tube double-Venturi* type, because the main air passage contains no air valve of any kind and is unrestricted except for the Venturi tubes. The fuel flows from the float chamber past the high-speed needle and assumes a similar level in the accelerating well. Referring to the illustration, at idling speeds the throttle butterfly is

practically closed.   Air comes through the small Venturi, passes through
the small opening controlled by the idling needle, picks up fuel from the
idling tube, and the resulting mixture enters the main passage just above
the edge of the throttle disk.   As the latter is opened for increased speeds,
the fuel is drawn from the accelerating well through a horizontal passage
and numerous small holes in the small Venturi tube.   A greater quantity
of mixture is thus produced in the large Venturi and flows to the cylinders.   The air bleeder is for the purpose of allowing a very small amount
of air to mix with the fuel and create an emulsifying action as it leaves
the accelerating well.

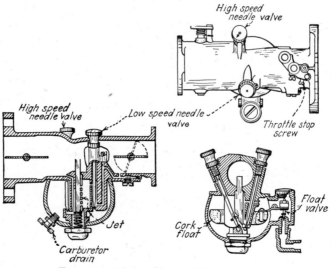

Fig. 100.—Schebler Model DLT carburetor.

It is thus observed that the high-speed needle must be open more or
less for both idling and high engine speeds, but is adjusted only for the
latter.   At low speeds the idling-adjustment needle controls the amount
of air entering the mixture and, therefore, is turned outward to produce a
leaner mixture and inward for a richer one.   In adjusting this carburetor,
the correct procedure is: (1) Open both needle valves about two turns.
(2) Close the throttle and adjust the idling needle valve until the engine
fires and operates smoothly.   (3) Open the throttle so as to increase the
engine speed and adjust the high-speed needle valve correctly.

110. Schebler Model DLT.—Referring to Fig. 100, it will be seen
that this carburetor does not have an air valve but is known as a plain-
tube type.   The mixture is controlled by two needle valves, one for
high speeds and loads, and one for low speed.

To adjust this carburetor, open the high-speed needle valve $1\frac{1}{4}$ turns
and the low-speed needle valve two turns, turning both counterclockwise.

Start the engine and warm it up. Then slow it down, partly retard the spark, and adjust the low-speed valve until the engine operates smoothly. Then speed up the engine, put it under load, and change the high-speed valve slightly, if necessary, until engine runs smoothly and gives its maximum power.

111. Ensign Model D.—The Ensign Model D carburetor (Fig. 101) differs somewhat in construction and operation from the usual types, in that the float chamber is located in the upper part of the device and the fuel is drawn downward through the suction tube before it enters the mixing chamber. The air in the centrifugal mixing chamber assumes a

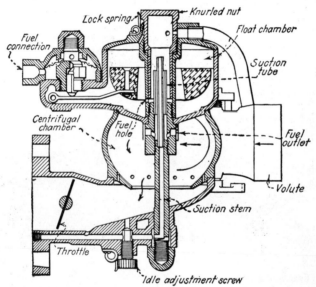

Fig. 101.—Ensign Model D carburetor.

whirling action drawing the fuel from the suction tube through the small openings as shown. This centrifugal effect on the fuel is said to aid in atomizing and vaporizing it.

The fuel supplied to the mixture is controlled by an adjustable sleeve attached to the knurled nut. Turning this sleeve in or downward reduces the fuel flow from the float chamber to the suction tube by closing up some small openings. This makes the mixture leaner. Turning this screw outward increases the fuel flow and makes the mixture richer. At very low or idling speeds, the mixture is adjusted by the idling adjustment screw. It controls the air rather than the fuel, however. Therefore, to make the mixture leaner, turn it out; and to enrich the idling mixture, screw it in. As a general rule, the high-speed adjustment should be open from one to one and one-half turns and the low-speed screw one-half to one turn.

112. Zenith Carburetor.—The Zenith carburetor (Figs. 102 and 103) is known as a *compound-nozzle* type. Fuel enters the float chamber through the screen and float valve and assumes a certain level as controlled by the float. The fuel flow from the main jet, which is directly connected with the fuel bowl, varies with the speed of the engine. In

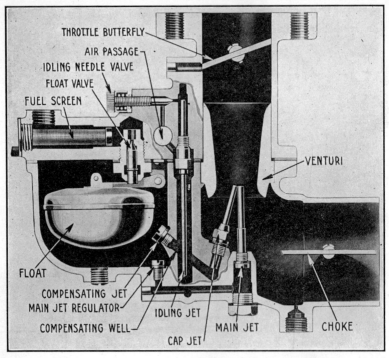

FIG. 102.—Sectional view of Zenith Model 90 carburetor.

fact the main jet supplies an increasing amount of fuel as the suction increases and the mixture grows richer. Its effect is most noticeable at high speeds. The fuel flow from the compensating jet is constant regardless of suction or engine speed, as it empties into a well open to the air. The cap jet connects with this well, discharging the fuel into the air stream. Since the compensator is very small, only a limited quantity of fuel issues from the cap jet, and the mixture produced by it grows poorer with increasing suction. But the excess flow from the main jet and the limited flow from the compensator and cap jet combine to form a mixture of correct proportions. Thus the two nozzles work together to supply a proper mixture regardless of engine speed.

Idling is taken care of, independently of the jets previously mentioned, by means of a special idling jet and air control. Since the throttle butterfly is practically closed, suction at the main and compensating jets is cut off. The fuel is drawn from the compensating well, up through the idling jet, and passes into the main mixture passage through a small hole at the edge of the closed throttle. In other words, the idling jet operates only when the throttle is open barely a crack. A needle valve regulates

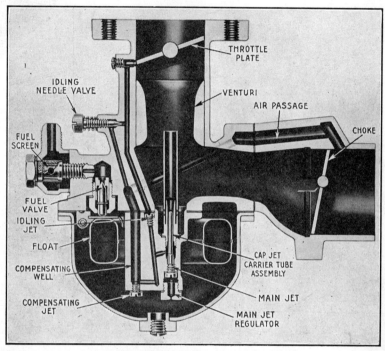

Fɪɢ. 103.—Zenith Model C6EV carburetor.

the idling-air supply and therefore the idling mixture. Turning this valve clockwise shuts off the air, making the mixture richer.

The idling needle valve is the only "quick" adjustment possible on the Zenith carburetor; that is, the adjustment is said to be of the fixed type. Usually this carburetor is properly fitted at the factory for use on a given engine. This consists of the selection of the correct size of Venturi, main jet, main-jet regulator, compensating jet, cap jet, and idling jet.

113. Zenith Model C6EV.—This carburetor (Fig. 103) is a special, heavy-duty device for tractor, truck, and marine service. Since the discharge and supplemental jets are located concentrically with respect to the float chamber, this carburetor is better adapted to engines operating at pronounced angles.

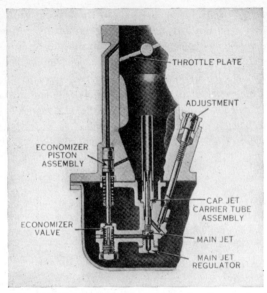

FIG. 104.—Zenith fuel economizer and accelerating mechanism.

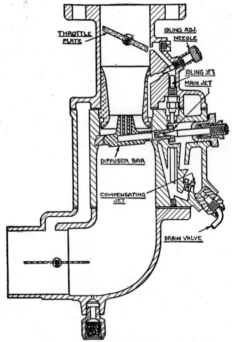

FIG. 105.—Zenith Model K5 carburetor.

Another feature is a special economizer and accelerating mechanism (Fig. 104) whose purpose is to supply instantly the necessarily richer mixture required for maximum power and proper acceleration. Its action is as follows:

Under part throttle, the suction or vacuum above the throttle is higher than when the throttle is open. This suction holds up the economizer and accelerating piston assembly. The economizer valve is closed, thus shutting off fuel from the power and accelerating jet.

When the throttle is opened, the suction falls and so does the piston. The piston pin hits the economizer valve, pushing it open, and an extra charge of fuel passes into the jet. This is the accelerating charge.

If the throttle is held open, the piston will remain at the bottom holding the economizer valve open. This allows fuel to continue flowing through the power and accelerating jet.

When the throttle is partly closed, the suction increases above it, the piston is drawn up to the top, the economizer valve closes, and only a very economical amount of fuel can be fed to the engine.

114. Zenith Model K5 Carburetor.—The Zenith Model K5 carburetor (Fig. 105) was specially designed for tractors burning heavy fuels and is found on several makes. It operates on the compound-jet principle and has the float mechanism built around the main jet to provide satisfactory operation on grades. The fuel from both jets discharges into the air stream through a diffuser bar, which permits better atomization of heavy fuel. This carburetor is equipped with an economizer device (Fig. 106) to provide closer mixture control at different loads. It operates as follows:

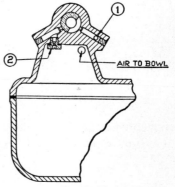

FIG. 106.—Zenith Model K5 carburetor fuel economizer.

The suction in the carburetor is transmitted through channel 1 but is not transmitted to the bowl because the notched collar around the throttle shaft closes the channel to the economizer jet 2. As the throttle is opened to a position of approximately one-quarter load, the notched collar forms a channel between channel 1 and economizer jet 2, thus transmitting the suction from the carburetor to the top of the float chamber. This retards the flow of fuel to the jets. Opening the throttle to a wide-open position rotates the notched collar so that it closes channel 1. This eliminates suction on the bowl and permits a free flow of fuel to the jets.

115. Downdraft Carburetors—I.H.C. Model A-10.—Downdraft carburetors, in which the device is elevated somewhat above the engine block and attached to the intake manifold in such a manner that the mixture flows downward, are now used exclusively on automobile engines, but only to a limited extent on tractors. The advantages claimed are (1) gravity assists the mixture in its flow to the cylinder, therefore effecting a saving in energy; (2) in case of a "flooded" carburetor or a leaky float, the liquid fuel runs directly into the manifold and not outside, thus lessening the fire hazard; and (3) the carburetor, being elevated, is usually more accessible and less exposed to dirt and dust.   Figure 107 shows the construction of the downdraft carburetor used on the Farmall F-12 tractor.   With the exception of the inverted construction, it is similar to other plain-tube carburetors with a single Venturi and both idling and high-speed adjustments.

116. Carburetion of Heavy Fuels.—As previously discussed, kerosene and distillate are not so volatile as gasoline; therefore they are more

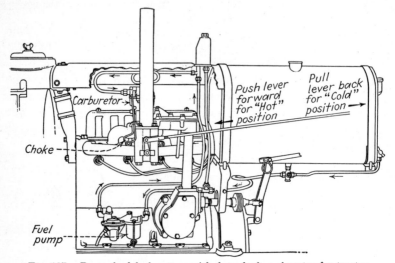

Fig. 107.—Pump-feed fuel system with downdraft carburetor for tractor.

difficult to vaporize when used as a gas-engine fuel.   They can be utilized successfully under certain conditions as follows:

1. The engine must operate at a uniform speed and under a medium to a heavy load.

2. Provision must be made for starting on gasoline with some convenient means of changing over when the engine is warmed up.   For this reason the use of such fuels is impractical if frequent starting and stopping are necessary.

3. A careful and precise adjustment of the fuel mixture is essential.

4. Multiple-cylinder engines having manifolds of any length require more or less preheating of the air-and-fuel mixture.

Obviously, the use of heavy fuels is confined largely to stationary, single-cylinder farm engines and farm tractors.   These engines, when in operation, run at a more or less uniform speed and load and, when once started, usually continue running for several hours.   On the other hand, their use in an automobile would result in constant trouble and inconvenience.

Even though a heavy-fuel-burning engine is operated continuously under uniform conditions, further provision for insuring complete vaporization and maximum efficiency is usually made by:

1. Heating the air, fuel, or mixture by means of the exhaust gases.

2. Controlling the cooling-water temperature according to weather and other operating conditions by means of a radiator shutter or curtain or a thermostatic valve.

Vaporization by Heat.—The utilization of heat to insure complete vaporization of the fuel is now a common practice in nearly all types of

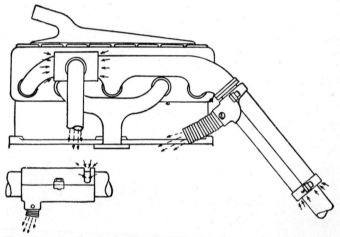

Fig. 108.—Devices for heating the air before it enters the carburetor.

engines burning either gasoline or kerosene. The heat of the exhaust gases, otherwise wasted, may be used to heat (1) the air entering the mixture, (2) the liquid fuel on its way to the carburetor, or (3) the mixture of air and partially vaporized fuel.

To heat the air, a metal jacket of some kind (Fig. 108) is usually constructed about the exhaust manifold or pipe at some point and connected to the air intake of the carburetor so that this air, as it is drawn into the latter, must pass through the hot jacket and become heated. The fuel coming in contact with the hot air will thus take up some of this heat and vaporize more readily.

Heating of the liquid fuel is seldom practiced. To accomplish this, the fuel line from the tank to the carburetor is connected to a coiled tube inclosed in a jacket about the exhaust manifold. The coil is thus kept at a high temperature and the fuel passing through it is heated and placed in a condition that insures its proper vaporization.

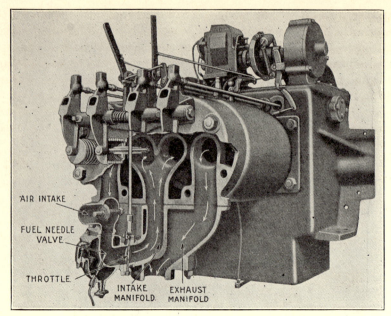

FIG. 109.—Intake and exhaust manifolds cast together for heating fuel mixture.

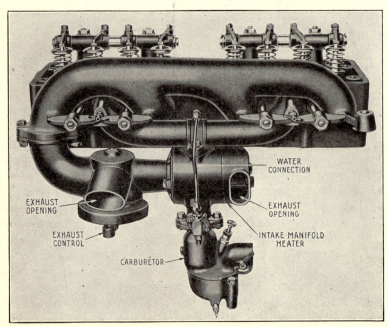

FIG. 110.—Intake manifold heater with heat-control device.

The application of exhaust heat to the fuel mixture, as it leaves the carburetor on its way to the cylinder, is perhaps the most common method of securing vaporization. The simplest arrangement consists merely of casting the intake and exhaust manifolds in one piece so that the iron wall separating the two passages becomes highly heated, and, in turn, heats and vaporizes the fuel. Figure 109 shows such an arrangement for a two-cylinder engine. The exhaust gases from both cylinders go out through the exhaust passages, which entirely surround the intake passage. The latter is thus kept at a high temperature.

It is undesirable to apply more than just enough heat to completely vaporize the fuel. Excessive heating causes the gas mixture to expand

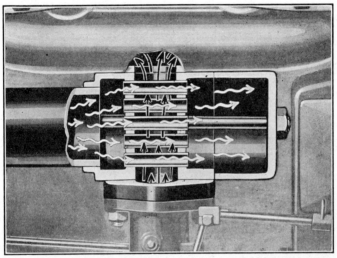

Fig. 111.—Sectional view of intake manifold heater.

in volume so that the cylinders do not receive the maximum fuel charge possible. Therefore, the maximum power output and efficiency are reduced. For this reason some preheating-type manifolds (Fig. 110) have dampers or shunt devices by which the heat applied to the mixture can be varied according to operating conditions. For example, in summer less preheating is necessary and the path of part or all of the exhaust gases can be deflected away from the intake manifold. Again, if gasoline is used instead of kerosene, less preheating is required and the exhaust can be by-passed and its heating effect reduced. In cold weather, when more preheating is necessary, the damper can be set for maximum heating effect as shown in detail by Fig. 111.

117. Carburetors for Heavy Fuels.—Heavy-fuel carburetors are constructed and operate in practically the same way as gasoline carburetors. In fact, any engine that burns such fuels successfully will also burn gasoline readily. Engines which are recommended to operate on gasoline only will likewise burn heavy fuels with the same carburetor, provided the load is fairly heavy and uniform, and the speed does not vary greatly. Any difference in the carburetors used on heavy-fuel engines will likely be due to the two provisions previously mentioned as being more or less essential to the successful carburetion of these fuels, namely, a means of supplying heat for complete vaporization, and a water connection and control valve.

118. Starting and Operation of Heavy-fuel-burning Engines.—A heavy-fuel-burning engine cannot be started easily on this fuel, even when warmed up or hot. In all cases some provision is made for starting on gasoline and then changing over either immediately or after running a few minutes. When the heavy fuel is first turned on, some gray-blue smoke may appear at the exhaust, but, when the engine receives the load and gets well warmed up, the exhaust should become clear. Great care should be taken in adjusting the kerosene needle valve. A small fraction of a turn will often produce misfiring and loss of power. Water should be turned on only when knocking occurs, and this adjustment likewise should be carefully made. The idea is often held that such fuels produce more carbon deposit in an engine than gasoline. This is untrue, however, unless the engine is allowed to operate on too rich a mixture for some time. Under such conditions, lack of power and other difficulties would soon develop. For best results, the ignition mechanism should function in the best possible manner at all times.

119. Idling-speed Adjustment.—All carburetors are equipped with an adjusting screw on the throttle butterfly lever arm (Fig. 100), for adjusting the idling speed of the engine. This screw, when correctly set, prevents the throttle from closing completely, even though the hand lever is in full-closed position. It should be set so that the engine runs as slowly as possible without load and still fires evenly and does not stop. To make the adjustment, start the engine and adjust the carburetor carefully and as nearly correctly as possible for both low and high speeds. Then slow the engine down, and with a screw driver turn the screw in or out, as the case may be, until the desired idling speed is secured with the hand lever in the closed position.

References

Commercial Carburetor Characteristics, *Purdue Univ., Eng. Exp. Sta., Bull.* 21.
Cornell Univ., N. Y. State Coll. Agr., Ext. Bull. 133.
Cornell Univ., N. Y. State Coll. Agr., Ext. Bull. 147.
DUELL: "The Motor Vehicle Manual."
"Dyke's Automobile and Gasoline Engine Encyclopedia."
ELLIOT and CONSOLIVER: "The Gasoline Automobile."
FRASER and JONES: "Motor Vehicles and Their Engines."
HELDT: "The Gasoline Automobile," Vol. IV.
Temperature Requirements of Hot Spot Manifolds, *Purdue Univ., Eng. Exp. Sta., Bull.* 15.

CHAPTER X

AIR CLEANERS

Early tractors were not equipped with any means or device for supplying clean air to the carburetor, mainly because the machines were so large that the air intake was high enough to be out of the dust-laden-air zone. Then, too, little thought was given to the harmful effects of even a very small quantity of fine dust. With the introduction of the smaller, light-weight tractors, it was soon discovered that, under most field conditions, enough dust and fine grit found its way into the cylinders through the carburetor to cause rather rapid wear. In fact, in some sections under certain conditions, enough damage could be done in even a day or two of operation to practically ruin the tractor.

METHODS OF SUPPLYING CLEAN AIR

Partial or complete elimination of engine troubles caused by foreign material entering the carburetor can be secured in the following ways:

1. By a periscope or long metal tube extending above the dust level.
2. By some form of air cleaner or filter.
3. By a combined air-cleaner and elevated-intake opening.

120. Elevated Air Intake.—The first idea of the manufacturers of small tractors was to attach a vertical tube to the carburetor air intake, extending it above the tractor far enough to obtain air that was comparatively free from the dust and grit stirred up by the machine itself. It was soon found that this arrangement, though simple, failed to do the work. Some device was needed that would prevent the finest dust as well as the more harmful coarse gritty material from getting into the engine. Then, too, these periscopes soon worked loose and fell off or were removed by the owner in order to permit the tractor being run under a shed. When once removed, they were seldom replaced, and the resulting trouble was blamed upon the tractor and its maker.

121. Air-cleaner Requirements.—The subject of air cleaners for tractors and trucks, and even for automobiles, has been given a great deal of study by research engineers and manufacturers. A number of reports of experimental work and experience with air cleaners have been published. These are listed at the end of the chapter, and the reader is referred to them for more complete information on the subject.

Any type of air cleaner to give maximum satisfaction must fulfill the following requirements:

119

1. It must have high cleaning efficiency at all engine speeds and under all operating conditions.

2. It should offer very little restriction to the air flow thus reducing the power or fuel economy, or interfere otherwise with the satisfactory operation of the engine.

3. It should require very little or only infrequent attention.

4. It should be compact, rigid, not too heavy, and rattleproof.

A large number of different air-cleaning devices, embodying various principles and ideas, have been developed and used, but it cannot be said that any particular one meets all the requirements enumerated. Thus far it might be said that all air cleaners are more or less a compromise between what it is desired to accomplish and the requirements of practical application to the engine.

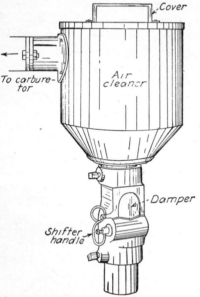

FIG. 112.—Dry- or filter-type air cleaner.

TYPES OF AIR CLEANERS

Air cleaners may be classified as follows:

1. Dry type.
  a. Filter type.
  b. Inertia type.
2. Water type.
3. Oil type.
4. Combination type.

122. Filter-type Cleaners.—The first tractor air cleaners were of the dry filter type (Fig. 112), consisting essentially of a sheet-metal closed container with an air opening in the side or bottom and an outlet in the top connected to the carburetor air intake. A filtering element, consisting of one or more layers of felt, eider-down cloth, wool, or similar fibrous material, was placed in the bottom of the cleaner and collected the dust as the air passed through. This type of filter soon proved unsatisfactory for several reasons. If it was dense or heavy enough to catch most of the dust, it offered too great restriction to the air flow. Also, it soon became clogged and required frequent cleaning or renewal and frequently got out of place owing to backfiring.

123. Inertia-type Cleaners.—The operation of the inertia-type air cleaner is based upon the same principle as the well-known centrifugal cream separator, namely, that if a liquid or a solid mixture of varying specific gravity is whirled violently, the heavier particles or portions will be thrown to the outside. Therefore, if the dust-laden air, on its way to the carburetor, is first made to assume a violent whirling action, the dust, grit, and other solid particles, being heavier, will be thrown to the outside and deposited or carried away, leaving the clean air to enter the carburetor at a small, central opening.

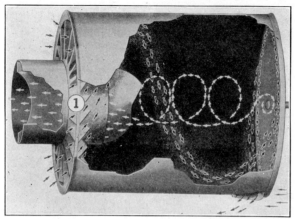

FIG. 113.—Dry, inertia-type air cleaner.

A simple inertia-type cleaner is illustrated by Fig. 113. It consists merely of a cylindrical sheet-metal device with louvered slots in the end where the air enters. The air, as it strikes the louvers, is given a whirling action that throws the dust to the out- side. Since the smaller carburetor opening is attached in the center of the larger cleaner body, only the clean dust-free air in the center of the whirl- ing mass enters the carburetor. The dust particles are discharged through a slotted opening in the lower part of the cleaner as shown.

Figure 114 illustrates another cleaner similar to the one just de- scribed, except that it is equipped with a small fan, which is rotated by the air stream. Attached to this fan and driven by it is a larger cone-shaped disk bearing some radial vanes. The air, as it enters the cleaner, is whirled, and the dust particles are thrown to the outside by the vanes.

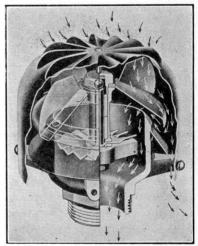

FIG. 114.—Inertia-type air cleaner with revolving fan.

Inertia and other dry types of air cleaners are not considered efficient and satisfactory for tractors, because the latter are usually operated under very dusty conditions.   The inertia cleaners offer little restriction to the air flow and therefore cannot affect the engine operation.   They are also convenient in that they seldom require attention.   However, in a farm tractor, the primary consideration is reliable and complete dust separation, and inertia cleaners are not so dependable in this respect as other types.

124. Water-type Air Cleaners.—Figure 115 shows the construction and operation of the water air cleaner or air washer.   The device consists of three principal parts: the body or bowl, the float, and central air-intake tube.   The float includes an inverted tube that fits over the outside of the air-intake passage.

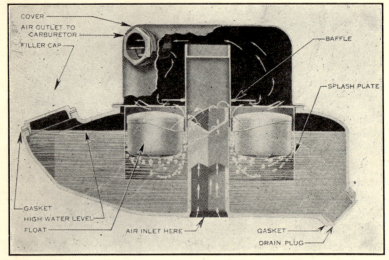

Fig. 115.—Water-type air cleaner or air washer.   (*Courtesy of Ford Motor Company.*)

Air enters the intake tube through an outside opening and passes upward through the upper end, where it meets the closed float tube and must turn and go downward between the two tubes.   Since the open end of the float tube is submerged in water, the air must pass through the water as shown, before it can reach the upper part of the chamber and enter the outlet connection.   The water takes up the dust and foreign matter that settles to the bottom of the chamber. This material must be removed periodically.   Deflector or baffle plates in the upper part of the cleaner body assist in preventing water or dirt particles being taken into the engine.   If the water level becomes too low, the float drops down until the float tube covers the air intake and chokes down the engine.

Water air cleaners have never been used extensively on tractors and are being superseded by the oil and combination-type cleaners for the following reasons: (1) they do not necessarily remove all the dust and dirt; (2) they require too frequent cleaning and replacement of the water, owing to a certain amount being taken up by the air and going into the engine; (3) there is considerable air-flow restriction and variation with the change in water level; (4) the metal parts corrode and leak or break; and (5) there is danger of freezing in cold weather.

125. Oil-type Air Cleaners.—Air filters and cleaners using oil as the dust-collecting medium have come into marked favor for tractor use in recent years. The cleaner shown in Fig. 116 consists essentially of two parts: the upper body and the lower oil cup, which fits tightly and is clamped to the body. This cup is kept about half filled with a light-grade oil or thin, drained crankcase oil.

The dust-laden air enters the so-called vortex chamber through a horizontal tangential inlet, which gives it a whirling movement as shown. Oil is picked up by the air and a certain amount carried with it into the upper part of the cleaner, which contains a

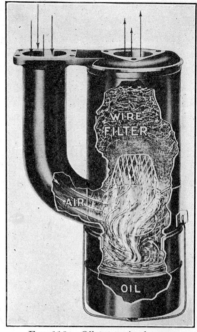

FIG. 116.—Oil-type air cleaner.

wire filtering material similar to steel wool. This filter, therefore, becomes saturated with oil and takes up the dust particles from the air passing through. The excess oil, dropping or running back into the lower cup, carries much of the dirt with it. The heavier particles are largely taken up by the oil before they enter the wire filter. A definite compact layer of sediment eventually forms in the bottom of the oil reservoir. It is recommended that this sediment be removed when the reservoir becomes one-fourth full or when the oil gets too thick with suspended dust particles to spray readily. At the same time fresh oil should be put in to the proper level. The frequency of cleaning will depend on how much the tractor is used and the dust conditions. It is claimed that with continued operation under very dusty conditions the air-flow restriction does not vary materially with this cleaner because the wire filter is being continually washed and relieved of the dirt.

The oil-type cleaner shown in Fig. 117 operates in a similar manner except that the air enters at the top and is drawn directly downward through a central tube. When the air reaches the lower end of this tube it comes in contact with the oil surface. At the same time its direction of flow is reversed and it passes upward through the filter, which consists of a number of circular sections of fine-wire screen. This filter, having become saturated with oil, readily takes up the dust, which finally works

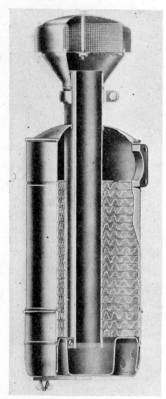

its way down into the oil container. These sections are removable for cleaning when necessary.

Another oil-type air cleaner (Fig. 118) consists of an outer hollow cylindrical metal shell and an inner cylindrical core having a thick fibrous wall. This core is the filtering element. The vegetable fiber is saturated with oil. Air enters the outer shell, passes through the core wall or filter, and into the carburetor. The oily fiber, therefore, acts as a strainer and retains the dust and dirt

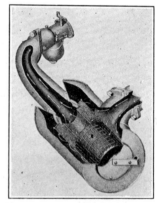

Fig. 117.—Oil-type air cleaner with removable wire filter sections.

Fig. 118.—Oil-filter-type air cleaner.

particles. In order to eliminate clogging and air-flow restriction, the filter core must be removed at least once daily, washed with clean gasoline or kerosene, saturated with light lubricating or crankcase oil, and replaced. Under unusually dusty conditions this should be done twice a day.

The Massey-Harris air cleaner (Fig. 119) operates in a somewhat different manner. The air enters at a top opening, passes downward through a narrow passage, around the end of a vertical partition, and then upward again and through the filter chamber filled with oily steel fiber. Near the lower end of the vertical plate is a small orifice from which oil is picked up, being fed from the oil chamber in the bottom of the cleaner.

The clean air, after leaving the filter, goes on to the carburetor as shown by the broken arrows. The oil collecting on the fine-wire filter retains the dust and eventually drops back into the lower chamber, as shown by the solid arrows. Provisions are made for cleaning the sediment out of the chamber and renewing the oil as frequently as necessary. The filter, likewise, must be periodically washed with gasoline or kerosene.

The principal objections to the oil-type cleaner are (1) possible variation and restriction of air flow as dust is taken up and (2) more or less frequent cleaning and attention. Neither of these is serious, however, and each is largely offset by the high cleaning efficiency.

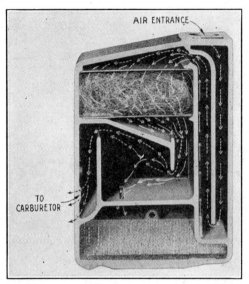

FIG. 119.—Oil-type air cleaner used on the Massey-Harris tractor.

126. Combination Cleaners.—A number of tractors are now equipped with so-called two-stage cleaners. For example, Fig. 120 shows an inertia-type cleaner combined with an oil filter. The first one separates out a large portion of the coarser dust and other material, leaving the second to take up the finer dust. Such an arrangement insures practically perfect cleaning of the air and also means less air-flow restriction and less frequent cleaning and attention.

Even where only a single cleaner is used, the air intake is usually extended considerably above the tractor, as shown by Fig. 121, in order to relieve the cleaner to a certain extent.

127. Breather Cleaners.—Any engine, such as a tractor motor with an enclosed crankcase, must have a breather opening and cap to permit the unequal pressure in the crankcase caused by the pumping effect of the pistons, to be balanced by the outside atmospheric pressure. The air drawn in by this action obviously carries considerable dust and dirt with it, which, when taken up by the lubricating oil, eventually causes considerable wear.   This trouble is prevented by using a special breather cap containing filtering material such as steel wool, or by connecting the breather opening to a special filter as shown by Fig. 319. [322]

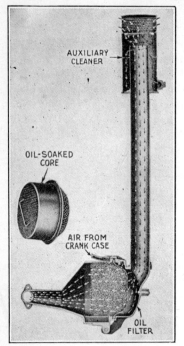

Fig. 120.—Combination inertia- and oil-type cleaner.

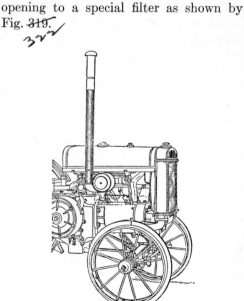

Fig. 121.—Air-cleaner extension.

HOW TO GET BEST RESULTS FROM AIR CLEANERS

Hoffman,[1] as a result of extensive studies and tests, offers the following advice and information concerning air cleaners:

128. No Leaky Connections.—No leaky connection should be permitted between air cleaner and carburetor or between parts of the air cleaner itself.   A loosely fitting slip joint or an ordinary flexible metal tube is almost sure to admit some air and a great deal of dust.   A piece of radiator hose, fitting tightly over the tubes to be connected, is very satisfactory if the connection is short.   Ordinary friction tape is a very satisfactory emergency material for stopping air leaks.   Filters, whether of cloth or fiber, cannot be efficient if they have holes in them.   The dusty air will go in largest quantity by the path of least resistance.

129. Breather Pipe Should Have Clean Air.—A special cleaner may be used or connection may be made to the carburetor air cleaner.   If clean air is not provided, the breather pipe itself acts as an oil-type air cleaner taking in breaths of dusty air, collecting the dust and sand on the oily walls, then blowing out puffs of clean air.   When dust goes in through the carburetor there is a chance that some

[1] Dust and the Tractor Engine, *Calif. Exp. Sta., Bull.* 362.

of it may escape through the exhaust. Not so with dust on the walls of the breather pipe. It remains until the oil washes it down into the crank case. When it is not feasible to connect the breather to the air cleaner, a small bag of eider-down blanketing or cotton flannel, lint side out, may be tied over the breather pipe.

130. Place Air Intake High.—Except in orchard work it is usually possible to use a periscope or high vertical extension for the air intake. The advantages are that the quantity of dust to be removed by the air cleaner is greatly reduced and that the coarser dust is avoided. Some air cleaners are regularly furnished with a jointed periscope. Part or all of it should be used whenever feasible. If a periscope is provided, it should be smooth inside and free from sharp turns. The inside diameter should be amply large (not less and preferably more than 2 in. for a 10–20 tractor), or the vacuum effect will be unnecessarily increased.

131. Give Needed Attention.—There is not on the market, so far as the writer is aware, any air cleaner that can be put on a tractor or automobile engine and forgotten, and yet, month after month, give adequate protection against dust. All require some attention and have their own peculiar troubles. Those with moving parts have troubles due to wear and to accumulation of dust and oil. Thus the float in some water-type cleaners may have holes worn which would permit dusty air to pass without going through water. Centrifugal types may become so encrusted with oil and dust that their action may almost entirely cease. Cleaners having small passages inside may clog up solid. Other kinds may so increase their vacuum effect due to accumulation of dust that the power of the engine may be greatly reduced. Nearly every air cleaner has its plate giving the manufacturer's directions for the care required. Some tractor operators may be able to improve upon these directions, but none may safely neglect them.

132. A Simple Efficiency Test for Air Cleaners.—Not what the air cleaner catches, but what it lets go by does the mischief in a tractor engine. Whether any appreciable amount of dust gets past an air cleaner may usually be known by disconnecting the air cleaner from the carburetor and wiping out the inside of the connecting tube. If, after a 10-hr. run under dusty conditions, only a trace of dust can be wiped out, the cleaner has probably done a first-rate job. The test may fail, if, as might possibly happen, so much water or oil should go over from the cleaner that the tube would be kept washed out.

References

Air Cleaners for Motor Vehicles, *Univ. Calif., Agr. Exp. Sta., Bull.* 499.
Air Cleaners on Trucks in Service, *S.A.E. Jour.*, Vol. 16, p. 249.
Air Filters, *S.A.E. Jour.*, Vol. 15, p. 66.
Best Location for Carburetor Intake, *S.A.E. Jour.*, Vol. 16, p. 501.
Does Your Carburetor Air Inlet Face Forwards? *Agr. Eng.*, Vol. 8, p. 13.
Dust and the Tractor Engine, *Univ. Calif., Agr. Exp. Sta., Bull.* 362.
Dusts Used for Testing Air Cleaner Efficiency, *Univ. Calif., Agr. Exp. Sta., Hilgardia*, Vol. 5, No. 2.
Efficiency Test of Radiator Fan-type Air Cleaners, *S.A.E. Jour.*, Vol. 21, p. 82.
HELDT: "The Gasoline Automobile," Vol. IV.
Mapping the Dust Concentration around Small Tractors, *Agr. Eng.*, Vol. 7, p. 12.
Report of 1924 California Air Cleaner Tests, *S.A.E. Jour.*, Vol. 16, p. 367.
Testing of Air Cleaners, *S.A.E. Jour.*, Vol. 15, p. 33.
The Physical Characteristics of Road and of Field Dust, *S.A.E. Jour.*, Vol. 16, p. 243.

CHAPTER XI

COOLING AND COOLING SYSTEMS

The temperature produced in the cylinder of a gas engine, at the instant the explosion takes place, varies from 2000 to 3000°F. In other words, there is an excess quantity of heat liberated that must be taken away rapidly. It is estimated that about one-third of this heat escapes through the cylinder wall. Consequently, the cylinder becomes very hot, particularly at the head and around the exhaust valve, and some means of conducting away the heat is necessary. If an engine were not equipped with some means of cooling, at least three troubles would arise as follows:

1. The piston and cylinder would expand to such an extent that the former would seize in the cylinder, injuring the latter and stopping the engine.
2. The lubricating qualities of the oil supplied to the cylinder and piston walls would be destroyed by the high temperatures existing.
3. Preignition of the fuel mixture would take place, resulting in knocking and loss of power.

All internal-combustion engines must operate at a certain temperature to produce the best results and seldom give the greatest efficiency unless the temperature around the cylinder varies from 160 to 200°F. Therefore, a cooling system that permits an excessive absorption of heat, resulting in a low operating temperature, is undesirable and indicates improper design.

The following is a classification of the common methods and systems of engine cooling:

1. Air.
2. Liquid.
   a. Open jacket or hopper.
   b. Thermosiphon.
   c. Forced circulation.
3. Combination of Air and Liquid.

132a. Air Cooling.—Cooling by air alone is not used extensively but is satisfactory for certain types of engines and under certain conditions. The cooling effect is produced usually by means of fins or projections on the walls of the cylinder as shown in Fig. 43. These fins may be placed transversely or longitudinally with respect to the cylinder, depending upon the use of the engine and the direction of the air flow past the cylinder. Such an arrangement of fins increases the radiating surface, and therefore the heat escapes faster than it would otherwise. Figure 122 illustrates an engine that has the cylinder and crankcase inclosed in a sheet-metal housing and the flywheel equipped with blades so that it will create a suction of air down through the cylinder, thus producing a greater cooling effect.

Fig. 122.—An air-cooled engine using the flywheel to produce circulation.

Air-cooled engines are usually of small bore and stroke, that is, they have small cylinders. Multiple-cylinder air-cooled engines have the cylinders cast individually rather than in pairs or in one block, so that the maximum cooling effect will be obtained. Common examples of air-cooled engines are those used in motorcycles and some small farm-lighting plants. A few automobile engines and certain types of airplane engines are likewise cooled by air. An air-cooled engine has the following advantages:

1. It is light in weight.
2. It is simpler in construction.
3. It is more convenient and less troublesome.
4. There is no danger of freezing in cold weather.

The principal disadvantage of air cooling is that it is difficult to maintain proper cooling under all conditions, and it is almost impossible to control the cylinder temperature. Air-cooled engines usually run a little hotter than water-cooled engines and require the use of heavier lubricating oil.

133. Cooling by Liquid.—Cooling systems using liquids, usually water, are used for all types of engines from the simple stationary farm engine to the most complicated multiple-cylinder high-speed types.

Water might be termed the universal cooling liquid for tractors, as well as for trucks and automobiles. It has certain important advantages, among which are the following:

1. It is plentiful and readily available nearly everywhere.
2. It absorbs heat well.
3. It circulates freely at all temperatures between the freezing and boiling points.
4. It is neither dangerous, harmful, nor disagreeable to use or handle.

The principal disadvantages of water for cooling are:

1. It has a high freezing point.
2. It may cause excessive corrosion of the radiator and certain metal parts of the engine. Clean, pure water such as rain water gives the best results.
3. It may cause troublesome deposits in the cylinder jackets.
4. Evaporation and boiling require frequent replenishing.

134. Oil for Cooling.—Oil instead of water has been used as the cooling liquid in some instances. The oil recommended is a crude oil product that is dark in color and comparatively thin. It should be a nonfreezing oil that has a minimum boiling temperature of about 400°F. Zero Black, Winter Black, Polar, and Arctic Ice Machine Oil are trade names of such oils.

The advantages of oil for cooling are:

1. It does not freeze in cold weather.
2. It is noncorrosive and does not injure the tank, radiator, and cylinder jacket in any way.
3. It evaporates slowly, requiring less attention.
4. It maintains a more uniform cylinder temperature.

The disadvantages of oil are:

1. It is not always available when needed.
2. Great care must be exercised in guarding against possible leakage because a leak would create a fire risk.
3. It destroys rubber-hose connections.
4. It may produce objectionable fumes and odors.
5. Its heat-absorbing capacity is low; therefore, larger cooling space, radiator, and connections and more liquid are necessary to maintain the same temperature.

Kerosene is sometimes used as a cooling medium, especially in automobiles in cold weather, but it has the disadvantage of destroying rubber-hose connections and thereby creating leakage and danger of fire. It likewise absorbs the heat more slowly than water and may cause the engine to overheat.

LIQUID COOLING SYSTEMS

135. Open-jacket or Hopper Cooling.—The simplest system of liquid cooling, known as the open-jacket or hopper system (Fig. 123), consists of a hollow space around the cylinder and cylinder head, known as the water jacket, and a cast-iron reservoir, usually cast with the cylinder and opening directly into this jacket. Such an arrangement is simple and convenient but is not adapted to portable engines because the cooling is dependent upon the heat taken up by the water as it evaporates and escapes from the hopper in the form of steam; therefore, if an automobile or tractor engine, for example, were equipped with this system, the water would naturally be thrown and splashed about. Another disadvantage is that more liquid is required. Consequently, the weight of the engine is greater. It is found only on stationary single-cylinder farm-type engines.

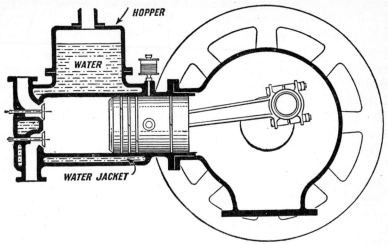

Fig. 123.—The open-jacket or hopper system of cooling. (*Courtesy of Standard Oil Company* [*Indiana*].)

136. Thermosiphon System.—The thermosiphon system of liquid cooling includes, in addition to the water jacket about the cylinder, a separate tank or reservoir connected at the top and bottom by a pipe

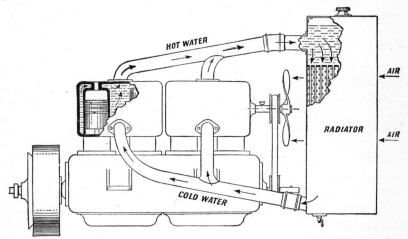

Fig. 124.—The thermosiphon system of cooling. (*Courtesy of Standard Oil Company* [*Indiana*].)

leading to the upper and lower parts of the cylinder, respectively. In tractors, this tank or reservoir is replaced by what is known as the radiator, this being nothing more than a tank made up of many fine passages through which the water flows and cools more rapidly than if held in one solid mass.

Figure 124 illustrates the construction and operation of this system. The operation is based upon the fact that, when water or any other liquid is heated, it expands and its weight per unit volume decreases; that is, the temperature of the water in the cylinder jacket increases as the engine warms up, and, therefore, the liquid expands and decreases in specific gravity and rises, being pushed out by the heavier cold water coming in from the reservoir through the lower pipe. A slow but continuous circulation is thus started and continues as long as the engine runs, the

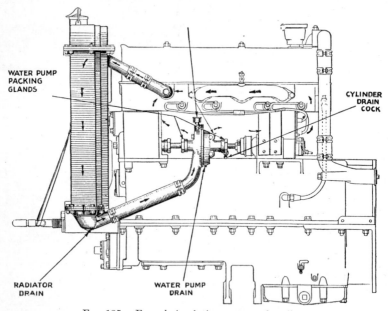

FIG. 125.—Forced-circulation system of cooling.

water passing upward through the upper pipe into the tank and reentering the cylinder from the tank by means of the lower connection. Since the circulation is somewhat slow or sluggish, large radiator connections and a slightly greater amount of liquid are necessary.

137. Forced-circulation System.—The forced-circulation system resembles the thermosiphon system with the exception that some sort of a pump is placed in the lower pipe leading to the cylinder, as shown by Fig. 125. This pump, usually of the centrifugal type, forces the water through the cylinder jacket and around to the reservoir, causing a more rapid circulation than that produced by the thermosiphon system. The advantage, therefore, of this system is that, since the water is circulated more rapidly, it is cooled faster, and less liquid is required to produce the same cooling effect. Another advantage is that, since the circulation is dependent upon the speed of the pump and therefore upon the speed of the engine, a more uniform temperature is apt to be maintained at all engine speeds and loads. The principal objections to the forced-circu-

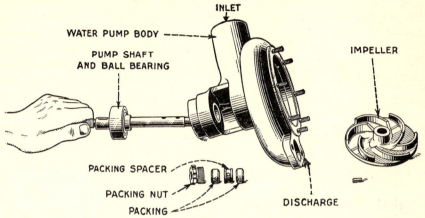

FIG. 126.—Water-circulating pump showing construction and parts.

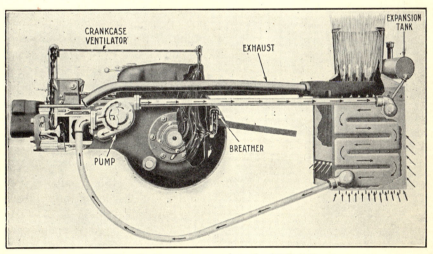

FIG. 127.—Cooling system used on the Oil Pull Tractor.

lation cooling system are the greater number of connections and the occasional leaking of the pump.

Occasional leaking, where the drive shaft enters the pump body, can always be expected in a centrifugal pump. This is usually overcome by tightening the packing nut or by repacking. In time, however, some difficulty may be encountered in stopping these leaks because of the shaft becoming rough due to corrosion.

138. Combined Air and Liquid Cooling.—The combination air and liquid cooling system is like that found in the automobile. A fan driven at high speed draws the air through the radiator, cooling the liquid rapidly, and also sends a blast of air past the cylinders driving the heat away from the engine.

139. Oil Cooling System.—A cooling system using oil is shown in Fig. 127. The radiator is specially constructed and differs materially from the usual type. It consists essentially of a number of thin, flat, rectangular sheet-metal sections, arranged vertically and connected together at the top and bottom by means of leak-proof fittings so that there is a vertical air space between the sections. A single section is made of two pieces of sheet metal, stamped out and put together with all four edges sealed. The complete section assembly is shown by Fig. 127. The hot oil from the engine enters the radiator from the upper pipe connection. As it passes slowly downward through the zigzagged section spaces, it gives up much of its heat, because both outer surfaces of each section are exposed to the air. By the time the oil has reached the lower pipe, it is sufficiently cooled to be conducted back around the cylinder jacket and again take up the heat. Circulation is produced by a centrifugal pump as shown.

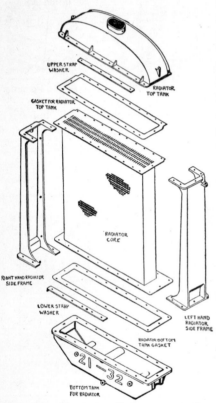

Fig. 128.—Tractor radiator construction and parts.

To insure the maintenance of a more nearly uniform cylinder temperature at any and all loads, the exhaust-discharge opening is placed directly above the radiator and is inclosed in the outer radiator housing, which terminates in a sort of exhaust or smokestack as shown. The exhaust outlet is in the form of a spiderlike device that spreads the exhaust out over a greater area and also makes it discharge vertically into the stack. Thus, the firing of the engine produces a suction or draft up through the radiator sections which increases the cooling effect. Furthermore, when the load is light, this suction will be weaker and the cooling effect less. With a heavy load, the exhaust will be very strong, the suction will be more intense, and the cooling action more pronounced. The effect, therefore, is to maintain the same cooling-liquid temperature at all engine loads.

140. Radiator Construction.—Water-cooled tractor engines are equipped with the conventional auto- or truck-type radiator (Fig. 128), consisting of the core, an upper and a lower reservoir, and the side members or frame pieces. Since tractors are subjected to considerable jarring and vibration, and surplus weight is of little consequence, the reservoirs and frame parts are made of cast iron rather than sheet metal.

Fig. 129.—Tubular radiator construction.

There are two general types of radiator cores, the tubular type with fins (Fig. 129) and the cellular or honeycomb type (Fig. 130). The former seems to predominate probably because of the lower manufacturing cost. The tubes are either round as in Fig. 129, or flat (Fig. 131). Horizontal, thin, metal fins fastened to the tubes increase the rate of heat radiation. The cellular radiator produces very effective cooling, but this does not offset its higher manufacturing cost.

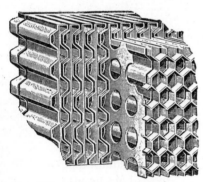

Fig. 130.—Cellular-type radiator construction.

Fig. 131.—Flat-tube type of radiator construction.

141. Circulating Pumps.—Engines equipped with the forced-circulation system of cooling use a centrifugal-type pump (Fig. 126) to produce this circulation. Such a pump consists of a cast-iron body, the rotating member or impeller with its curved blades, the drive shaft, and the packing nut or gland and packing. The water enters the body near the center and is discharged at a tangential opening by the centrifugal action of the impeller.

The pump may be located between the lower radiator connection and the cylinder block as in Fig. 125, or directly behind the fan (Fig.

132). In the first case, it is driven by a separate shaft as shown; in the second, it is driven directly by the fan shaft and drive.

142. Fans and Fan Drives.—The cooling fan in all tractors is located just behind the radiator (Fig. 133). It does its work by drawing the air through the radiator at a high velocity and by maintaining rapid movement of the air around the engine. This action obviously takes the heat away very rapidly.

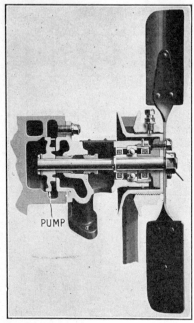

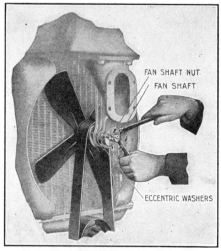

FIG. 132.—Water pump and fan on same shaft.

FIG. 133.—Eccentric device for tightening fan belt. (*Courtesy of Ford Motor Company.*)

Fans may have two, four, or six blades, depending upon the size of the engine and the fan speed. The sheet-metal housing attached to the rear of the radiator and partly surrounding the fan is known as the fan shroud.

Some typical fan mountings are shown in Figs. 134 and 135. It will be noted that antifriction bearings are used in all cases, because they are more durable, easier to lubricate, and give little power loss.

Two types of fan drives are found in tractors, namely, belt (Fig. 133) and gear (Fig. 135). The belt drive with either a flat belt or a V-belt is probably the more used of the two. Some provision must be made for tightening the belt as shown by Figs. 133 and 134. The gear-driven fan requires practically no attention except lubrication and cannot fail to operate at regular speed unless the gear becomes loose, which is not likely to occur.

FIG. 134.—Belt-driven fan with tapered roller bearings.

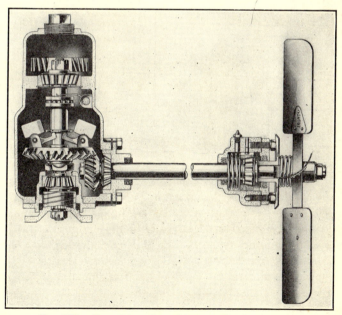

FIG. 135.—Gear-driven fan with tapered roller bearings.

Some fans are not attached rigidly to the drive shaft but are driven from this shaft through a simple friction-clutch device, as shown in Fig. 135. A heavy spring maintains the proper tension, which can be adjusted if necessary. The purpose of this device is to overcome the wear and loosening tendency owing to shocks when the engine starts and stops.   Since a fan has a relatively high speed, the sudden starting or stopping of the engine subjects the fan to a rather sudden jerk which might gradually loosen a rigid, keyed connection.

143. Engine-heat Control.—Since tractors are operated under a great variety of weather and load conditions, most designers provide some means of control by which a fairly uniform engine-operating temperature may be maintained regardless of the varying factors mentioned.   The best operating temperature is between 170 and   190°F. If the cooling liquid can be maintained within this temperature range at light as well as heavy loads, or in cold as well as warm weather, better engine performance will be secured.

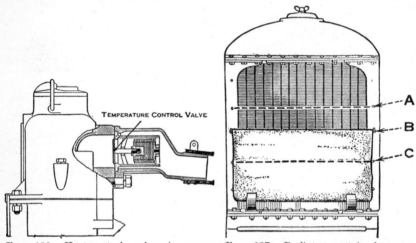

TEMPERATURE CONTROL VALVE

A

B

C

FIG.   136.—Heat-control   valve   in   upper   FIG.   137.—Radiator   curtain   for   con-
             radiator connection.   trolling engine temperature.

One such method of heat control is shown by Fig. 136.   When the engine is cold the thermostatic valve in the upper radiator connection is contracted and the passage closed so that no circulation takes place. When the liquid gets hot enough, the copper bellows expands and opens the valve, and circulation begins.   As long as the engine is operating at the proper temperature, the valve remains open.   Should the temperature start to drop, the valve closes enough to restrict the circulation and thereby prevents any appreciable temperature change.

Another heat-control device consists of an adjustable canvas curtain attached to the front of the radiator (Fig. 137).   This curtain can be placed at different heights so as to cover a certain portion of the radiator depending upon the conditions.   At very light loads or in cold weather about one-half to three-fourths of the radiator could be covered.   In medium-cool weather it should be covered about one-third, and so on. Observation and experience will best indicate the setting of the curtain.

Some tractor radiators are covered with a wire screen for preventing grass, weeds, leaves, and similar material from clogging the radiator and interfering with its operation. Occasional cleaning of the screen is usually necessary but this is easier than cleaning the radiator passages.

144. Antifreezing Cooling Mixtures.—In certain sections, where freezing temperatures exist for several days or even months at a time, it is often advantageous to replace the water in a cooling system with some solution that will not freeze readily. Freezing of the cooling solution usually results in one or more troubles as follows:

1. It may crack the cylinder head or block and produce either an internal or an external leak.

2. It may weaken the radiator and connections or create a leak in these parts, which is often difficult to repair.

3. Freezing at a certain point in the cooling system during operation of the engine may interfere with the proper circulation of the cooling liquid and permit the engine to run too hot.

A number of liquid materials, used either alone or mixed with water, can be utilized to prevent these troubles. Only a very few, however, meet the usual requirements of a satisfactory antifreeze solution which are as follows:[1]

1. The ingredients used should be easily obtainable in operating localities.

2. The possibility of freezing should be negligible.

3. The solution should not be injurious to either the engine or the radiator through corrosion or electrolytic action, or to rubber-hose connections.

4. It should not lose its nonfreezing and noncongealing properties after continued use.

5. The possibility of fire hazard should be a minimum.

6. The boiling point of the solution should not differ materially from that of water.

7. The viscosity should be as constant as possible through the entire temperature range involved, and the solution should remain perfectly fluid and not tend to stop up any small openings in the system.

8. It should be able to conduct heat away as rapidly as possible.

As previously mentioned, an oil, such as kerosene, might be used but is not recommended for the reasons stated. A solution of calcium chloride and water, usually about $3\frac{1}{2}$ lb. of calcium chloride to 1 gal. of water, will not freeze except at very low temperatures; but unless the calcium chloride is pure, it may cause excessive corrosion or deposits on the inside of the cylinder jacket and water tank or radiator.

A mixture of denatured alcohol and water is one of the best antifreezing solutions. It has no destructive action of any kind and is not

[1] *Lubrication,* Vol. 13, No. 11, publ. by The Texas Company, New York.

dangerous. Its principal disadvantages are that it has a low boiling point and the alcohol evaporates rather rapidly and must be replenished frequently. The mixture used depends upon how low the temperature may go. Under some conditions a solution of 60 per cent alcohol and 40 per cent water by volume may be necessary; but ordinarily a solution containing 40 per cent alcohol and 60 per cent water by volume will not freeze at 0°F. or above. In replacing evaporation, more alcohol than water must be replaced because the alcohol evaporates faster.

Antifreezing solutions containing glycerine are rapidly coming into favor. The first cost of the glycerine is greater than that of alcohol and other materials, but since there is little or no evaporation the total cost over a long period may be even less. If the pure 95 per cent glycerine is used, it should be mixed with water to form a 50 per cent mixture by volume. A special radiator glycerine, produced by certain manufacturers and marketed under various trade names, is likewise proving satisfactory. This product is usually about 60 per cent glycerine by weight. Therefore, it is recommended that 3 to 4 parts by volume be mixed with 1 part of water.

Table IX gives some ready information concerning the most common antifreezing solutions and their freezing points.

TABLE IX.—ANTIFREEZING MIXTURES FOR TEMPERATURES DOWN TO 0°F.[1]

Material	Quality, per cent	Quantity, per gallon of water	Sp. gr. of mixture at 60°F.	Boiling point of mixture, degrees Fahrenheit
Denatured alcohol...........	90 (by vol.)	2.5 qt.	0.958	187
Wood alcohol..............	97 (by vol.)	1.6 qt.	0.966	185
Pure glycerine.............	95 (by wt.)	2.7 qt.	1.112	221
Radiator glycerine..........	60 (by wt.)	8.5 qt.	1.112	221
Ethylene glycol............	95 (by wt.)	1.9 qt.	1.048	219
Calcium chloride...........	75	3.0 lb.	1.178	221
Honey....................	....	8.5 qt.	1.296	225

[1] *Journal S.A.E.*, Vol. 19, No. 1, p. 99.

For a temperature of −20°F., add the following quantities per gallon of water:

Antifreezing mixture	Quarts
Denatured alcohol...	4.5
Wood alcohol...	2.7
Pure glycerine..	5.0
Ethylene glycol...	3.3

145. Miscellaneous Considerations.—In selecting an engine from the standpoint of the cooling system, consideration should be given to the purpose for which the engine is to be used, the type of engine, and the construction of the cooling system itself. A stationary farm gas engine or tractor, which usually operates for long periods of time under heavy loads, should be equipped with a system that will give efficient and thorough cooling at all times. Where high speed and light weight are necessary, an air-cooling system may prove satisfactory. In the selection of the system itself, observation should be made of the cooling space provided around the valves, and there should be a good drain at one or more of the lowest points of the system so that the liquid can be drained out readily when necessary.

CHAPTER XII

GOVERNING AND GOVERNING SYSTEMS

Nearly all stationary farm machines, such as threshers, ensilage cutters, feed mills, and so on, must operate at a certain uniform speed. The power-supplying device must, therefore, maintain this speed whether the machine is running empty or at its maximum capacity. In other words, it must instantly adjust its power output to the power required, at the same time maintaining uniform speed. For this reason, the stationary farm gas engine or farm tractor, when used for belt work particularly, must be equipped with a governor, a mechanical device that instantly adjusts the power output to the power requirements of the driven machine in such a way as to maintain practically a uniform speed.

Suppose a wood saw is being operated by a stationary engine. The load on the latter fluctuates constantly: that is, when the saw is cutting, the load is a maximum; and when the stick is shifted for the next cut, the load becomes a minimum. If the power output of the engine remained at the maximum required by the saw when cutting, the engine would run away or race under lighter loads, and wear rapidly or fly to pieces. A governor on the engine would vary its power output to meet the power required by the saw and thus keep the speed of the latter practically constant.

On the other hand, it is desirable and necessary to vary the engine speed in such machines as automobiles, trucks, and airplanes in order to obtain different traveling speeds. When such a machine is in use, the operator of course must be present to guide it; therefore, the manual control of the speed involves no great difficulty. The speed of a tractor engine operating a stationary belt-driven machine might be held fairly constant by the manual operation of a convenient hand lever, but the use of a governor would eliminate the need of the extra man and soon pay for itself.

146. Systems of Governing.—For stationary farm engines two systems of governing are used. Some engines that burn gasoline are equipped with what is known as the *hit-and-miss* system. Those farm engines that are made to operate on kerosene, as well as all farm tractors and some trucks, are equipped with the *throttle* system of governing.

147. Hit-and-miss System.—The fundamental principle involved in the hit-and-miss system of governing is to keep all explosions alike and at the maximum intensity, but to vary the number per time interval depending upon the power output required.

Example.—A single-cylinder four-stroke-cycle engine running at 500 r.p.m. and developing its maximum power would produce a maximum of 250 explosions per minute. If the belt suddenly slipped off, relieving the engine of its load, there would be an excessive increase in speed. If, on the other hand, the number of explosions per minute were reduced by some mechanical means to, say 50, the power output would be correspondingly less and the engine would continue to run at its normal speed of 500 r.p.m.

This system of governing, then, varies the number of explosions or working cycles per time interval so as to maintain a reasonably constant engine speed at any and all loads, the same maximum mixture charge being taken into the cylinder for each explosion.

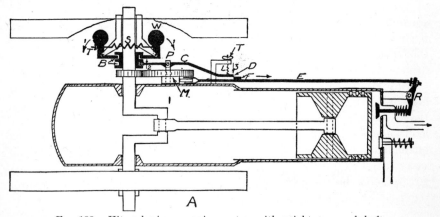

Fig. 138.—Hit-and-miss governing system with weights on crankshaft.

The governing mechanism (Fig. 138) consists of the following parts: weights *W*, sliding collar *B*, detent arm or finger *C*, and notch or catch block *F* on exhaust-valve push rod *E*. The weights, usually located on one flywheel on the crankshaft, are held together by springs, but when rotated are drawn apart against the spring tension by centrifugal force. The greater the engine speed, therefore, the farther they separate. This expansion of the weights slides the grooved collar on the crankshaft, and this sliding movement of the collar reacting on one end of the pivoted detent arm causes the opposite end to catch in a notch on the exhaust push rod, holding the valve open. This valve being open, a fuel charge will not be drawn in and compressed and fired on the next cycle. Consequently, the engine will lose speed, and the springs will draw the weights together owing to the decreasing centrifugal force, thus sliding the collar back and releasing the detent arm and push rod, and permitting another explosion. Such an engine, when carrying light loads, operates at a more or less uneven or fluctuating speed, but when running at full load fires steadily and runs uniformly.

The governor weights, in some cases, as shown by Fig. 139, are not located on the crankshaft but are mounted separately and driven by a small spur pinion meshing with the cam gear.

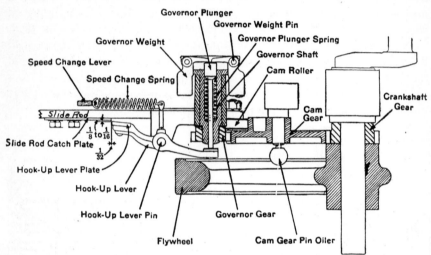

Fig. 139.—Hit-and-miss governing system with gear-driven weights.

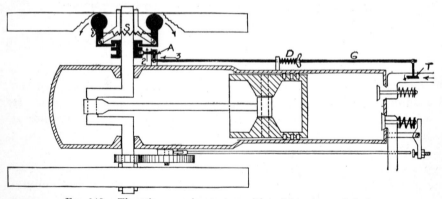

Fig. 140.—Throttle governing system with weights on crankshaft.

As previously stated, the hit-and-miss system of governing is not used on kerosene-burning engines because the interval between explosions during which the exhaust valve is held open permits the cylinder to become cooled rather than to remain at a uniform temperature. Therefore, this uniformity of conditions, so essential for the vaporization and combustion of kerosene, does not prevail.

148. Throttle System.—The principle of the throttle system of governing is to permit the engine to fire the maximum number of times, that is, on every fourth piston stroke, regardless of the load, but to vary the fuel charge per cycle and the resultant explosion intensity as the load varies. That is, an engine having a normal speed of 500 r.p.m. would fire 250 times per minute at any load, the governor functioning in such a manner as to reduce the quantity of fuel mixture entering the cylinder on each intake stroke at light loads and increasing it with an increasing load.

The mechanism (Figs. 140 and 141) consists of (1) set of weights, held together by springs and located on the crankshaft or driven by a special pinion, (2) a sliding collar, (3) a throttle connecting rod or arm, and (4) a throttle butterfly. As the engine speed increases, the weights fly apart moving the collar and actuating the throttle connecting rod, thus partly

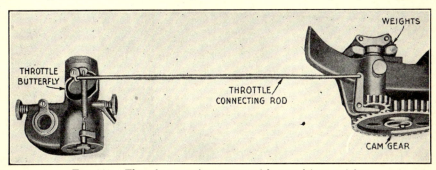

Fig. 141.—Throttle governing system with gear-driven weights.

closing the throttle butterfly and cutting off the fuel mixture. If the engine speed decreases with an increased load, the springs counteract the centrifugal force on the weights, drawing them together and sliding the collar so as to open the throttle. If the mechanism is well designed and operates freely, the power output of the engine will be adjusted instantly to the power required and a uniform speed produced and maintained at all loads.

TRACTOR GOVERNING

A governor is necessary on a tractor engine to maintain a uniform speed with varying loads, particularly when operating belt-driven machines. Again a governor protects the engine against operation at excessive speeds when doing either belt or drawbar work.

149. Governing Mechanism.—All tractors are equipped with the throttle system of governing, the mechanism being actuated and controlled by centrifugal force by means of rotating weights. A typical layout is shown by Fig. 142. Referring to this figure, the operation is as follows: Two weights in the gear chamber behind gear *A* are rotated by this gear and are held by and swivel about the two pins B. These weights fly out when the motor speeds up, moving governor shifter S outward. This action presses the ball-bearing thrust contained in ball-bearing housing D outward, moving lever E. This lever swivels on pin F, which is fixed in the gear cover. The movement of E causes a movement of rod G and closes butterfly valve H.

The angle at which the butterfly valve stands can be changed by disconnecting ball joint P from lever E and giving the ball joint one or more turns on rod G. Turning it to the right will cause H to stand more nearly vertical; and to the left will cause it to become more horizontal.

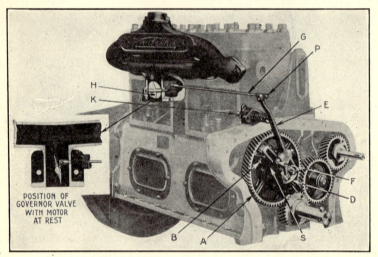

POSITION OF
GOVERNOR VALVE
WITH MOTOR
AT REST

Fig. 142.—Tractor governor showing parts and construction.

The adjustment for speed is made by turning screw K. Turning it to the right speeds up the motor, while turning it to the left slows down the engine.

Figure 143 illustrates another tractor-governing mechanism. It will be observed that the general construction, operation, and adjustment are practically the same. The weights are located in the front gear housing of the engine and rotated by a gear from the crankshaft. All mechanism, with the exception of the lever rods leading to the throttle butterfly, is enclosed and oiled largely by a mist from the crankcase. This eliminates the possibility of poor governing due to wear and lost motion, or sticking and hanging. Any tractor governor in order to function properly must work freely but with as little lost motion in the connections as possible.

It is possible to adjust a governor to maintain a higher or a lower operating speed. In some cases this is done by means of a specially provided hand screw, as shown in Fig. 143, which changes the tension on the governor spring. Others do not have this provision but the speed can be changed by shortening or lengthening the throttle rod. As a rule, such practice is not recommended. Every tractor is designed by the manufacturer to operate within a certain speed range, and any changes in this respect are seldom necessary or advisable.

The engine speed of any tractor can be controlled, to a certain extent, by means of the hand throttle. This is connected to the butterfly valve indirectly through the governor spring, lever arm, and throttle connecting rod. When in the closed position, it holds the throttle butterfly nearly

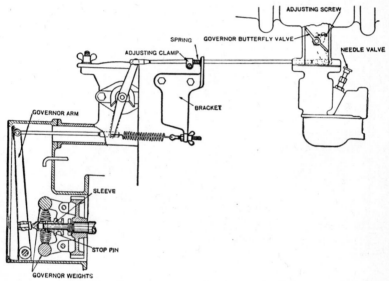

Fig. 143.—Tractor governor showing speed adjustment.

closed and counteracts any tendency of the contracted weights to open the valve. As the hand lever is moved toward the open position, the control of the valve is gradually shifted to the governor until, in the wide-open position, the governor has complete control.

150. Automatic or Vacuum-type Governor.—Motor trucks, busses, and automobiles are sometimes equipped with a governor for the purpose, primarily, of maintaining a set maximum road speed that may be considerably below the maximum speed possible without such a device. An automatic or vacuum-type governor (Fig. 144) is used for this purpose. It is a simple device located between the carburetor and the intake manifold and has no mechanical connection to any other parts of the engine.

Referring to Fig. 144, the device consists of a housing, a throttle butterfly valve mounted off center and connected to a spring-controlled cam-and-lever mechanism. With the butterfly mounted off center, the longer portion is subjected to the pressure reaction of the gas mixture passing through the device. Thus as the engine speed and resultant suction increase, the reaction tends to close the valve and thereby to maintain a certain maximum speed. As the engine speed and suction decrease, the spring opens the valve. The desired speed range is maintained by a simple adjustment of the spring tension. The successful operation of this type of governor depends largely upon precision of construction, freedom of movement of the parts, and proper installation, adjustment, and adaptability to the engine.

151. Speed Variation.—A perfectly functioning governing mechanism should not permit any appreciable variation in the engine speed between

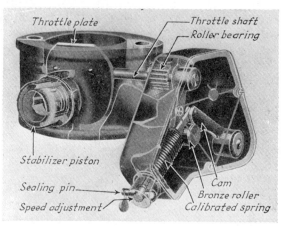

FIG. 144.—Vacuum-type governor.

no load and full load. Obviously the natural action is for an engine to run faster without load, with the speed dropping off as the load increases. The total variation will depend largely upon the rated engine speed. If it is a high-speed engine with a rated r.p.m. of 1,200, for example, a variation of 75 to 100 r.p.m. would not be excessive or unreasonable. A low-speed engine whose rated r.p.m. is 750, say, should not show over 50 r.p.m. variation. Poor governor control is usually due to faulty or imperfect design, but may be caused by lack of good lubrication of all working parts, by paint in the lever connections, by a bent, broken, or poorly fitted throttle butterfly, or by incorrect adjustment. The speed of an engine can be checked readily by means of a speed indicator (Fig. 381). The tip is inserted in the end of the crankshaft and the indicator held firmly with one hand with a watch in the other hand. One complete turn of the indicator disk is 100 revolutions. By counting the number of disk r.p.m. and multiplying by 100, the total r.p.m. is secured.

152. Governor Hunting.—Frequently when an engine is first started or after it is warmed up and is working under load, particularly in belt work, its speed will become uneven or irregular. It speeds up quickly, the governor suddenly responds, the speed drops quickly, the governor responds again and the action is repeated. This is known as *hunting*. It is usually caused by an incorrect carburetor adjustment and can be corrected by making the mixture either slightly leaner or slightly richer. It is possible also for the governor itself to cause hunting by being too stiff or striking or binding at some point, so that it fails to act freely.

CHAPTER XIII

IGNITION AND IGNITION METHODS

In any internal-combustion engine, the fuel mixture must be ignited or "set on fire," so to speak, before the explosion can take place and power be generated. This ignition always takes place near the end or during the latter part of the compression stroke. In the development of the internal-combustion engine, four different methods of ignition have been devised as follows:

1. Open-flame method.
2. Hot-tube, hot-bulb, or hot-bolt method.
3. Ignition by the heat of compression.
4. Electric-spark ignition.

Of these four the last three named are still in use, but by far the greatest number of engines utilize the electric-spark method in some form or another.

153. Open-flame Ignition.—The open-flame arrangement was perhaps the first successful means of ignition. However, it was soon displaced by the other methods mentioned, as they proved more satisfactory for several reasons. This system consisted of two gas jets, one burning continuously on the outside of the engine and a second jet that alternately communicated with the first jet and the combustion chamber. This second jet was extinguished each time by the explosion and, therefore, had to be relighted. It was placed on the inside of a hollow, rotating valve whose one port or opening first registered with an opening near the first jet. This lighted the second jet. Further movement of the valve permitted its port to register with an opening leading to the combustion space just as compression was nearly completed. Thus, the charge coming in contact with the flame was ignited, and an explosion produced. As previously stated, this system of ignition survived only a short time, owing to the immediate development of other more reliable ones.

154. Hot-tube and Hot-bulb Ignition.—The hot-tube system of ignition originally consisted of a tubelike projection on the end or side of the cylinder, as shown in Fig. 145. The outer end of the tube was closed and the other end opened into the combustion space. It was heated by a torch or flame and, as the charge was compressed, a certain portion entered the tube and was ignited by contact with the hot inner surface. The ignition of the charge in the tube then spread or propagated through the mixture in the cylinder and produced an explosion. The

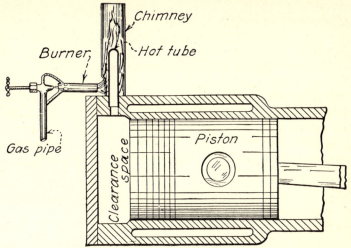

FIG. 145.—Simple hot-tube method of ignition.

tube was sometimes covered by an asbestos-lined jacket to retain the heat. In some cases it was necessary to keep the torch burning as long as the engine was in operation. In others, the flame was taken away

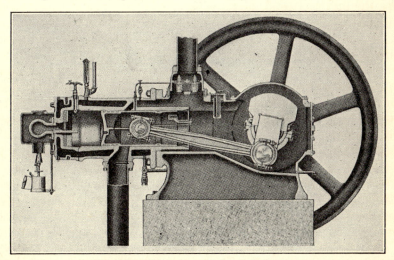

FIG. 146.—Oil engine with hot-bulb ignition.

once the engine was started and the hot exhaust gases which remained in the tube following the explosion served to keep it hot and ignite the next charge.

The original hot-tube arrangement just described has been superseded by a modification known as the hot bulb (Fig. 146).   The bulblike projection on the cylinder head is heated by a torch until the engine starts.   In most cases this torch is then removed or extinguished, the hot exhaust gases continuing to keep the bulb sufficiently hot to produce ignition.

Some heavy-duty, stationary oil engines are equipped with a bolt or pin in the cylinder head that projects into the combustion space. This bolt, when heated by some external means, provides the ignition for starting the engine.   Once the engine is in operation the heating torch

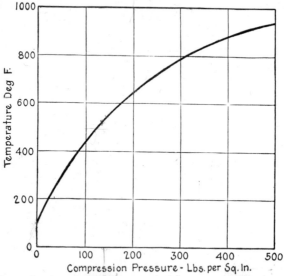

Fig. 147.—Curve showing relation of compression pressure to cylinder temperature.

is removed and ignition is produced by what is known as high compression. This hot-bolt or hot-pin arrangement, however, may be considered as another modification of the hot tube.

Hot-tube and hot-bulb devices for producing ignition have certain distinct disadvantages that make their use impractical except for certain types of engines, namely, heavy-duty, one-speed, stationary oil engines, such as are used for operating large pumping outfits, electric generators, cotton gins, and the like.   Some of these disadvantages are:

1. Several minutes are required to heat the tube or bulb before the engine can be started.

2. Instant change of time of ignition, so essential for multiple-cylinder variable-speed engines, is impossible.

3. Unless made of a special alloy, or high-grade material, the tube or bulb oxidizes rapidly, on account of the excessive heat, and must be replaced frequently.

155. Ignition by High Compression.—When air or gaseous material is compressed rapidly and consequently reduced in volume, work is required that is instantly converted into heat, thereby causing a correspondingly rapid rise in the temperature of the gas compressed. This physical action is utilized in producing ignition in certain so-called high-compression engines; that is, the compression pressure is great enough to produce a temperature high enough to ignite the fuel mixture at the proper time without the assistance or use of any special ignition device.

The effect of rapid compression on temperature is shown by the curve in Fig. 147. In preparing this curve, an initial temperature of 60°F. has been assumed. However, this initial cylinder temperature will be considerably higher after an engine has been running a few minutes. Table VIII shows that the ignition temperature of the common internal-combustion-engine fuels varies perhaps from 500 to 900°F. Consequently, a compression pressure of 300 to 500 lb. per square inch will readily ignite any hydrocarbon fuel mixture. In other words, compression-ignition engines require a compression ratio of 1:15 or higher, while 1:5 or 6 is the average compression ratio of ordinary electric-ignition, carbureting-type engines. Since Dr. Rudolph Diesel, a German, first demonstrated the possibilities of high compression in producing ignition in internal-combustion engines, all such engines are now usually known as Diesel engines.

156. Electric-spark Ignition.—Electric-spark-ignition systems vary greatly in construction and operation, depending upon the type of engine. Therefore they will be discussed fully in later chapters.

References

CONSOLIVER and BURLING: "Automotive Electricity."
HELDT: "The Gasoline Automobile," Vol. III.
STONE: "Electricity and Its Application to Automotive Vehicles."
STREETER and LICHTY: "Internal-combustion Engines."

CHAPTER XIV

DIESEL-ENGINE CONSTRUCTION AND OPERATION

For a number of years following the efforts of Dr. Rudolph Diesel in successfully developing the high-compression-ignition engine in 1898, only a limited number of heavy-duty, slow-speed stationary engines were manufactured. Eventually, however, the possibilities of the utilization of the Diesel principle in smaller, higher-speed, single- and multiple-cylinder engines of both the stationary and automotive type were given consideration. During the past decade there has been a marked expansion of the application of the Diesel principle of ignition and operation to various types of internal-combustion engines.

PRINCIPLES OF OPERATION

As explained in Chap. XIII, the Diesel engine differs from a carbureting-type engine, primarily, in two ways, namely, (1) only air is taken in on the intake stroke of the piston, the liquid fuel being injected directly into the combustion chamber at the end of the compression stroke; and (2) the fuel mixture is ignited by high compression, and no special ignition device or mechanism is needed. On the other hand, all Diesel engines operate on either the two- or the four-stroke-cycle principle like other internal-combustion engines.

157. Two-stroke-cycle Diesel.—The principles of operation and events involved for a two-stroke-cycle engine are shown by Fig. 148. The cycle begins with the upward movement of the piston from its C.D.C. position. The intake and exhaust ports are closed and the charge of fresh air is compressed to approximately 500 lb. per square inch. At the same time the crankcase volume is increasing and air enters the crankcase through an automatic suction valve. This air is sometimes called *scavenging* air. When the piston reaches H.D.C., a charge of fuel is injected into the combustion space. The high temperature existing ignites the mixture of atomized fuel and air, and combustion takes place in such a manner that a constant pressure equal to the compression pressure is maintained as the piston moves downward on the power stroke. Expansion continues until the exhaust port is opened and the burned gases are released. The intake port is likewise uncovered immediately after the exhaust port, and the air in the crankcase, which is now under pressure, by-passes into the cylinder, thus completing the cycle. Figure 149 shows the construction of such an engine.

154

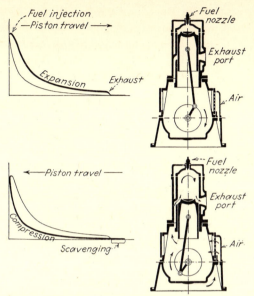

Fig. 148.—Two-stroke Diesel-engine cycle.

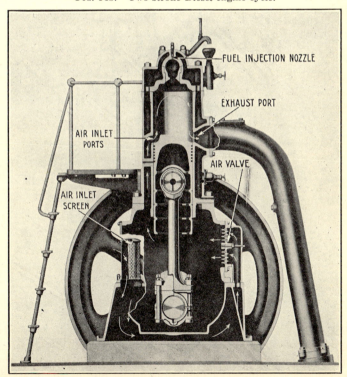

Fig. 149.—Cross-section view of a two-stroke cycle, Diesel-type oil engine

158. Four-stroke-cycle Diesel.—Figures 150 and 151 show the principles of operation of a four-stroke-cycle Diesel engine. It will be observed that the events take place in the same manner as in the ordinary

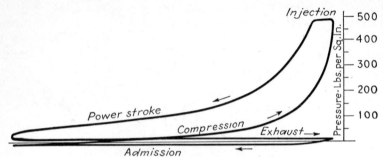

Fig. 150.—Pressure diagram of four-stroke Diesel-engine operation.

four-stroke-cycle carbureting-type engine (see Chap. IV) with the exception that air alone is drawn in on the intake stroke and the liquid fuel is injected into the cylinder at or near the end of the compression stroke.

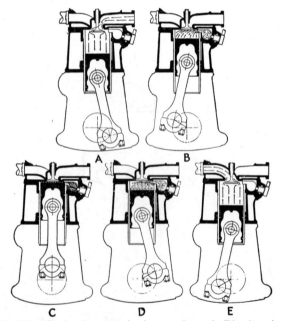

Fig. 151.—Cycle of events of a four-stroke-cycle Diesel engine.

Obviously, two valves, intake and exhaust, operated by a gear, camshaft and the usual mechanism, are necessary, as shown by Fig. 152.

159. Fuel Injection.—The proper injection of the fuel into the combustion chamber against the high pressure is one of the most difficult problems encountered by the Diesel-engine designer.   Since the mechanism involved must supply a fuel charge sufficient only for a single explosion, it is obvious that it must be carefully designed to operate with the utmost precision.   The principal requirements of a Diesel fuel supply and injection mechanism are (1) that it positively supply a correct fuel charge to each cylinder according to the engine load and speed; (2) that it inject the fuel at the correct time in the cycle; (3) that it facilitate

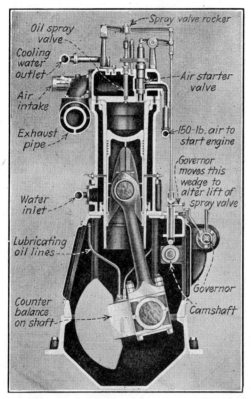

Fig. 152.—Four-stroke-cycle, Diesel-type oil engine.

efficient fuel utilization by atomizing the charge at the time of injection; and (4) that it should not be subject to undue wear or require frequent adjustment or servicing.   Considering these factors, it is obvious that for engines having limited piston displacement and high speed the utilization of the Diesel principle becomes increasingly difficult.

The two common systems of fuel injection are *air injection* and *direct* or *solid injection*.   The former utilizes a stream of air under high pressure (1,000 to 1,200 lb. per square inch) to force the fuel from the injector nozzle (Fig. 153) into the combustion chamber.   Direct or solid injection involves the direct application of high pressure to the liquid fuel by a

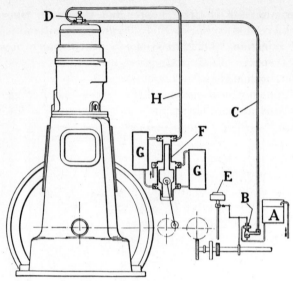

FIG. 153.—Diesel-engine fuel injection using air injection.   *A*, fuel tank; *B*, fuel pump; *C*, fuel line; *D*, injection nozzle; *E*, governor; *F*, three-stage air compressor; *G*, air tanks; *H*, air line.

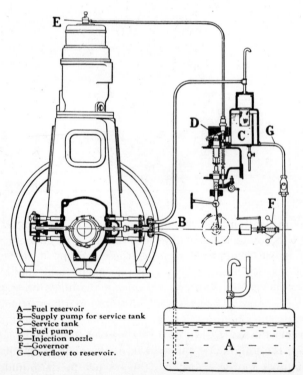

A—Fuel reservoir
B—Supply pump for service tank
C—Service tank
D—Fuel pump
E—Injection nozzle
F—Governor
G—Overflow to reservoir.

FIG. 154.—Diesel-engine fuel injection using solid injection.

pump to force it into the fuel nozzle and thence to the combustion chamber (see Fig. 154).

160. Air Injection.—Figure 153 shows the complete layout for an air-injection system.   An air compressor, usually of the three-stage type, pumps air for both starting and injection into storage tanks, a pressure of 1,000 to 1,500 lb. per square inch being maintained. When the engine is in operation, this high-pressure air passes through a line to the fuel-injection valve, which also receives the fuel under pressure from the fuel-oil pump.   As the fuel is forced by the injection air through the fuel-injector nozzle into the combustion chamber, it is highly atomized.   The injection timing is controlled by a suitable mechanism that opens the injector needle valve just as the piston reaches compression dead center.

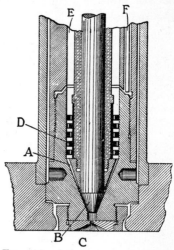

Figure 155 shows the construction and operation of a fuel-injection valve and nozzle for air injection.   The fuel enters at *F* and fills the annular space *E*, which

Fig. 155.—Injection nozzle for air injection.

contains air at a pressure higher than the compression pressure.   When valve *B* is opened, the air in space *E* forces the fuel through the atomizer disks *D* and spray nozzle *C* into the combustion chamber.

The air-injection system is used primarily on very large, stationary, four-stroke-cycle engines.   For the smaller sizes of Diesel power units of

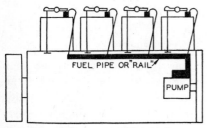

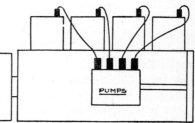

Fig. 156.—"Common rail" injection system.

Fig. 157.—Individual pump injection system.

both the stationary and automotive type, solid injection is preferable because of its simplicity and the elimination of heavy, bulky air tanks, air compressor, and other parts.

161. Direct or Solid Injection.—In the solid-injection system (Figs. 154 and 157), the fuel is first pumped from the main supply tank to an auxiliary chamber by means of the transfer pump.   The high-pressure injection pumps receive the fuel from the auxiliary chamber and force the proper charge to the injector in the cylinder head. The timing of the charge is controlled by timing the stroke of the injector pump with the crankshaft and piston position.

For multiple-cylinder Diesel engines, two systems of solid injection are used, namely, (1) "common rail" (Fig. 156) and (2) individual pumps (Fig. 157). In the former, a single pump supplies fuel to all cylinders

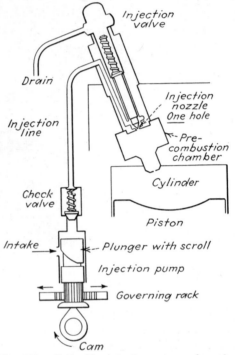

FIG. 158.—Caterpillar injection pump and nozzle.

by means of a pipe or "rail" that carries fuel constantly under high pressure. The fuel is fed from this pipe into each cylinder by a mechanically operated valve. Although comparatively simple, the "common rail" method is seldom used because it is difficult to adjust the mechanism so that each cylinder will receive a correct fuel charge, particularly at light loads or idling speeds.

In the individual pump type, as the name indicates, a separate fuel-injection pump and line are used for each cylinder.

Figure 158 shows the construction and operation of the injection valve for a solid-injection system. The valve is held to its seat by a spring. The fuel pressure developed by the injection pump overcomes the spring pressure and lifts the valve. Most Diesel engines of this type use the precombustion chamber type of injector as shown. As the fuel is sprayed into this small chamber the highly-heated air causes a small portion of the fuel to ignite. As injection continues, the fuel is enveloped by flame, becomes gaseous, and, owing to the high pressure developed, rushes at high velocity into the main combustion chamber where complete combustion takes place. The use of a precombustion chamber rather than direct injection into the main combustion chamber

does not require so high an injection pressure. Furthermore, it permits the use of an injector having a large single jet rather than several small jets which would clog more readily.

162. Diesel Governing.—T h e principle involved in governing Diesel engines is not unlike that used in other types; that is, the reaction of centrifugal force produced by revolving weights is utilized to control the quantity of fuel injected according to the engine load. However, no attempt is made to vary the air charge, and it remains constant at all loads. Figure 159 shows how such a governor operates. The governor rack 1 is connected to and actuated by the

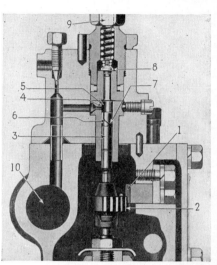

Fig. 159.—Caterpillar Diesel governing mechanism.

governor. This rack revolves the pump gear 2, which is attached to the bottom of the pump plunger 3. As the gear is turned by the rack, the plunger is turned and lifted so that it closes the inlet port 4 to which fuel is fed from manifold 10. This operation traps the fuel in the compartment 5, above the plunger, in the groove 6 and in the space below the scroll 7. Here a pressure is created and as the plunger is raised by the camshaft the fuel is forced through the check valve 8 and through the line 9 into the injection valve. As the scroll edge 7 reaches the fuel-inlet port 4, pressure is relieved, causing check valve 8 to close. Thus the position of the scroll with respect to the inlet port determines the size of the fuel charge trapped above the plunger and consequently forced to the injector.

MODIFIED DIESEL AND HEAVY-FUEL ENGINES

163. Hvid-type Engine.—An engine made in small sizes, particularly for farm purposes, and operating more or less upon the Diesel principle, is illustrated in Fig. 160. It is known as the Hvid-type engine. It is a four-stroke-cycle engine and depends entirely upon high compression for ignition. However, the introduction of the fuel charge is unique

Fig. 160.—Cross-section view of a Hvid-type engine.

and strikingly different from that of the Diesel. Referring to Fig. 160, the fuel is pumped from the tank to the fuel injector, located in the

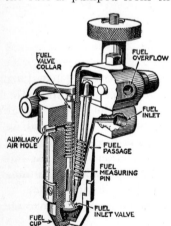

Fig. 161.—Fuel injector for Hvid-type engine.

cylinder head as shown, the excess fuel returning to the tank through the overflow pipe. On the intake or suction stroke of the piston, air only is drawn into the cylinder through the mechanically operated air-inlet valve (not shown). At the same time, the fuel-inlet valve opens (see Fig. 161) and permits a charge of fuel to enter the fuel cup. The amount admitted is controlled by the fuel-measuring pin, which in turn is controlled by the governor. At the same time that the fuel is being admitted into the cup, a small amount of fresh air is drawn through the auxiliary air hole, down past the fuel-inlet valve into the fuel cup. At the end of the suction stroke, the fuel-inlet valve and air valve close. The air in the cylinder is now compressed to about 450 lb. per square inch and its temperature raised to between 800 and 1000°F. This hot air rushes into the fuel cup through three small holes and vaporizes and ignites the lighter portions of the fuel charge, creating a primary explosion in the cup and forcing the balance of the fuel out into the combustion chamber where it is completely vaporized and ignited and the major explosion produced. The resulting expansion sends the piston out on the power stroke, near the end of which the exhaust valve opens, permitting the products of combustion to escape or be pushed out by the piston as it moves back through the exhaust stroke. At the end of this stroke the valve closes and the cycle is completed.

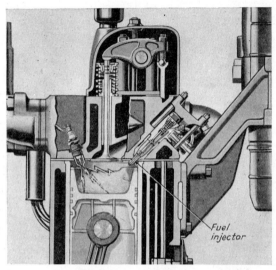

Fig. 162.—Spark-ignition and fuel-injection-type engine.

The operation of the Hvid-type engine may be said to be based upon the fact that every fuel, no matter how heavy it may be, contains some light hydrocarbons that will vaporize at a fairly low temperature. The explosion of the lighter fractions of the fuel charge in the cup acts as a propellant to force the remaining heavy portion of the charge out into the combustion chamber. Kerosene is recommended as the most satisfactory fuel for the farm-type Hvid engine. By making certain minor changes, it can be made to burn heavier fuels.

164. Hesselman Engine.—The Hesselman-type engine is one that burns heavy fuels and uses electric-spark ignition in combination with mechanical fuel injection (Fig. 162). The compression ratio is about 1:7.5, which means that the compression pressure is somewhat higher than for carbureting-type engines, but considerably below that of the Diesel. Therefore, the pressure is insufficient to produce ignition, and an electric-spark system is necessary. On the other hand, the relatively high compression pressure combined with the use of a low-cost fuel provides a rather low fuel consumption and cost characteristic. The principal advantage of this type of engine over the Diesel is that the smaller sizes can be cranked by hand, and a special mechanical starting device is unnecessary. Gasoline placed in a small auxiliary tank is used for starting.

165. Diesel Starting Methods.—One of the chief disadvantages of a Diesel engine, regardless of size or number of cylinders, is the energy required to crank it when starting, owing to the very high compression pressure. Hand cranking is usually out of the question, and, in most cases, some type of mechanical starter is used.

Fig. 163.—McCormick-Deering Diesel-engine construction and starting mechanism.

The following methods are used for cranking Diesel engines: (1) hand cranking by temporarily releasing or reducing the compression pressure (adaptable only to engines of limited displacement such as automotive-type Diesels), (2) use of compressed air from a high-pressure air-storage tank, (3) use of a small auxiliary starting engine, and (4) use of a storage battery and electric starting motor.

A simple application of the hand-cranking method is used for the Hvid-type stationary engine previously described. The intake valve is locked open, relieving the pressure, the engine is rotated as fast as possible by hand and the valve lock is released. The inertia attained will usually be sufficient to carry the piston over a compression stroke and firing will begin.

The McCormick-Deering Diesel engine (Fig. 163) is an example of a hand-started automotive-type Diesel engine. It is equipped with an auxiliary combustion chamber that is connected to the main combustion chamber by a special valve. Likewise, there is a high-tension magneto with spark plugs and a gasoline carburetor and manifold. To start this engine, a lever 1 is given one-fourth of a turn. This opens the valve 3 to the auxiliary combustion chamber 4 and reduces the compression pressure. At the same time the magneto is engaged, and the gasoline carburetor and manifold are connected to the auxiliary chamber by the valve 5. The engine is now cranked in the ordinary manner and started on gasoline. When it has made about 700 revolutions (run about 1 min.) rod 2 releases the shaft turned by lever 1. This permits valve 3 to close,

Fig. 164.—Caterpillar Diesel starting engine.

which cuts out the auxiliary chamber 4 and isolates the spark plug. The same operation disengages the magneto and closes the valve 5. The engine immediately begins operating as a full Diesel.

Compressed-air starting (Fig. 153) is used only for stationary Diesel engines. Air under high pressure is stored in a tank that is connected by a pipe to the cylinder head. Opening a valve permits the air to act on the piston and turn the engine until firing begins. Such engines must be equipped with an air compressor to replenish the air supply.

Caterpillar Diesel tractor engines are equipped with a small high-speed auxiliary gasoline-burning starting engine (Fig. 164). This engine is cranked by hand and then engaged with the tractor-engine flywheel by a special starter-drive coupling. The drive is by means of a small spur gear that meshes with teeth on the flywheel. In starting the tractor engine, the compression pressure is released by locking the exhaust valves open. As soon as the engine is turning over properly, the compression release lever is disengaged, the fuel-injection pumps are engaged, and firing begins, after which the starter engine is disengaged and stopped.

Some automotive-type Diesel engines are equipped with a storage battery and electric starting motor for cranking. Usually a 12- or a 24-volt system is necessary, particularly for large engines. Starting is accomplished by closing a switch connecting the battery to the starting motor. As the latter begins to rotate, a small spur pinion on the armature slides into engagement with the teeth on the flywheel, and the engine turns. When firing begins, the gears automatically disengage and the starter switch is opened. A generator is also needed to keep the battery charged.

166. Diesel-engine Design and Construction.—Owing to the higher pressures and strains to which certain parts of a Diesel engine are subjected, consideration must be given to designing these parts heavier and stronger or of special materials. Likewise, extreme precision in manufacturing and assembly is important. Some concrete examples relative to this problem are (1) special tie-rods, (Fig. 165) extending through the cylinder block to tie the main bearings securely to the cylinder head and provide greater security against compression and combustion stresses, (2) heavier, special carbon-steel crankshafts, (3) large precision-fitted bearings of special quality bearing metal, (4) correctly designed and fitted pistons with a sufficient number of rings to retain compression, (5) alloy-steel heat-treated valves, carefully ground and fitted to eliminate leakage and withstand high temperatures, (6) positive lubrication of all moving parts, (7) efficient cooling under all operating conditions, and (8) positive and reliable fuel-supply and injection-system and fuel-filtering devices. As a rule Diesel-type engines are somewhat heavier and cost more per horsepower than spark-ignition engines of the same type.

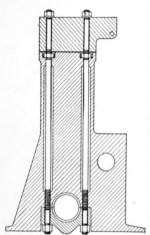

Fig. 165.—Special tie-rods for Diesel-tractor cylinder head and bearings.

The outstanding operating characteristic of any Diesel engine as compared with an electric-ignition engine is its lower fuel consumption per horsepower-hour. Referring to Fig. 394 it will be observed that a Diesel engine uses from 0.50 lb. per horsepower-hour at full load to approximately 0.90 lb. at one-fourth load. A similar electric-ignition engine burning gasoline consumes from 0.70 lb. per horsepower-hour at full load to 1.40 lb. at one-fourth load. This lower fuel consumption combined with the lower first cost of the Diesel fuel itself means a pronounced saving in operating costs. For example, the cost of the fuel per hour for a 50-hp. Diesel engine, using fuel costing 6 cents and weighing 7 lb. per gallon, would be $\dfrac{50 \times 0.5 \times 6}{7} = 21.4$ cents.

For an engine of the same size, using gasoline costing 12 cents and weighing 6.15 lb. per gallon, the fuel cost per hour would be

$$\frac{50 \times 0.7 \times 12}{6.15} = 68.3 \text{ cents.}$$

Multiple-cylinder automotive-type Diesel engines show a somewhat different torque characteristic as compared with similar electric-ignition engines. The difference in this respect is such that a Diesel engine is said to have better "lugging" ability, that is, it hangs to the load and

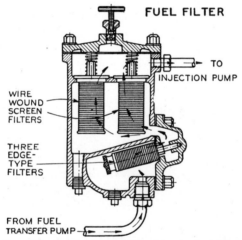

Fig. 166.—Special type of Diesel-tractor fuel filter.

continues to pull even though the speed drops off perceptibly. This characteristic is probably brought about by the time and duration of the fuel injection and the fact that the subsequent combustion process is somewhat slower and lacks the more spontaneous explosive action characteristic of the electric-spark-ignition engine.

167. Diesel Fuels.—Diesel fuels and their characteristics are discussed in Chap. VIII. A Diesel engine, to give satisfactory service, must not only operate on a fuel of proper specifications, but also must be equipped with efficient strainers and protective devices to prevent the slightest trace of foreign matter of any kind reaching the fuel-injection mechanism. This mechanism, being extremely delicate and sensitive in construction and operation, might have its proper performance noticeably affected by a minute particle of dirt, moisture, or other apparently harmless material.

Figure 166 shows one type of two-stage filtering device. The fuel enters as indicated and first passes through three metal edge-type filters, which remove the bulk of the foreign material. It is then forced through two wire-wound screen filters with still finer openings, designed to remove silt, fibrous, and flat materials. The filtering elements should be readily removable and thoroughly cleaned at frequent intervals. Other necessary precautions are to put only clean fuel in the fuel tank, keep the tank closed, keep all fuel lines tight and see that the air cleaner is given proper attention and kept in good working condition.

References

A Diesel Engine for Tractor Service, *Agr. Eng.*, Vol. 13, No. 1.
ADAMS: "Elements of Diesel Engineering."
HELDT: "High Speed Diesel Engines."

CHAPTER XV

ELECTRICAL IGNITION—SOURCES OF ELECTRICITY

Any electrical ignition system is made up of certain parts or units according to the following outline:

1. Source of current.
   a. Chemical—dry cells and storage cells.
   b. Mechanical—magnetos and similar electric generating devices.
2. Coil.
   a. Low tension or make and break.
   b. High tension or jump spark.
      (1) Vibrating.
      (2) Nonvibrating.
3. Timing mechanism.
4. Sparking device.
   a. Spark plug.
   b. Igniter.
5. Switch and wire connections.

168. Chemical Generation of Electricity.—Electricity for ignition purposes may be generated in two ways, namely, by chemical means or by a mechanical device. Chemical devices for generating an electric current are known as cells or batteries. Correctly speaking, a cell is a single unit, and a battery consists of two or more cells connected together. Such cells for generating electricity are made up of four fundamental parts: (1) the positive material, (2) the negative material, (3) the electrolyte, and (4) the container. That is, a simple cell (Fig. 167) consists essentially of two dissimilar materials immersed in a solution called the electrolyte. If these materials are connected externally by a good con-

TABLE X.—CHEMICAL COMBINATIONS USED IN DIFFERENT TYPES OF CELLS

	(A) Daniell cell	(B) Leclanché cell or dry cell	(C) Lead-acid storage cell
Positive material............	Copper (Cu)	Carbon (C)	Lead dioxide (PbO_2)
Negative material..........	Zinc (Zn)	Zinc (Zn)	Lead (Pb)
Electrolyte................	Sulphuric acid (H_2SO_4)	Ammonium chloride (NH_4Cl)	Sulphuric acid (H_2SO_4)
Voltage...................	1.0	1.5	2.0

ductor of electricity, such as a copper wire, chemical action takes place between the solution and these materials, and an electric current flows through the wire, that is, chemical energy is converted into electrical energy.

Only certain combinations of materials will generate electricity in this manner. Some of the more common ones and the voltage produced are given in Table X.

Combinations *A* and *B* (Table X) form what are known as primary cells. Combination *C* is the one used in the ordinary lead-type automobile and farm light-plant batteries and forms what is known as a secondary cell or battery.

169. Primary and Secondary Cells.—A *primary cell* is one in which the chemical action, going on as the cell discharges, changes one or more of the active materials— particularly the negative element and the electrolyte—in such a way that when the cell is completely discharged or "dead," it can be restored to its original condition only *by renewing the materials that have been so changed*. For example, in Fig. 167 the zinc plate is gradually consumed, the zinc replacing the hydrogen in the sulphuric acid. This hydrogen

Fɪɢ. 167.—A simple cell for generating electricity.

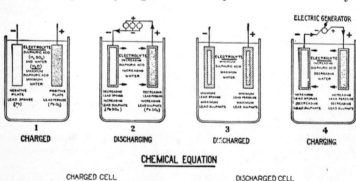

Fɪɢ. 168.—Action taking place in a lead-acid storage cell during a cycle of discharge and charge.

collects on the positive copper plate which remains unchanged. Consequently, the negative zinc and the electrolyte are gradually broken down and eventually must be replaced.

In a *secondary cell* (Fig. 168), the active materials likewise undergo a chemical change during the discharging process, but when the cell becomes completely discharged, a replacement of these materials is unnecessary and it can be restored to its original condition *by sending an electric current through it in a direction opposite to that of discharge.* This process is commonly known as recharging. Of course, this charging and discharging action cannot be carried on indefinitely because these secondary or so-called storage cells gradually lose their strength and efficiency for other reasons. Their life varies from 2 to 10 years or more depending upon their type, construction, quality of materials, use, care, and other factors.

170. Conductors and Insulators.—Certain materials, largely metals, transmit an electric current readily and are known, therefore, as good *conductors;* that is, they are said to offer a low resistance to the flow of electricity through them. Some of these substances are: silver, copper, aluminum, zinc, brass, platinum, iron, nickel, tin, and lead. On the other hand, many materials are very poor conductors, or apparently do not conduct an electric current at all and are known as *insulators.* Some common insulating materials are porcelain, mica, rubber, glass, fiber, and Bakelite.

171. Electrical Units and Measurement.—The flow of an electric current through a conductor can be compared to the flow of water through a pipe. The quantity of water flowing or rate of flow, usually expressed in gallons per minute, is dependent upon the pressure in pounds per square inch or head, and the resistance offered by the pipe according to its size and length. Likewise the rate of flow of an electric current measured in amperes is determined by the electrical pressure in volts and the resistance of the conductor according to its size and length, and the material used.

The usual definitions of the common electrical units are as follows:

Volt.—The unit of electrical pressure or electromotive force (e.m.f.). It is the pressure required to send a current of 1 amp. through a circuit whose resistance is 1 ohm.

Ampere.—The unit of the rate of flow of an electric current. It is that quantity of electricity that is made to flow by a pressure of 1 volt through a circuit whose resistance is 1 ohm.

Ohm.—The unit of electrical resistance. It is the resistance offered to the flow of 1 amp. under a pressure of 1 volt.

Ampere-hour.—The quantity of current flowing in amperes for a period of 1 hr., or

$$\text{Ampere-hours} = \text{amperes} \times \text{hours.}$$

Watt.—The unit of electrical power or the rate at which work is performed by 1 amp. of current flowing under a pressure of 1 volt, or

$$\text{Watts} = \text{volts} \times \text{amperes}$$

Watt-hour.—The power consumed when 1 watt is used for 1 hr., or

$$\text{Watt-hours} = \text{volts} \times \text{amperes} \times \text{hours}$$

Kilowatt.—One thousand watts.

Kilowatt-hour.—The power consumed when 1 kw. of power is consumed for 1 hr., or

$$\text{Kilowatt-hours} = \text{kilowatts} \times \text{hours}$$
$$= \frac{\text{amperes} \times \text{volts} \times \text{hours}}{1,000}$$

172. Ohm's Law.—As previously stated, the current flowing in an electrical circuit is dependent upon the resistance of the circuit and the pressure or voltage.   In fact, there is a definite relationship between the current flowing in amperes, the pressure in volts, and the resistance in ohms.   This relation, known as Ohm's law, is expressed as follows:

$$\text{Current in amperes } (I) = \frac{\text{pressure in volts } (E)}{\text{resistance in ohms } (R)}$$

or using the common symbols

$$I = \frac{E}{R}$$

Likewise,

$$E = IR$$

and

$$R = \frac{E}{I}$$

This law is of inestimable value, particularly to the electrical engineer, for calculating the voltage required to transmit a certain current a given distance, for determining correct wire sizes, and so on.

173. The Dry Cell.—The common dry cell is nothing more than a primary cell whose principal active materials are carbon, zinc, and ammonium chloride.   However, it is made up in such a manner that the electrolyte is in a so-called nonspillable form so that the cell or battery can be carried about or placed in any convenient position without injury to its contents or hindrance to its action.

The dry cell serves as a source of electricity for a number of purposes, including gas-engine ignition, flashlights, doorbells, telephones, radio receivers, and so on.   It is made up in different forms and sizes according to the purposes for which it is to be used.   For ignition purposes, it is cylindrical in shape, the standard size being $2\frac{1}{2}$ in. in diameter and 6 in. high.

Referring to the cross-sectional view (Fig. 169), the construction of a dry cell is as follows:

1. The outer pasteboard carton serves as a protector from moisture and possible short circuits.

2. The zinc cup serves as a container and as the negative material.

3. The carbon stick acts as a conductor to conduct the current from the cell, but does not serve exclusively as the positive material although it is often erroneously considered as such. This carbon stick has the terminal fastened to it, although the entire stick might be considered as the positive terminal.

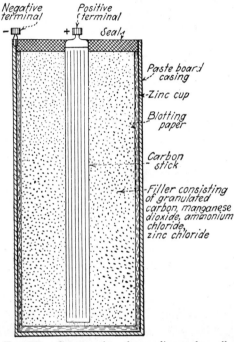

FIG. 169.—Construction of an ordinary dry cell.

4. The filler is a mixture of (*a*) granulated carbon or graphite, which serves as the active positive material; (*b*) manganese dioxide (MnO_2), which is called the depolarizer; (*c*) ammonium chloride (NH_4Cl), in solution, which is the electrolyte; (*d*) zinc chloride ($ZnCl_2$), which reduces the chemical action between the zinc and ammonium chloride when the cell is not being used.

5. The blotting paper which completely lines the zinc cup acts as an insulator between it and the positive carbon. This blotting paper, however, permits the electrolyte to permeate it and come in contact with the zinc, thus producing the chemical action and generating an electric current.

6. The top is sealed airtight with some hard waterproof material, usually a pitch or tar compound.

174. Polarization.—The chemical action taking place in a dry cell, as it discharges, results in the liberation of some of the hydrogen of the electrolyte. This hydrogen collects in the form of bubbles on the carbon, which serves as the positive electrode, and has a tendency to act as an insulator, preventing the carbon from receiving its positive charge of electricity and functioning as it should. The current output, or rate of discharge, is thus gradually reduced. This action is known as polarization and is more pronounced if the cell is discharged at a high or excessive rate.

The manganese dioxide, mixed with the carbon and electrolyte, acts as a depolarizer; that is, it liberates some of its oxygen, which unites with the hydrogen to form water. The manganese dioxide is probably reduced to a lower oxide.

Fig. 170.—Testing a dry cell with a pocket ammeter.

175. Testing Dry Cells.—The most practical method of determining the condition of a dry cell is to use a common pocket voltammeter (Fig. 170). This instrument is equipped with two stationary contacts labeled "amps" and "volts," a flexible contact, and a double scale. To read the current in amperes, the stationary "amps" contact is placed on the positive terminal of the cell and the negative terminal is touched with the flexible contact. The reading in amperes will be indicated by the hand on the upper 0 to 30 scale. It is important in testing for amperes that the reading be made as quickly as possible when the indicator comes to rest. If the cell shows 20 to 30 amp., it is in good condition. If it shows only 5 to 10 amp., it is practically exhausted and of little value for ignition or other purposes.

To test the voltage of a dry cell, the "volts" terminal of the instrument is placed on the positive terminal of the cell and the flexible contact touched to the negative terminal. The hand will indicate the voltage on the 0 to 12 or "volts" scale. For a good cell, the reading will be 1.5 and will not drop perceptibly until it is nearly worn out. That is, the amperage of a dry cell or battery decreases gradually with use and age but the voltage remains practically constant. Therefore, the condition of a cell should be determined by testing for amperes rather than for volts.

To obtain the best results and maximum service from dry cells they should be stored in a cool place when not in use and kept free from dampness at all times.

176. Connecting Dry Cells.—Ignition systems using a battery as a source of electricity require it to have a pressure of at least 6 volts. Since the voltage of the common dry cell is only 1.5 and cannot be increased by increasing the size of the cell, several units must be connected in such a way as to produce this higher pressure. Figure 171 illustrates the *series* method of connecting cells, in which the carbon or positive of one is connected to the zinc or negative of the next, and so on. If a voltmeter were placed in the circuit, it would register about 6 volts. On the other

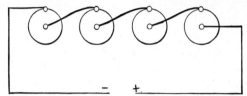

FIG. 171.—Series method of cell connection.

hand, if the current reading in amperes were taken, it would show little or no increase over that of the individual cell; that is, the series connection produces a total voltage equal to the sum of the voltages

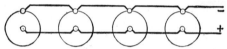

FIG. 172.—Parallel or multiple method of cell connection.

of the cells and a total amperage equal, approximately, to the average amperages of the individual cells.

Another arrangement (Fig. 172), in which all of the positive terminals are connected by one wire and all of the negatives by another, is known as the *parallel* or *multiple* method. In this case, the voltage of the set would be about 1.5 or the average of the voltages of the individual cells, but the total amperage would be equal to the sum of the amperages of the separate cells. This method is impractical and seldom used.

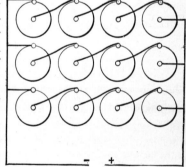

FIG. 173.—Parallel- or multiple-series method of cell connection.

Figure 173 illustrates a third method of connecting cells, which is known as *parallel series* or *multiple series*. It will be observed that the individual cells are connected in series and the sets in parallel. Such an arrangement is advantageous, particularly under two conditions, namely: (1) when an engine using a dry-cell battery for ignition is operated almost continuously every day as on a concrete mixer, for example, and (2) when a supply of good dry cells is not available and a number of weak ones must be used.

Dry cells deteriorate and lose their strength even when allowed to stand unused, their maximum life seldom exceeding 1 year under any conditions.   Therefore, if an engine is operated intermittently for a few hours or a day or two a week, it is advisable to buy only four or five fresh cells and connect them in series.

On the other hand, if an engine is in constant operation every day it is usually more economical to purchase eight or ten cells and connect them in multiple series, for the reason that the discharge rate in amperes per cell is reduced one-half.   For example, referring to Fig. 209, which is a diagram of a simple ignition circuit, when the points close, the circuit is complete and an electric current flows.   The rate of flow in amperes, according to Ohm's law, depends upon the voltage and the resistance of the entire circuit in ohms.   Since there are four dry cells in series, the voltage will be about 6.   Suppose the total resistance is 1 ohm.   Then,

$$I = \frac{E}{R} = \frac{6}{1} = 6 \text{ amp.}$$

which is the rate of current flow when the points are closed.   Now suppose, instead of four cells in series, eight cells in multiple series are

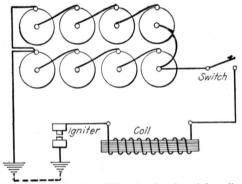

Fig. 174.—A simple make-and-break ignition circuit using eight cells in parallel series.

connected to the same ignition circuit, as shown by Fig. 174.   The voltage and resistance remain the same; consequently the current flow in amperes must remain the same.   But this current is now supplied by eight cells connected in multiple series instead of by four cells in series.   Therefore, each set of four cells is supplying 3 amp. to the circuit, and the discharge rate per cell has been reduced one-half, as compared to the rate when only four cells in series are used.

Tests have shown that if a cell of any kind is required to generate an electric current continuously for hours or days at a time, the lower this current drain per cell, the greater the life of the cell or battery.   In fact, eight or ten cells, connected in multiple series, will last considerably longer under continuous service than eight or ten cells used four or five at a time in series.

Frequently a dry-cell ignition battery becomes weak and it is found impossible to secure new cells at once.   In such an emergency it is often possible to relieve the situation by connecting eight, ten, or even twelve weak dry cells in multiple series.   Suppose for example, 12 cells testing 1.4 volts and 8 amp. each are connected as in Fig. 173.   The maximum voltage and amperage of the battery would be 5.6 and 24, respectively, which is nearly the same as that of a set of four or five good cells connected in series.

References

CONSOLIVER and BURLING: "Automotive Electricity."
DUELL: "The Motor Vehicle Manual."
"Dyke's Automobile and Gasoline Engine Encyclopedia."
HELDT: "The Gasoline Automobile," Vol. III.
STONE: "Electricity and Its Application to Automotive Vehicles."
The Electrical Characteristics and Testing of Dry Cells, *U. S. Bur. Standards, Circ.* 79.

CHAPTER XVI

SECONDARY OR STORAGE CELLS AND BATTERIES

Storage cells or batteries are used extensively for many purposes, some of which are:

1. Automobile and truck ignition, starting, and lighting.
2. Farm lighting.
3. Train lighting.
4. Radio-receiver operation.
5. Reserve and auxiliary lighting and power service in central power stations.
6. Motor power for vehicles, trucks, mine locomotives, and so on.
7. Railway-signal and switch operation.

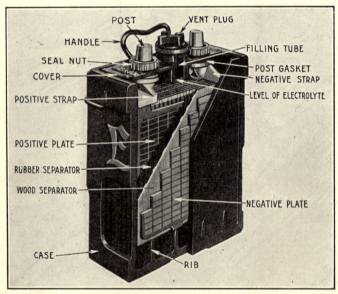

FIG. 175.—Lead-type storage-cell construction.

177. Kinds of Storage Cells.—The two types of secondary cells are the lead-acid cell (Fig. 175) and the Edison or nickel-iron alkaline cell (Fig. 176). The essential parts of the former are: (1) positive plates of lead dioxide (PbO_2), (2) negative plates of sponge lead (Pb), (3) electrolyte of dilute sulphuric acid (H_2SO_4), (4) wood or threaded rubber separators or insulators, and (5) rubber or glass container.

178

The Edison cell consists of (1) positive plates of nickel oxide (Ni₂O₃), (2) negative plates of iron (Fe), (3) electrolyte of potassium hydroxide (KOH), (4) hard-rubber insulators, and (5) nickel-plated steel container.

178. The Edison Cell.—The Edison storage cell is not used so extensively as the lead-acid type for the following reasons:

1. It has a low discharge voltage (1.2 volts); therefore, more cells are necessary to obtain a desired circuit voltage.

2. Its current output and efficiency are greatly reduced at low temperatures.

3. Its maximum, short-interval discharge rate is comparatively low for a battery of given voltage and capacity.

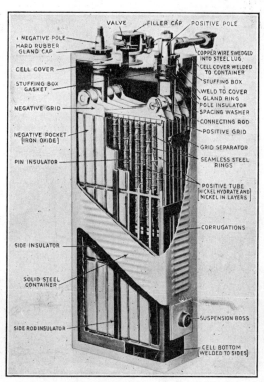

Fig. 176.—Edison-type storage-cell construction.

Some advantages of the Edison cell are:

1. It is not injured by remaining in a charged or discharged condition for a long time.

2. It is rugged in construction and will withstand severe vibration.

3. It does not give off corrosive gases.

4. It is light in weight.

5. It is not injured by excessive charging and discharging rates.

LEAD-TYPE BATTERY CONSTRUCTION AND OPERATION

179. Plates.—The active positive and negative materials in the lead-acid cell are in the form of rectangular plates. A number of positive plates are connected together to form what is known as a *positive group*

Fig. 177. Fig. 178.
Fig. 177.—Positive-plate group for lead-type cell.
Fig. 178.—Negative-plate group for lead-type cell.

or element (Fig. 177). In a similar manner, a *negative group* is made up of several negative plates (Fig. 178). The two groups are then placed together (Fig. 179), so that the positive and negative plates alternate. Since it is necessary to have a negative on each side of a positive, there must be one more of the former, and the total plates per cell will be an

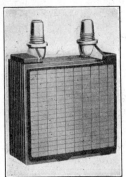
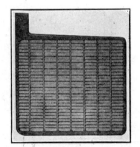

Fig. 179. Fig. 180.
Fig. 179.—Complete plate assembly for lead-type cell.
Fig. 180.—A plate grid for a Faure-type plate.

odd number. Separators or insulators, as described later, are inserted between all plates.

180. Plate Construction and Manufacture.—Since the active materials in both the positive and negative plates are of a brittle nature and have little mechanical strength, a framework made of some neutral metals is necessary to hold them in place. This cellular frame (Fig. 180), made of a lead-antimony alloy, is known as a grid.   There is considerable variation in the grid construction, as used by different manufacturers, but in all cases the purpose is the same, namely, to prevent the active materials from cracking and falling out of the plate and thereby reducing the cell output.

Two different methods or processes are used in manufacturing plates and the latter are designated accordingly as either Planté or Faure, after the name of the man inventing the process.   The Planté was the first one evolved.   It consists of filling the grid structure with pure lead and then subjecting the plate to an electrochemical action that converts a thin surface layer into lead dioxide (PbO_2).

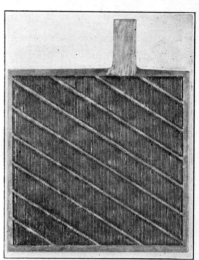

This being the active positive material, the plate, if it is to be a positive one, is ready for use.   Negative plates are formed from these positives by reducing the lead dioxide to sponge lead.

Although only the surface layer of a Planté-type positive plate has been converted to lead dioxide when it is first put into use, the remaining unchanged material that is still plain lead is gradually converted as this outer layer sheds off.   In order to secure the desired current-generating capacity, as well as a reasonable length of life, the plates are made unusually thick and with a roughened

FIG. 181.—A Planté-type plate.

or grooved surface to secure the greatest possible exposure.   Figure 181 illustrates a Planté-type plate used in a farm lighting battery.

In manufacturing Faure plates the active material, in paste form, is pressed into the grid and allowed to harden.   The materials as pasted into the grids are red lead (Pb_3O_4) for the positive, and litharge or lead monoxide (PbO) for the negative, mixed with sulphuric acid and water into a stiff paste.   When pressed into place the paste hardens like cement and adheres to the ribbed structure.   At this stage the positive plate will be bright red in color and the negative a yellowish gray.

The plates are then subjected to an electrochemical process known as *forming,* which consists of submerging them in sulphuric acid and water and sending a direct current through them, the current flowing in at the positive plate and out at the negative.   This process converts the active material of the positive plate into lead dioxide, which is chocolate-brown in color and that of the negative into gray spongy lead.

Faure-type plates are used more extensively than the Planté type for the reason that they can be made thinner. Therefore, the cell or battery will be smaller and lighter. Vehicle-type storage batteries are of the Faure type exclusively. The batteries supplied with some farm electric-lighting plants have Planté-type positive plates and Faure negative plates. The former can be identified by their extreme thickness which is often as much as $\frac{7}{16}$ in. The Faure negatives are about $\frac{3}{16}$ in. thick.

The size, thickness, and number of plates per cell vary considerably according to the purpose for which the battery is to be used. Further information concerning these items is given in the discussion of Battery Ratings and Capacity.

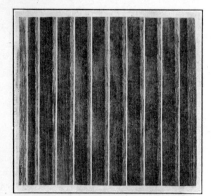

Fig. 182.   Fig. 183.

Fig. 182.—A wood separator for a lead-type cell.

Fig. 183.—A threaded-rubber separator for a lead-type cell.

181. Separators.—The purpose of the separators is to act as insulators between the plates, preventing them from coming in contact with each other and creating a so-called internal short circuit. At the same time, they permit the electrolyte to penetrate or pass through and come in contact with the entire plate surface and to circulate freely. Separators are usually thin wood sheets cut slightly larger than the plates and grooved on one side (Fig. 182). Cedar, cypress, redwood, fir, and poplar give the best results. In addition to being made of clear, high-grade material, wood separators are carefully treated with certain chemicals to remove any impurities and to increase their porosity.

A so-called threaded-rubber separator (Fig. 183) used by one manu-facturer consists of a thin sheet of rubber-composition material, which is finely perforated. Each perforation contains a fine, short thread which acts as a wick and thus permits thorough diffusion of the electrolyte. These separators are of about the same thickness as the wood type and likewise grooved on one side. This corrugated surface is always placed next to the positive plate with the grooves running vertically. The purpose of the grooves is to permit the sediment forming on the positive-plate surfaces, during the charging and discharging action, to settle to the bottom of the cell container. The advantage claimed for the threaded-rubber separator is that it lasts longer than the wooden separator for the reason that the acid electrolyte does not attack and weaken it so readily. These rubber separators should not be confused with the thin, perforated, rubber sheets, known as retaining walls (Fig. 184), which are often used with wood separators and are placed next to the positive plate. They are about $\frac{1}{64}$ in. in thickness and serve to prevent the active material of the plate from dropping out.

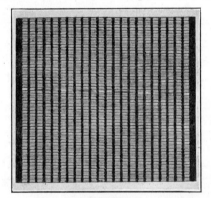 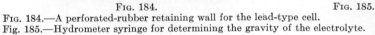

FIG. 184.   FIG. 185.

FIG. 184.—A perforated-rubber retaining wall for the lead-type cell.
Fig. 185.—Hydrometer syringe for determining the gravity of the electrolyte.

182. The Electrolyte.—The electrolyte used in the lead-acid cell consists of a mixture of about 2 parts of chemically pure, concentrated sulphuric acid (H_2SO_4) to 5 parts of distilled water by volume. The concentrated acid will have a specific gravity of 1.835 but the diluted mixture of the above proportions will drop to about 1.300 at 70°F. Some types of storage batteries, particularly farm light-plant batteries, use even a weaker solution than this, its specific gravity, when the cells are fully charged, varying from 1.220 to 1.250.

The usual method of determining the strength of the solution is to test it by means of a hydrometer syringe, as shown in Fig. 185. The hydrometer itself is placed inside the syringe so that it is only necessary to insert the latter in the cell, draw out some of the liquid, make the reading, and squirt the liquid back into the cell without removing the syringe nozzle. The ordinary battery hydrometer is graduated from 1.100 to 1.300, which is the maximum range of variation of the electrolyte between a completely discharged and a fully charged condition. Table

XI gives the conditions present in a battery at different stages of charge, assuming the surrounding temperature to be 70°F.

TABLE XI.—CHARACTERISTICS OF LEAD-ACID-TYPE STORAGE CELLS AT DIFFERENT STAGES OF CHARGE

Condition of battery	Specific gravity	Cell voltage	Freezing point, °F.
Fully charged.................................	1.230 to 1.300	2.2	−90
Three-quarters charged.......................	1.200 to 1.250	2.1	−60
One-half charged.............................	1.190 to 1.220	2.0	−20
One-quarter charged..........................	1.175 to 1.190	1.9	Zero
Completely discharged........................	1.150 or less	1.8 or less	−20

183. Container.—The container for the lead-acid cell is made either of a hard-rubber composition material (Fig. 175) or of glass (Fig. 186). For automobile and radio work the former is preferred because it will with-

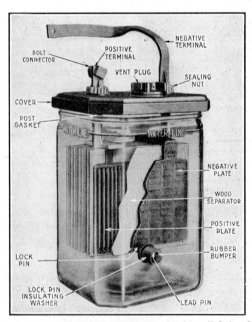

FIG. 186.—Lead-type storage-cell construction for farm-lighting battery.

stand jars and light shocks when carried about. Glass is preferable for light-plant batteries, which are not moved about and thus subjected to possible cracking and breakage. The glass jar has the advantage of permitting better observation of the cell contents and its condition.

The tops of both the rubber and the glass-jar cells are covered with a rubber-composition material and sealed with a tarry, acid and waterproof compound. This cover has a vent or opening that permits testing of the electrolyte and refilling the cell. The vent plug is provided with one or more small holes, which allow the escape of gas formed during the charging and discharging action.

184. Chemical Action in the Lead-acid Cell.—As already stated, the active materials in a lead-acid cell, when fully charged, consist of the lead dioxide (PbO_2) or positive material, the sponge lead (Pb) or negative material, and the dilute sulphuric acid solution, or electrolyte (H_2SO_4). If the terminals of such a cell are connected by a conductor so that an electric current will flow, the cell is said to be discharging, and the chemical action takes place as follows:

$$PbO_2 + Pb + 2H_2SO_4 = 2PbSO_4 + 2H_2O$$
(Lead dioxide + lead + sulphuric acid = lead sulphate + water)

Thus, during discharge, the active materials in the plates are changed to lead sulphate, a light-gray compound of somewhat greater volume, and, consequently, the plates swell and tighten.

Likewise, some of the acid is consumed in the formation of the sulphate, and there is an increase in the water content of the electrolyte so that the latter becomes weaker and lower in specific gravity. It is not possible for a cell to discharge until all the active materials in the plates are converted to sulphate; the action is confined largely to the surface layer.

The initial voltage of a fully charged lead-acid cell in good condition is about 2.2, while the average cell will show a voltage of about 2. The minimum permissible voltage for a discharged cell is 1.7. Although a further discharge is possible, the current output is low and the life of the cell is apt to be shortened.

The variation in voltage of the lead-acid cell, according to its condition of charge, might seem to offer a convenient means of determining the stage of charge, but such is not the case for the reason that the change is merely a fraction of a volt. Obviously, a very delicate and extremely accurate testing instrument would be required. However, this voltage variation does provide a possible check on the general condition of a cell; that is, the electrolyte, when tested for specific gravity, may indicate that the cell is in a fully or partly charged condition, but, if the voltage is checked and found to be below normal, something else is wrong with the cell. Nevertheless, since there is a marked and uniform drop in the specific gravity of the electrolyte, owing to the formation of water and the absorption of the acid during the discharging process as already described, the measurement of the specific gravity serves as the most convenient, accurate, and satisfactory means of determining the state of charge of a cell or battery.

When a battery is charged, the chemical reaction is opposite to that during discharge, as shown by the following equation:

$$2PbSO_4 + 2H_2O = PbO_2 + Pb + 2H_2SO_4$$

(Lead sulphate + water = lead dioxide + lead + sulphuric acid)

Thus, the electrolyte gradually assumes its former concentration and the stage of charge, whether half charged, full charged, and so on, can be readily determined by means of the battery hydrometer (Fig. 185). When a fully charged condition is reached, the specific gravity of the electrolyte will not increase further, and noticeable or violent gassing will take place in the cell due to the breaking down of some of the water into hydrogen and oxygen. Excessive gassing and bubbling in a cell or battery are a positive indication that it is completely charged, provided it is in good condition in other respects.

185. Effect of Temperature on Specific Gravity of Electrolyte.—In determining the specific gravity of the electrolyte in any storage cell, its temperature should be about 70°F., or else a correction should be made, according to the existing temperature, for the reason that the liquid will expand or increase in volume and have a lower specific gravity at high temperatures; or will contract and have a higher specific gravity at low temperatures. For the lead-acid battery, the specific-gravity readings for cells in various stages of charge, as given in Table XI, are always assumed to be taken at 70°F. A variation of 3°F. will produce a variation of 0.001 in the hydrometer reading. The usual rule is: For each 3°F. above 70°F., add 0.001 to the observed hydrometer reading, or for each 3°F. below 70°F. subtract 0.001. The result will be the specific-gravity reading if the temperature of the electrolyte were 70°F. For example, if the temperature of the electrolyte is 85°F. and its specific gravity 1.190, the latter corrected to 70°F. would be

$$1.190 + \frac{85 - 70}{3} \times 0.001 = 1.195$$

186. Size, Rating, and Capacity.—Storage batteries are rated according to (1) voltage and (2) ampere-hour capacity. As already explained, the voltage of either the Edison or the lead-acid cell varies only slightly with its condition of charge, the average discharge voltage of the former being considered as 1.2 and of the latter 2.0 volts. The total voltage rating of a battery, therefore, depends only upon the type of cell and the method used in connecting them; that is, for cells connected in series it would be:

Number of cells $\times$ 2 (for lead-acid battery).

Number of cells $\times$ 1.2 (for Edison battery).

For a number of cells connected in parallel, the voltage reading would be about 2.0 for the lead-acid battery or 1.2 for the Edison type, regardless of the number of cells in the battery. Figure 187 shows 32 lead-acid cells connected in multiple series; that is, each of the two sets *A* and *B* consists of 16 cells connected in series, and, therefore, has a voltage of 32. Since *A* and *B* are connected in parallel, the total battery voltage is likewise 32.

187. Ampere-hour Rating or Capacity.—It is frequently convenient and necessary to designate in some manner the size and capacity of storage batteries used for various purposes. This is done by giving

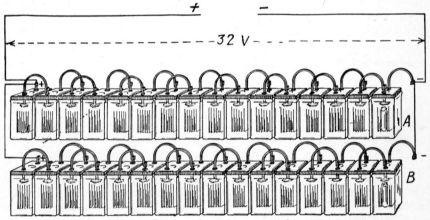

Fig. 187.—Two 16-cell storage batteries connected in multiple series.

them a certain ampere-hour rating. For example, suppose that a given battery is rated at 60 amp.-hr. This means, theoretically, that when in good condition and fully charged, it will deliver 60 amp. for 1 hr., or 30 amp. for 2 hr., or 20 amp. for 3 hr., or 10 amp. for 6 hr., or 7.5 amp. for 8 hr., or 5 amp. for 12 hr., and so on, when it will be completely discharged. In other words, its 1-hr. discharge rate would be 60 amp.; its 2-hr. rate, 30 amp.; its 3-hr. rate 20 amp., and so on. Practically, however, the battery would not deliver 60 amp. for 1 hr. or 30 amp. for 2 hr. or even 10 amp. for 6 hr. In other words, the ampere-hour capacity of any storage battery actually varies greatly, depending upon the rate of discharge in amperes per hour. The lower the rate becomes, the longer the period during which it will discharge at that rate and the greater its total ampere-hour capacity up to a certain limit. This is best explained by reference to the curves (Fig. 188).

It is the usual practice to base the ampere-hour rating of a battery upon the 8-hr. discharge rate; that is, a 60-amp.-hr. battery, when fully charged, should supply a current of 7.5 amp. for 8 hr.; an 80-amp.-hr. battery, 10 amp.; a 160-amp.-hr. battery, 20 amp.; and so on, when discharged to a voltage of 1.75 volts per cell.   As already explained, a battery may be discharged at a rate greater than the 8-hr. rate but such a discharge rate should not be maintained for more than a few minutes at a time for reasons that will be explained later.

Another method of rating storage batteries used with farm light plants is known as the 72-hr. intermittent rating.   It is based upon a discharge rate equal to one-twenty-fourth of the battery rating and a final voltage of 1.75 volts per cell.   The discharge is intermittent, involves

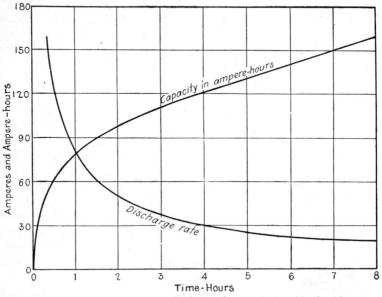

Fig. 188.—Curves showing capacity and discharge characteristics of lead-acid storage cell.

three rest periods, and covers a total period of 72 hr. as follows: discharge for 4 hr., rest for 16 hr.; discharge for 8 hr., rest for 16 hr.; discharge for 8 hr., rest for 16 hr.; discharge for 4 hr.   The intermittent rating of a given battery is usually 25 to 35 per cent higher than the 8-hr. rating.

The watt-hour rating as used by some manufacturers is obtained by multiplying the ampere-hour rating by the battery voltage.

Since nearly all storage batteries consist of a number of cells connected in series, the ampere-hour capacity and rating of the battery are the same as the capacity and rating of the individual cells.   It is primarily dependent, therefore, upon the size and number of plates per cell, because the greater the size or number of plates or both, the greater the total plate surface and the quantity of active materials available for generating electricity.   This is best explained by reference to Table XII, which lists the various plate dimensions and sizes of batteries for farm-lighting plants as supplied by one manufacturer. Other factors affecting the ampere-

TABLE XII.—NUMBER AND SIZE OF PLATES AND RATING OF FARM LIGHT-PLANT BATTERIES

Type[1]	Number of plates per cell	Capacity, ampere-hours		Capacity, watt-hours for 16 cells		Size of plates, inches		Thickness of plates, inches		Weight per cell, pounds
		Intermittent rate	Eight-hour rate	Intermittent rate	Eight-hour rate	Height	Width	Positive	Negative	
XLT-5	5	65	44	2,080	1,408	5¾	5¾	$\frac{7}{32}$	$\frac{3}{16}$	20
XLT-7	7	95	66	3,040	2,112	5¾	5¾	$\frac{7}{32}$	$\frac{3}{16}$	31
XLT-9	9	128	88	4,096	2,816	5¾	5¾	$\frac{7}{32}$	$\frac{3}{16}$	41
XLT-11	11	160	110	5,120	3,520	5¾	5¾	$\frac{7}{32}$	$\frac{3}{16}$	43
FLT-9	9	205	140	6,560	4,480	8	6½	¼	$\frac{3}{16}$	59
FLT-11	11	255	175	8,160	5,600	8	6½	¼	$\frac{3}{16}$	62
FLT-13	13	305	210	9,760	6,720	8	6½	¼	$\frac{3}{16}$	75
FLT-15	15	355	245	11,360	7,840	8	6½	¼	$\frac{3}{16}$	79
FLT-17	17	405	280	12,960	8,960	8	6½	¼	$\frac{3}{16}$	83

[1] Manufacturer's designation.

hour capacity are thickness of plates, porosity of plates and separators, quantity and density of electrolyte, and temperature.

188. Automobile Battery Sizes and Ratings.—Automobile batteries, in most cases, consist of three cells and, therefore, have a voltage rating of 6. The plates are $\frac{5}{64}$ to $\frac{7}{32}$ in. in thickness, somewhat thinner than those used for farm lighting service, for the reason that an automobile requires a battery having a comparatively low capacity but capable of supplying a high discharge rate for very short periods, namely, when operating the starting motor.

The size is determined by the number of plates per cell, which varies from 13 to 21, and the number of plates in turn determines the capacity in ampere-hours. The present standard method of rating automotive-type batteries as established by the Society of Automotive Engineers is as follows:

Batteries for combined starting and lighting service shall have two ratings except as noted. The first rating shall indicate the lighting ability and shall be the capacity in ampere-hours of the battery when it is discharged continuously to an average final terminal voltage equivalent to 1.75 per cell at the 20-hr. rate for passenger car and motor truck service and at the 6-hr. rate for motorcoach service. The temperature of the battery at the beginning of such discharge shall be exactly 80°F., and an average temperature of 80°F. shall be maintained during discharge with a maximum variation of ± 5°F. New batteries shall meet these ratings on or before the third discharge when discharged at a rate in amperes obtained by dividing the rated capacity in ampere-hours by the number of hours at which the rating is specified.   The second rating shall apply only to batteries used in passenger car and motor truck starting and lighting service.   This rating shall indicate the cranking ability of the battery at low temperatures and shall be (1) the time in minutes when the battery is discharged continuously at 300 amp. to a final average terminal voltage equivalent to 1.0 volt per cell, the temperature of the battery at the beginning of such discharge being 0°F.; and (2) the terminal battery voltage 5 sec. after beginning such discharge.   New batteries shall be prepared for this test by giving them a complete charge followed by an initial discharge for 5 hr. at a rate in amperes equal to one-sixth of the 20-hr. rate in ampere-hours specified for the respective sizes and types of batteries. This discharge shall be made while the battery is at a temperature of 80°F. following which the batteries shall be fully charged and then placed in an atmosphere of 0°F. ± 1°F. for not less than 24 consecutive hours prior to discharge at 300 amp. at 0°F.

189. Battery Efficiency.—The efficiency of a storage battery, like that of any similar chemical or mechanical energy-producing device, is the ratio of the energy output of the battery when discharged to the energy input when charged.   This energy output and input may be expressed either in ampere-hours or in watt-hours; that is,

$$\text{Ampere-hour efficiency} = \frac{\text{output in ampere-hours}}{\text{input in ampere-hours}}$$

or

$$\text{Watt-hour efficiency} = \frac{\text{output in watt-hours}}{\text{input in watt-hours}}$$

The efficiency of the lead-acid battery depends upon several factors, such as (1) rate of discharge, (2) condition of plates, (3) temperature, and (4) internal resistance.   For a battery in good condition and discharged at a normal rate and temperature, it may be 80 to 85 per cent.   High discharge rates and low temperatures tend to lower the efficiency.

190. Battery Charging.—As previously stated, a secondary cell or battery is so designated because, when discharged, it can be restored to its original condition or recharged by sending an electric current through it in a direction opposite to that of the current flow during discharge. Also, as already explained, the charging process is an electrochemical action that converts the active materials into their original form.   To charge any battery the following conditions must be observed:

1. Direct current only can be used.   This may be supplied in a number of ways as described later.
2. The flow of the current through the battery in charging must be opposite to that during discharge.   Therefore, the polarity of the charging circuit must be known, and its positive and negative terminals connected to the positive and negative terminals, respectively, of the battery.

3. The voltage of the charging circuit must be greater than that of the battery, but not excessive. A charging voltage equal to about 2.5 volts per cell is recommended to obtain the correct charging rate in amperes. If the voltage of the charging circuit is too low, the rate will be decreased and a longer time required to fully charge the battery. A charging voltage less than the battery voltage will obviously discharge rather than charge the battery. A charging voltage greater than 2.5 to 3.0 volts per cell will charge the battery at too high a rate and result in heating and permanent injury to the materials.

191. Methods of Charging Batteries.—Some of the usual methods and devices used for charging storage batteries are as follows:

1. By a direct-current generator driven by an internal-combustion engine. Examples: (1) All automobiles are equipped with a small direct-current generator, which charges the battery whenever the engine is running beyond a certain speed. (2) The

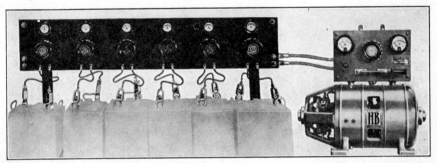

Fig. 189.—A constant-potential battery-charging outfit.

common battery-type farm lighting plant usually consists of a small high-speed engine directly connected to a direct-current generator, which may be operated whenever the battery needs charging.

2. By a special battery-charging outfit or motor generator set consisting of an alternating-current electric motor driven by current from a central power station and connected directly to a direct-current generator, which generates and supplies the necessary charging current. Example: The common battery-charging sets (Fig. 189) as used in automobile battery service stations.

3. By means of a rectifier in connection with an ordinary alternating-current supply line. The rectifier converts the alternating to direct current. There are a number of different types available. Their use is confined largely to "home charging" of radio batteries.

4. By means of a high voltage (110 or 220 volts) direct-current supply line, from a large power plant. This high-voltage direct-current service, being seldom available, is consequently little used.

192. Charging Rate.—The rate in amperes at which the current flows when a storage battery is being charged is known as the charging rate. Obviously, the higher the charging rate for a given battery, the shorter the time required to charge it. On the other hand, an excessively high charging rate may cause permanent injury because of the heating effect on the plates and cell contents.

During the early stages of the charging process, a higher rate of charge is permissible. As the battery becomes partly charged and bubbling or gassing is observed, the rate should be reduced until, near the end and during the last few hours, it is less than one-half of the initial rate. A safe rule to follow in the case of a completely discharged battery, which is in good condition otherwise, is to determine or estimate its ampere-hour rating and start charging it at a rate equal to one-eighth of this capacity, continuing until gassing begins and then reducing the rate about one-half and completing the charge. This method will require from 12 to 24 hr. of continuous charging.

A method of charging automobile batteries, which is now in extensive use in battery service stations, is known as the constant-potential method and permits the complete charging of a battery in about 8 hr. A constant voltage equal to about 2.5 volts per cell is applied to the battery. Since the resistance of the discharged battery is low, the initial charging rate will be high but no harmful effects are produced. As the battery charges and the resistance increases, the applied voltage remaining constant, the charging rate drops or tapers off during the final hour or two of the charging period. A typical layout for constant-potential charging is illustrated by Fig. 189.

193. Equalizing Charge.—After a battery has been given a complete charge, it may be found that the specific gravity of the electrolyte in one or more cells is considerably lower or higher than that of the others. This might be caused by (1) excessive evaporation of the electrolyte,

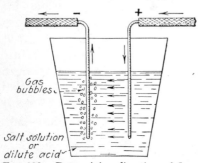

Fig. 190.—Determining direction of flow of a battery current.

(2) spilling or leaking out of the electrolyte, (3) acid added at some time instead of water to replace evaporation, (4) excessive sulphation, and (5) incomplete charge.

If a cell tests too low, the level of the electrolyte should be made to correspond with the others and the complete battery given an overcharge at a low rate for a few hours until the cell or cells in question reach the same specific gravity. This is known as an *equalizing charge*. If there is no change then, it is advisable to replace the electrolyte with a fresh mixture of the correct strength. High-voltage batteries consisting of a large number of cells such as farm lighting-plant batteries should be tested at least once a month, and, if necessary, given an equalizing charge. This will insure a maximum period of service and less trouble.

194. Hints on Battery Charging.—When a battery is connected to a charging circuit, care must be taken to connect it correctly, that is, the positive terminal of the circuit to the positive of the battery, and negative to negative. If the battery terminals are not marked or otherwise distinguishable, wires may be attached to them and the other ends immersed in a glass of water (Fig. 190) containing a small amount of common salt in solution. The ends of the wire should not touch. Bubbles will be seen to form on the end of the wire attached to the negative terminal of the battery. If a battery is not properly connected for charging, it will discharge further and the plates will be seriously injured.

It is not advisable to add water to the electrolyte in a completely discharged battery unless the electrolyte is below the top edge of the plates and they are exposed. The charging action increases the volume of the electrolyte; therefore, if too much water is added, the solution may bubble up into the filler opening or even run over.

Each cell in a battery should be tested by means of a hydrometer every 3 to 5 hr. to determine the progress of the charging action. Little change may occur during the first few hours, but, if the cell is taking the charge, bubbling will eventually take place and, near the end, excessive bubbling or gassing should develop. When gassing begins, the charging rate should be low and the charge should not be continued more than 1 or 2 hr. longer, unless one or more cells have failed to come up with the others. Continued charging, after gassing begins, lowers the level of the electrolyte, causes heating of the cell, and may injure the plates. A battery is fully charged when all cells are gassing freely and the specific gravity and voltage have reached a high value and show no further change.

BATTERY TROUBLES

195. Sulphation.—The life of any storage battery is dependent upon the care and attention it receives. Maximum service should not be expected from a storage battery of any kind if it is not inspected and checked up periodically. Perhaps the most common cause of short life is sulphation, which results from a protracted discharged condition. As already mentioned, during the normal discharging process, the active material in the plates is changed to lead sulphate ($PbSO_4$) which, in turn, is converted back into lead dioxide and lead when the cell is recharged. However, if the cell or battery is permitted to remain in a discharged condition for any length of time, this sulphate gradually hardens and is not readily converted to its former state. The longer the cell remains uncharged, the more difficult it is to recharge it. A badly sulphated cell or battery can be recharged only with difficulty, if at all, and then seldom regains its original effectiveness.

196. High Charging and Discharging Rates.—Excessive charging or discharging rates for periods exceeding a few seconds or minutes are injurious to storage batteries in that heat is generated, which causes warping or buckling of the plates and thereby loosens the active material and reduces the life and efficiency of the battery. As previously stated, it is important that a battery be charged or discharged only at a certain rate based upon its capacity in ampere-hours.

197. Other Causes of Battery Troubles.—Other common causes of battery troubles are: (1) failure to keep electrolyte at proper level and over tops of plates; (2) use of undistilled or impure water for replenishing electrolyte; (3) dirty, badly corroded, or loose connections and terminals; and (4) use of too strong acid for replacement of electrolyte.

198. Care of Batteries When Not in Use.—The care of a battery that is not to be used for some time depends upon the length of the period of disuse. If it is only for a few months or a year, it may be stored "wet," so to speak, that is, it should be given a full charge at the beginning of the idle period and recharged at a low rate at the end of each month, if possible.

If the period of disuse is a year or more, "dry" storage is recommended. This consists of the following steps: (1) give the battery a good full charge; (2) empty electrolyte out of all cells and replace immediately with some pure water and allow to stand 5 hr.; (3) empty out the water and disassemble cells, removing plate groups and separators; (4) pour water over plates to clean them thoroughly; (5) allow plates to dry and then lay them away until needed.

When the battery is to be used again, the following procedure is recommended: (1) Reassemble plate groups, using new separators if possible; (2) add fresh electrolyte of same strength as that discarded; (3) allow battery to stand from 5 to 10 hr. and then charge at a very low rate until it gasses freely.

References

CONSOLIVER and BURLING: "Automotive Electricity."

JANSKY and WOOD: "Elements of Storage Batteries."

Operation and Care of Vehicle Type Batteries, *U. S. Bur. Standards, Circ.* 92.

The Design, Manufacture, Operation, and Care of Lead Storage Batteries for Farm Lighting Plants, *Trans. Amer. Soc. Agr. Eng.*, Vol. 12.

WRIGHT: "Automotive Repair," Vol. III.

CHAPTER XVII

MAGNETS AND MAGNETISM—INDUCTION

It has been found that under certain conditions a piece of iron or steel or some iron alloy possesses the property of attracting or repelling other pieces of iron or materials containing this metal. This peculiar action or form of energy is known as *magnetism*, and the metal possessing it is called a *magnet*. Other metals and nonferrous alloys, such as aluminum, copper, brass, bronze, tin, lead, zinc, and so on, do not have this property and are therefore termed *nonmagnetic*.

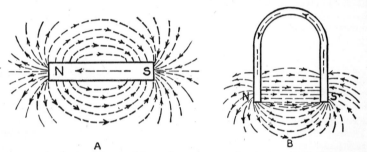

FIG. 191.—Steel-bar and horseshoe magnets showing magnetic field.

199. Nature of Magnetism.—Soft iron of a more or less pure form absorbs or takes up magnetism readily and in greater quantity, so to speak, but retains this property only while the magnetizing force or medium is very near or acting upon it. In other words, soft or pure iron forms what is known as a *temporary magnet*. Steel and alloys containing iron are not magnetized so readily, but when once so treated, retain this property to a certain degree even after the magnetizing medium has been removed. The common bar and horseshoe magnets (Fig. 191) are, therefore, made of hard steel and are known as *permanent magnets*. The well-known magneto magnets are likewise permanent magnets and are made of a very high grade steel alloy.

200. Characteristics of Magnetism.—If a steel-bar magnet is held under a piece of cardboard (Fig. 192), on which has been placed a thin layer of iron filings, and the cardboard is tapped gently, the filings will assume a rather definite arrangement and form more or less distinct lines, which radiate from one end of the magnet to the other. The attraction appears to be greater at the ends or poles of the magnet but also exists

in the space around it between the poles.   This space about a magnet in
which there exists more or less attraction for other pieces of iron is known
as the *magnetic field*.   When iron filings are held over a U-shaped or

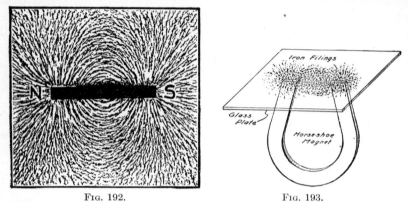

Fɪɢ. 192.   Fɪɢ. 193.

Fɪɢ. 192.—Lines of magnetic force around a bar magnet as shown by iron filings.
Fɪɢ. 193.—Effect produced by a magnet held under a glass plate which is covered with
iron filings.

horseshoe magnet (Fig. 193), a similar effect is produced, and a magnetic
field likewise exists.

If a pocket compass is brought near an ordinary bar magnet, the end
of the needle, which ordinarily points north, will be attracted by one pole

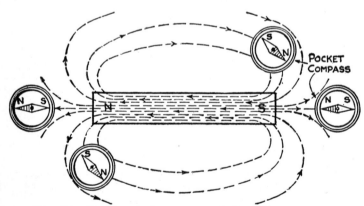

Fɪɢ. 194.—Polarity of a bar magnet as indicated by a compass.

of this magnet but will be repelled by the other pole, as shown in Fig. 194.
The end of the magnet attracting the north end of the needle is called the
south pole and the end repelling the north end of the needle is known as
the *north pole* of the magnet.   It is likewise observed that the lines of
force about the magnet flow from north to south.

If the north poles of two steel magnets are brought close to each other (Fig. 195), and iron filings are sprinkled on a cardboard held over them, the filings will arrange themselves as shown, indicating that there is a repulsion or repulsing effect.   On the other hand, if a north pole of one magnet is brought close to the south pole of the other (Fig. 196), there is a very strong attraction and the filings arrange themselves differently.   In

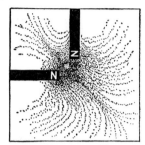 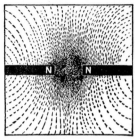

Fig. 195.—The repelling effect of like poles.

other words, it can be said that like poles repel, and unlike poles attract each other.

If a piece of soft iron is brought near to a permanent steel magnet, the soft iron itself becomes a magnet and the lines of force tend to concentrate themselves in it.   If the soft iron is removed, however, it does not retain its magnetism.   If, instead of a piece of soft iron, a piece of unmagnetized soft steel is placed near a permanent steel magnet, as previously

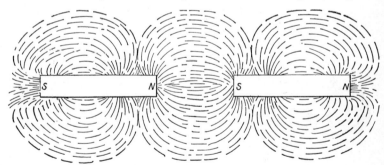

Fig. 196.—The magnetic attraction produced by unlike poles.

described, the piece of soft steel, upon removal from the magnet will, perhaps, retain a small amount of magnetism, known as *residual magnetism*.   The ability to retain this residual magnetism is known as *retentivity*.

A piece of very hard steel, unmagnetized, when brought close to a magnetized piece of steel or a strong magnet, will not take the magnetism so readily as soft iron, but once it becomes magnetized, it retains a considerable amount and is said to have better retentivity than soft iron or soft steel. These characteristics of soft iron, soft steel, and hard steel are very important and useful in connection with the successful

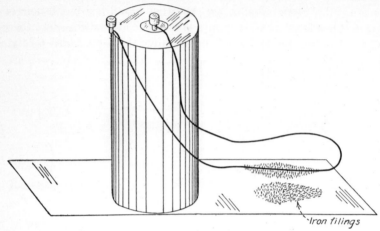

Fig. 197.—The magnetic field about a conductor of electricity as indicated by iron filings.

operation of numerous electrical devices such as coils, magnetos, and electric generators.

201. Electromagnetism.—Although magnetism and electricity are two entirely different forms of energy, they are more or less connected or related. For example, if an electric current is passed through a piece of copper or iron wire, and a portion of the latter dipped in iron filings (Fig. 197), they will be attracted to it and the wire will have the properties of a magnet. When the circuit is broken, the filings drop off. This shows that the conductor of electricity has a magnetic field about it. If an ordinary pocket compass is placed in different positions about a wire carrying an electric current, and the action of the needle observed in these different positions, it will be found that the lines of force form concentric circles about the wire, as shown in Fig. 198. Likewise, it has been found that the direction taken by the lines of force is determined by the direction of the current through the wire, as shown in Fig. 199.

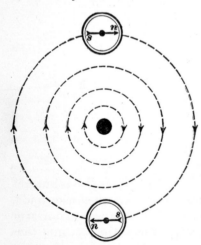

Fig. 198.—Direction of lines of force about a conductor of electricity as indicated by a compass.

If a certain length of wire is made into a coil, as in Fig. 200, and the current passed through it, the lines of force about each turn of wire in the coil will combine to form a concentrated or strong magnetic field about the entire coil. Such a coil of wire through which an electric current is flowing is known as a *solenoid* and possesses the same properties as a piece of magnetized steel; that is, it has a north and a south pole and a magnetic field around it. If the current is suddenly broken or ceases to flow, the magnetism likewise disappears.

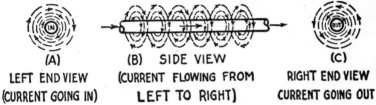

(A)
LEFT END VIEW
(CURRENT GOING IN)

(B) SIDE VIEW
(CURRENT FLOWING FROM
LEFT TO RIGHT)

(C)
RIGHT END VIEW
CURRENT GOING OUT

Fig. 199.—Effect of direction of flow of current on lines of force.

The polarity of a solenoid depends upon the direction of flow of the current through the wire, and may be determined as follows if the direction of flow is known: Grasp the coil with the right hand, with the fingers pointing around the coil in the direction of the current flow. The thumb will then point toward the north pole. The polarity can likewise be determined by means of an ordinary compass in the same manner that the polarity of the steel magnet is determined (Fig. 194).

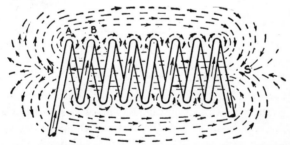

Fig. 200.—The magnetic field about a coil of wire carrying an electric current.

202. The Electromagnet.—If the end of a piece of soft iron is brought near a coil or solenoid, consisting of many turns of insulated wire, and a current is passed through the coil, the iron will be attracted and drawn or sucked into the center of the solenoid (Fig. 201). The soft iron thus becomes magnetized and is known as an electromagnet and possesses the same characteristics as an ordinary steel-bar magnet. Owing to the fact that the soft iron is extremely magnetic and readily magnetized, the lines of force tend to concentrate in this iron core, and the coil becomes much stronger, magnetically, than without the iron core. If the core of soft iron is withdrawn, this iron will lose its magnetic properties. A steel core would not produce so strong an electromagnet because steel does not take up magnetism as readily as soft iron.

The strength of such an electromagnet depends upon two factors, namely, the number of turns of wire around the core making up the coil, and the quantity of current flowing in the wire in amperes; that is, the greater the number of turns of wire and the greater the current flow in amperes, the stronger the electromagnetic field and the electromagnet produced.

203. Induction.—In the preceding discussion it has been shown that a conductor of an electric current has a magnetic field existing about it as

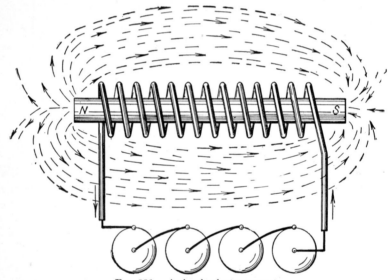

Fig. 201.—A simple electromagnet.

long as the current flows. Now, if the action is reversed and, in some manner, a magnetic field is suddenly set up about a conductor of electricity, an electric current will flow in the conductor just at the instant the field builds up. Again, if a magnetic field existing about a conductor is suddenly reduced in strength or completely broken down, a current will flow in the conductor at the instant the change occurs. This peculiar relationship or interaction of electricity and magnetism is known as *induction*.

It should not be assumed that if a conductor is merely held in a magnetic field of uniform strength a current will flow. Such a current is generated only by a change in the number of lines of force about the conductor. This change may be from a low to a high field strength or from a high to a low field strength.

204. Electromagnetic Induction.—The phenomenon of induction is well demonstrated and illustrated by means of an ordinary single-winding coil (Fig. 202). If an ordinary 6-volt battery is connected to such a coil

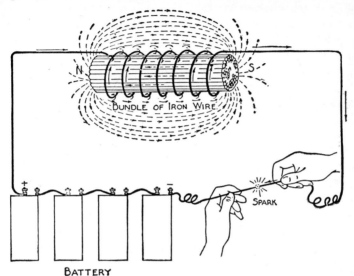

Fig. 202.—Action of a simple low-voltage induction coil.

as shown and the circuit is closed, a strong magnetic field is built up and exists as long as the current flows. Some of the electrical energy of the battery becomes magnetic energy. Now, if the circuit is quickly broken

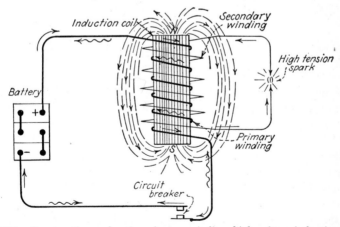

Fig. 203.—Construction and action of a two-winding, high-voltage induction coil.

at some point, an electric spark of considerable size is observed where the break occurs. Bearing in mind that an electric spark is electricity flowing through air and that the latter is a very poor conductor, it is evident that this sudden breakdown of the magnetic field about the coil sets up a high voltage in the circuit, which produces an instantaneous flow of current. The magnetic energy is thus dissipated and converted into electrical energy again and this action is known as *electromagnetic induction.*

The double-winding coil (Fig. 203) demonstrates this action in a somewhat more conclusive manner, perhaps, than the single-winding coil. Suppose that the winding attached to the battery, ordinarily known as the primary, is made of reasonably coarse wire and has fewer turns than the other or secondary winding, which is finer and has many turns. If the primary circuit is closed, current flows through it and sets up a magnetic field that also surrounds the secondary winding, since it is on the same iron core. Now, if the primary circuit is suddenly broken, a spark will jump the small gap left between the ends of the secondary. Thus the magnetic field which existed about this winding, although created by the primary winding, generates for an instant a current flow in the secondary winding, due to its sudden breakdown or dissipation. This again is called an *induced current.*

The phenomenon of induction, produced in a somewhat different manner than previously described but fundamentally the same in

IF A COIL OF WIRE IS MOVED UP AND DOWN A CURRENT WILL BE GENERATED TO FLOW IN THE WIRE

FIG. 204.—Showing principle involved in generating an electric current by induction.

principle, is shown in Fig. 204. If the permanent steel horseshoe magnet is first held so that the coil of insulated copper wire is between its poles, a magnetic field exists about the coil. Now, if the coil is jerked away in the direction indicated, a voltage and resultant current flow will be generated for an instant as indicated by a sensitive galvanometer. Likewise, if the coil is quickly returned to the first position, a current will flow just at the instant the coil enters the field. In this device it will be observed that the movement of the magnet and its field away from the coil, or the movement of the coil away from the magnet, results apparently in the lines of force being cut or broken. In fact, the action is often so expressed, and the maximum voltage and current flow occur when the greatest number of lines of force are cut.

The voltage and amount of current induced in an electric circuit depend upon three factors, namely, (1) the strength of the magnetic field about the conductor, (2) the number of turns making up the conductor involved, and (3) the speed or rate at which the magnetic field changes or breaks down. In other words, the stronger the magnetic field, the greater the number of turns of wire in the field, and the quicker the field strength changes with respect to the conductor, the stronger the voltage and current induced. The principles of electromagnetic induction, as just described, are important and fundamental to a clear understanding of the action and operation of all ignition coils, magnetos, electric generators of all kinds, transformers, doorbells, and alarms and many similar devices.

205. Direction of Flow of Induced Current.—Just as there is a definite relation between the direction of the current in a solenoid or coil

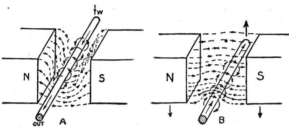

FIG. 205.—Effect of polarity and direction of motion of conductor on direction of current flow.

and the polarity of the electromagnet so formed, so there is also a definite relation between the arrangement of the poles, the direction of movement of the conductor through the magnetic field, and the direction of flow of the induced current. For example, referring to Fig. 205*A*, it will be observed that if the conductor is moved downward through the field between the poles of the magnet, the lines of force are bent or distorted and assume a more or less circular form about the conductor as the latter passes through the field. The direction of the induced current is thus determined by the direction of the circular lines of force about the conductor. In Fig. 205*B* the conductor is shown moving upward through the field, so that the lines of force pass around it in the opposite direction; therefore the induced current flows the other way.

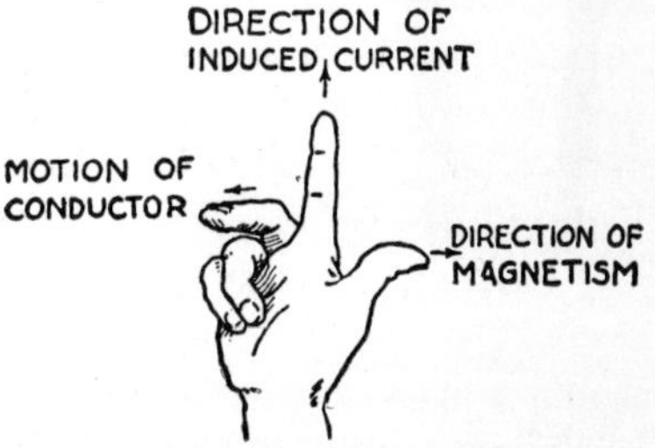

FIG. 206.—Right-hand, three-finger method for determining direction of induced current.

A rule for determining any one of the three conditions involved, namely, the direction of flow of the magnetic lines of force, the direction of motion of the conductor, and the direction of flow of the induced current, when two of them are known, is called the *right-hand three-finger rule* and is illustrated by Fig. 206.

References

CONSOLIVER and BURLING: "Automotive Electricity."
HELDT: "The Gasoline Automobile," Vol. III.
STONE: "Electricity and Its Application to Automotive Vehicles."

CHAPTER XVIII

BATTERY IGNITION SYSTEMS

All electrical ignition systems for internal-combustion engines are either one or the other of two types, namely, (1) the *make-and-break* or *low-tension* system and (2) the *jump-spark* or *high-tension* system. The current for either system may be generated chemically, that is, by means of a battery of some kind; or mechanically, by means of a magneto or similar electric generator.

The essential functions of any electrical ignition system are, first, the generation of a large, hot spark in the cylinder; and, second, the production of this spark at the right instant in the travel of the piston. In other words, if a good spark is produced in the cylinder at the right time, the combustible mixture of fuel and air should be properly ignited.

206. Ignition Voltage and Requirements.—An electric spark is nothing more or less than an electric current passing through air. In any electrical ignition system such a spark is produced by making the current flow between two points separated by a small fraction of an inch and forming a part of an otherwise complete electrical circuit. Since air and other gases are very poor conductors of electricity, an extremely high voltage is necessary to produce the desired spark in the cylinder of any engine. Furthermore, at the time that the spark is required in the cycle, the gaseous mixture surrounding the sparking points is under compression. Consequently, this mixture is considerably more resistant to the passage of an electric current than if it were at a lower or ordinary atmospheric pressure. In fact, referring to Fig. 207, it is observed that the voltage required to produce a satisfactory spark varies almost directly as the compression pressure in pounds per square inch.

Again, the greater the width of the spark gap, the greater the voltage necessary to make the current flow across the gap and create a desirable spark. This is clearly shown by reference to the curve in Fig. 207. Therefore, to obtain the best results, it is important to use as narrow a gap as possible in the spark plug or igniter and to maintain the highest possible voltage in the circuit.

It has already been explained that the voltage of a single dry cell, a storage cell, or any similar chemical generator of electricity seldom exceeds 2.0 volts. In view of this fact, and since voltages ranging from a few hundred to several thousand volts are required if any electrical ignition system is to function properly, it would seem impractical to attempt to utilize this source of electric energy for ignition purposes. That is, to do so, would apparently require a large number of cells connected in series which would mean excessive cost, weight, and depreciation and, perhaps, considerable trouble. However, since the spark is needed in the cylinder at a certain instant only and need not continue flowing for any length of time, even for a second or less, the source of the electrical energy is not required to supply a steady flow of current, and it is possible to utilize the

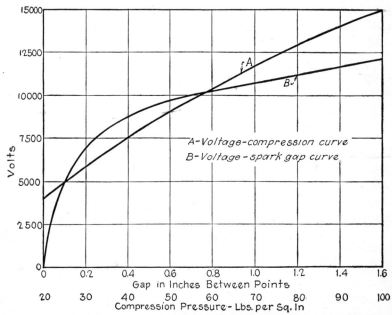

Fig. 207.—Curves showing voltage requirements for different compression pressures and spark-gap widths.

principle of electromagnetic induction, as previously explained, to increase or "step up" the low voltage of an ordinary three- or four-cell battery to the desired degree.

207. Make-and-break System.—If four dry cells are connected in series and the ends of the two terminal wires are rubbed together, only a very fine spark will be observed. In fact, the current will not flow across even a very minute gap, unless the circuit is first completed and then broken. Now, if this battery is connected properly to a comparatively simple and inexpensive induction coil (Fig. 202), a voltage is set up that will produce a much fatter and brighter spark. It is evident, therefore, that the introduction of the coil and the sudden breaking of the electrical circuit result in the generation for an instant of a very high voltage.

It might be said that this spark is produced indirectly and not directly

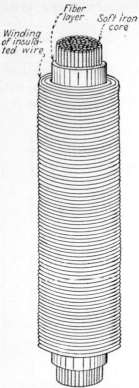

Fiber layer
Soft iron core
Winding of insulated wire,

by the battery; that is, the battery current forms the electromagnet and the sudden or instantaneous collapse of the magnetic field generates a second or induced current of much higher voltage. The quicker the circuit is broken, the larger the spark, for the reason, as previously discussed, that the voltage of an induced current is dependent upon three factors, one of which is the rate at which the magnetic field is cut or broken down.

The make-and-break system of ignition (Fig. 209), using a battery as the source of current, operates more or less in the manner just described. The coil (Fig. 208) is very simple in construction, consisting essentially of a soft-iron core, made up of a bundle of soft-iron wires rather than a solid piece of iron, so that the magnetizing and demagnetizing action will take place as rapidly as possible; a layer of insulating material to bind the iron wires together and separate them from the winding; a single winding consisting of many turns of insulated copper wire; a metal or fiber protective case; and the two terminals that are the ends of the winding.

Fig. 208.—Construction of a low-tension or make-and-break ignition coil.

The device for breaking the circuit and producing the spark in the cylinder is called an igniter. As shown in Fig. 210, it consists of a cast-iron block or frame,

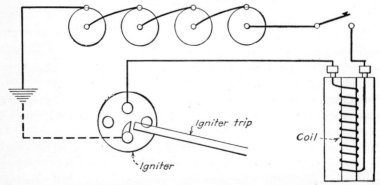

Igniter trip

Coil

Igniter

Fig. 209.—Wiring diagram for the make-and-break ignition system.

flanged and drilled for bolting to the engine; a stationary pin or electrode, insulated by mica from the igniter block and other parts; and a movable pin, equipped with springs to permit it to snap away or cause a quick separation of the points.

The complete system for a single-cylinder engine (Fig. 209) consists of a battery, coil, igniter, switch, and the igniter-tripping device on the engine, operated by the cam gear and attached, usually, to the exhaust-valve push rod. The parts are connected to each other in series as shown. If the switch is closed the circuit will apparently be complete. However, the igniter points are held open a fraction of an inch during a greater part of the cycle in order to prevent undue waste of the battery current. Just before the spark is needed in the cylinder, the igniter trip arm (Fig. 209) pushes the movable point against the stationary insulated point, and the current flows and builds up an electromagnetic field about the coil. Then, at the instant the spark should occur, the trip arm releases the movable electrode, and a large bright spark is produced at the points as they separate. A quick separation of the points is brought about by springs on the movable electrode. The body of the igniter and the engine frame serve as part of the

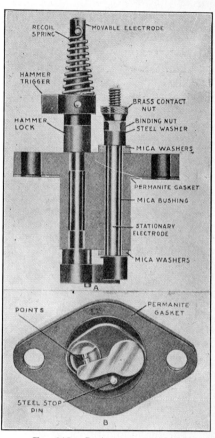

Fig. 210.—Igniter construction.

electric circuit. Therefore, one terminal of the battery is grounded at any convenient metal part of the engine.

The igniter points were formerly made of platinum, but, owing to the high cost of this metal, alloys of nickel, chromium, or tungsten, which are very hard, are now used entirely.

Most engines equipped with the make-and-break system of ignition have a small hand lever of some kind by which the spark can be retarded for starting. This device functions so as to slightly delay the tripping of the igniter until the piston reaches compression dead center. There is also some simple adjustment on the tripping mechanism by which the time of ignition can be changed a small amount, or reset, should it get out of time. Figure 211 illustrates the spark-advance and retard lever and the timing adjustment as used on many engines.

208. Adaptability and Advantages of Make-and-break System.—The use of the make-and-break system of ignition is confined principally to farm type stationary low-speed engines. It is simple from an electrical standpoint and gives a good hot spark, making it particularly well adapted to kerosene-burning engines. On the other hand, it has certain disadvantages that limit its utility for multiple-cylinder high-speed engines. These are (1) The igniter-tripping mechanism creates more or less noise. (2) Frequent adjustment for wear is usually necessary, other-

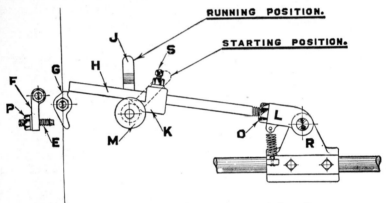

Fig. 211.—Igniter-tripping mechanism showing timing adjustment.

wise the spark gets out of time. (3) The igniter requires a large opening in the cylinder necessitating the use of a gasket to prevent loss of compression. (4) The igniter cannot be removed so conveniently and easily as a spark plug and is more expensive to replace.

JUMP-SPARK SYSTEM

The make-and-break system of ignition just d cribed is known as a low-tension or low-voltage system because the circuit is first completely closed and then suddenly broken at the igniter points and a spark produced. The momentarily induced voltage is only 300 or 400 volts, which is insufficient to make the current jump the gap of a spark plug, the sparking device used in the jump-spark system. That is, in the jump-spark system a permanent air gap exists between the points of the plug and, with the high compression pressure, a voltage of 5,000 to 10,000 is necessary to make the current flow across this gap and ignite the charge. To set up this high voltage, the principles of electromagnetic induction are utilized as in the make-and-break system, but a somewhat different and more complicated type of coil is necessary.

The jump-spark system with battery may be subdivided into (1) the *nonvibrating system,* and (2) the *vibrating system,* depending upon whether the coil used is of the nonvibrating or the vibrating type, respectively. The fundamental principles of each are more or less identical.

209. Nonvibrating System.—Referring to Fig. 212, a nonvibrating coil is made up of a soft-iron core consisting of a bundle of soft-iron wires, a primary winding of about No. 16 insulated copper wire, and a secondary winding consisting of many thousands of turns of very fine insulated copper wire. In the manufacture of such a coil, the wire itself that makes up the winding is well insulated. Likewise the different turns of wire are insulated from each other and the layers forming each winding are well insulated one from the other. This thorough insulation is necessary in order to prevent the excessive voltage generated in the coil winding from breaking through and creating an internal short circuit. In addition to thorough insulation, the coil is surrounded by a moisture- and waterproof compound and incased in a metal or fiber casing. The ends of the windings are brought through the case and fastened to the terminal pieces or posts.

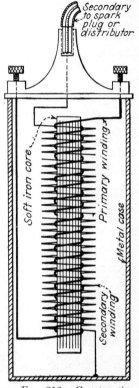

Fig. 212.—Construction of a typical nonvibrating coil for jump-spark ignition.

The operation of such a coil is as follows: If a battery is connected in the primary circuit, as shown in Fig. 203, the current flows and forms an electromagnet, as previously described. If this battery current is now suddenly broken, the electromagnetism dies out very quickly and thereby induces a current in the primary winding, and also in the secondary winding, for the reason that it likewise is in the magnetic field. In other words, it is not necessary that the secondary winding have an actual electrical connection with the primary winding to have a current induced in it by the breakdown of the magnetic field. Since the secondary winding consists of many more turns than the primary winding, the voltage induced is very high, and the current flows across the gap and produces a spark. Therefore, if a coil of this kind can be connected to a battery and to the spark plug of the engine, and some means can be devised by which the primary circuit can be broken just at the instant the spark is needed in the cylinder, the fuel charge will be ignited.

The device (Figs. 214 and 215) that is used to interrupt or break the primary circuit is known as a *breaker* or *interrupter mechanism.* It consists essentially of three parts (1) a stationary insulated point, (2) a

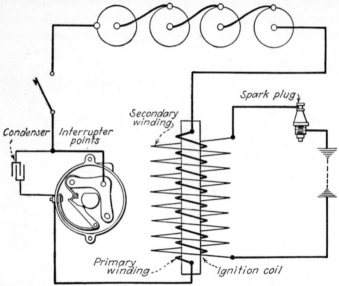

Fig. 213.—Wiring diagram for a single-cylinder engine equipped with the jump-spark system with nonvibrating coil.

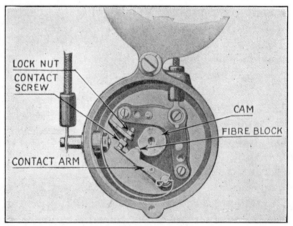

Fig. 214.—Single-cylinder-engine breaker mechanism for nonvibrating coil system.

movable, noninsulated point, and (3) the rotating cam. The cam strikes the movable point and separates it from the stationary one at the instant the spark is desired.

A diagram of the wiring for a single-cylinder engine equipped with the nonvibrating jump-spark system is shown in Fig. 213. The primary circuit consists of the battery, the primary winding of the coil, and the breaker mechanism. The secondary circuit consists only of the secondary winding of the coil and the spark plug. It is observed that the cam operating the breaker points is so constructed that the points are held open a greater part of the time. They close only long enough to just complete the primary circuit and then open again. This prevents needless flow of the battery current.

210. The Condenser.—When the breaker points open, a current is induced in the primary as well as in the secondary winding, and a spark will occur at the points, just as it did at the igniter points in the make-and-break system. Since this sparking will soon pit or roughen the points and eventually render them inefficient or inoperative, it is desirable that it be eliminated. This is done by means of a condenser (Fig. 216), which consists of alternate layers of tin foil separated and insulated from each other by waxed paper. As shown, it is not

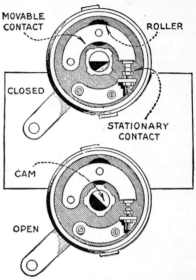

FIG. 215.—Four-cylinder breaker mechanism for nonvibrating coil system.

possible for an electric current to flow through a condenser as it would through a wire or a coil. The condenser is connected in the circuit as shown in Fig. 213. That is, it might be said that it is connected across the breaker points or bridges them.

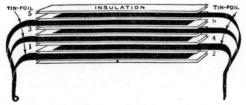

FIG. 216.—Construction of a condenser.

The condenser action is as follows: The current induced in the primary circuit at the time of the separation of the points, instead of dissipating itself in the form of a spark at the points, surges into the condenser but not through it, then surges outward into the circuit again in the opposite direction; then reverses again and surges backward and continues until it finally dies out. This action not only eliminates the arcing almost entirely but, in addition, helps to demagnetize the iron core and break down the electromagnetic field much more rapidly than it is broken down without the condenser. Consequently, the current induced in the secondary circuit is of a much higher voltage, and a better spark is produced. In other words, we might say that the condenser protects the points by eliminating the arcing and intensifies the current in the secondary winding, producing a better spark at the spark-plug gap. The condenser may be built in the coil unit or it may be found near the breaker-point mechanism.

211. The Breaker Mechanism.—The breaker mechanism (Figs. 214 and 215) used with the nonvibrating coil system is usually a rather small delicate mechanism with the stationary point carefully insulated from the other parts. A small spring of some kind holds the movable point against the stationary one when contact is desired. The contact points themselves are usually made of some hard and high-heat-resistant material, such as a platinum or tungsten alloy. The breaker cam for a single-cylinder engine, or for a four-stroke-cycle engine with any number of cylinders for that matter, rotates at camshaft speed or one-half crankshaft speed. That is, it is obvious that in a single-cylinder, four-stroke-cycle engine, the points must open once in two revolutions of the crankshaft to produce the correct number of sparks. Therefore, if the cam is in time, that is, opens the points at the right time in the compression stroke, the ignition system will function correctly.

For multiple-cylinder engines the breaker cam, as shown in Fig. 219, has as many faces or projections as there are cylinders and opens the points that number of times per revolution, thereby producing the required number of sparks without increasing its speed with reference to the crankshaft speed. The advancing or retarding of the spark is usually produced by shifting the breaker points by means of the projecting arm (Fig. 215), which is attached to the advance and retard lever of the engine.

212. Multiple-breaker Mechanisms.—Engines having six or more cylinders and operating at a speed of 2,000 r.p.m. or more are sometimes equipped with multiple or double breaker-point mechanisms, in order to provide a longer closed interval and therefore permit the coil to build up fully and give a strong spark. For example, an eight-cylinder engine running at 3,000 r.p.m. fires 12,000 times per minute or 200 times per second. This means that if a single set of points were used the coil would have a $\frac{1}{200}$-sec. time interval to build up its field and then break down and produce a spark.

Two distinct arrangements involving a double-breaker mechanism are used as follows:

1. Two sets of interrupter points connected in parallel with half as many cam lobes as cylinders, each set of points firing one-half the number of cylinders (Fig. 217).

2. Two sets of interrupter points connected in parallel with the same number of cam lobes as cylinders, one set serving only to close the circuit after the other set has broken it to produce a spark (Fig. 218).

The ignition system, in other respects is the same as when a single set of points is used. In either system, if one set of points should remain

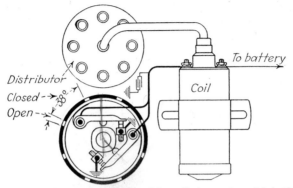

Fig. 217.—Multiple-breaker mechanism for eight-cylinder engine with half as many cam lobes as cylinders.

closed or become short-circuited, the circuit would not be interrupted by the other set of points and no spark would be produced. On the other hand, in system 1 if one set of points failed to close, only half of the cylinders would fire. In system 2, in case of failure of either set of

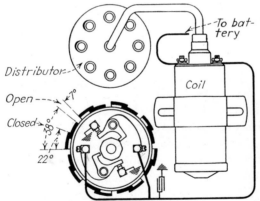

Fig. 218.—Multiple-breaker mechanism for eight-cylinder engine with same number of cam lobes as cylinders.

points to close, all cylinders would still fire. However, if the one set of points that acts only to close the circuit fired the cylinders, owing to the other set being held open, then the spark would be out of time.

The proper functioning of an ignition system using a double-breaker mechanism depends upon the proper opening and closing of the points with respect to each other—this being called *synchronization*—as well as the correct timing of the points with respect to the piston travel. To synchronize system 1, the gaps are first adjusted correctly and equalized. Then one set of points is adjusted or moved with respect to the cam until the correct open and closed interval relationship is obtained. A special synchronizing tool of some kind is necessary in making this adjustment. Synchronizing system 2 merely involves setting and equalizing the gaps for the two sets of points.

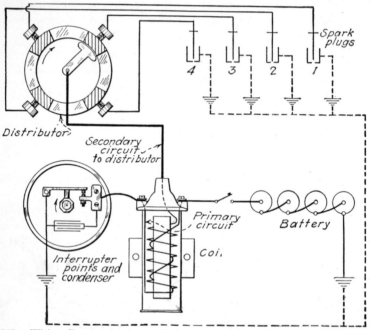

Fig. 219.—Wiring diagram for four-cylinder engine equipped with jump-spark ignition system with nonvibrating coil.

213. Vibrating Coil System.—The vibrating type of induction coil consists of (1) a soft-iron core made up of a bundle of soft-iron wires, (2) a primary winding made of about No. 16 insulated copper wire and relatively few turns, (3) a secondary winding of very fine insulated wire and many hundreds of turns, (4) the vibrator points, and (5) a condenser.

Figure 220 shows the general layout of such a coil and the circuits involved. Referring to this figure, it is observed that the battery current flows from the positive terminal to one of the vibrator points, through these points to the primary winding, and back to the negative side of the battery. The circuit being complete, an electromagnetic field is set up and the core is magnetized and attracts the flexible vibrator point, the other point being stationary. The circuit is thus broken at the points, the magnetic field dies out, and a current of very high voltage is induced in the secondary winding, producing a spark at the gap. However, at the instant that the vibrator points separate, the iron core loses its magnetism and the flexible vibrator point is released and again makes contact with the stationary point. This completes the circuit, and the action becomes continuous. The rapid vibration induces a series of high-voltage currents in the secondary circuit and a shower of sparks occurs at the gap. Here we observe that the vibrator points, operated by magnetic attraction, take the place of the mechanical interrupter as used in the nonvibrating coil; that is, the vibrator points become the inter-

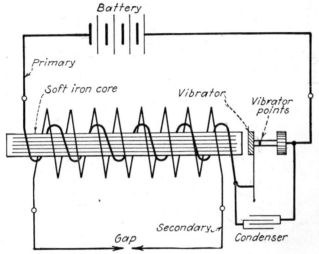

Fig. 220.—Construction and wiring of a vibrating-type jump-spark coil.

rupting or breaker mechanism for breaking down the primary circuit and the magnetic field, thereby creating an induced current in the secondary circuit. A condenser is necessary to eliminate the arcing at the vibrator points and keep them from becoming rough. It is connected across the points as shown and acts in the same manner as previously described in connection with the nonvibrating coil.

If the battery and coil are now connected to a spark plug, and some mechanism is placed in the primary circuit that will permit the current to flow only at the time that the spark is needed, the ignition system will be complete. In other words, some mechanism is needed to close the primary circuit just at the time the spark is to occur in the cylinder and then open it so that the current will not flow continuously. This continuous flow of current would produce a constant vibration of the vibrator, thus not only wasting the battery current, but also creating a spark in the cylinder continuously throughout the cycle, which would

interfere with the proper operation of the engine. This mechanism is
known as a timer (Fig. 222) and consists essentially of an insulated

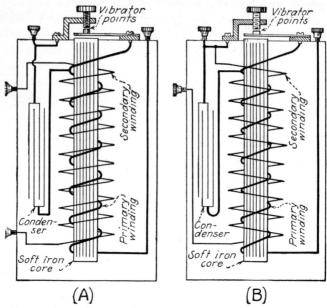

FIG. 221.—Construction and wiring of vibrator coil. *A*. Four-terminal type. *B*. Three-
terminal type.

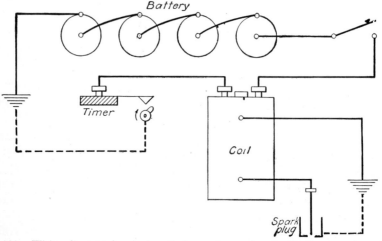

FIG. 222.—Wiring diagram for single-cylinder engine equipped with jump-spark system
with four-terminal vibrating coil.

stationary projection on the engine and a grounded rotating part that
makes contact at the right time in the cycle.

214. Four-terminal Jump-spark Vibrating Coil.—Figure 221*A* illustrates one type of vibrating coil and shows four terminals, two of which are the ends of the primary winding and are known as the *primary terminals;* and two of which are the ends of the secondary winding and are known as the *secondary terminals.* The usual method of connecting such a coil to an engine for jump-spark ignition is shown by the wiring diagram (Fig. 222). The primary circuit consists of the battery, switch, vibrator points, primary winding, and timer. The secondary circuit consists only of the secondary winding and spark plug. The condenser

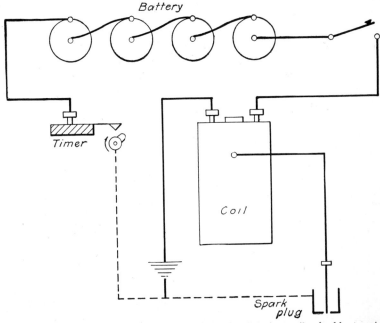

Fɪɢ. 223.—Single-cylinder, jump-spark system with vibrating coil—double terminal of coil grounded.

is on the inside of the coil and is protected by the case. It is always properly and permanently connected when the coil is manufactured.

215. Three-terminal Vibrating Coil.—Vibrating coils that are now on the market have only three terminals (Fig. 221*B*). That is, one primary end and one secondary end have been fastened together on the inside of the coil and brought to one common terminal, so that we may say that the coil has one primary terminal, a double terminal, and one secondary terminal. Figure 223 illustrates the wiring of a single-cylinder engine using a coil of this type. Tracing the circuit, the current passes from the positive of the battery through the switch to the primary terminal; through the primary winding to the double terminal; through the ground wire, and then through the engine block, to the timer; and back to the battery. The secondary circuit flows to the spark plug, jumps the gap, passes through the engine block and back through the ground wire to the double terminal of the coil.

216. Vibrator-coil System for Multiple-cylinder Engine.—Multiple-cylinder engines using jump-spark ignition with vibrator coils require as many coil units as there are cylinders, as shown in Fig. 224.   Likewise, the timer will have as many contacts as there are cylinders.   In other words, it might be said that a multiple-cylinder engine using this system

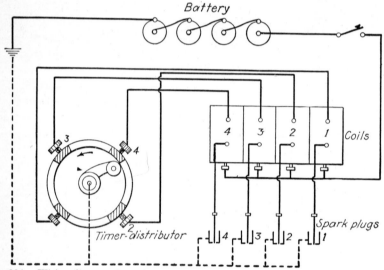

FIG. 224.—Wiring diagram for a four-cylinder engine using the jump-spark system with vibrating coils.

is merely so many single-cylinder engines connected together.   The timer not only completes the circuit for the different cylinders, but also acts as a distributor to send the sparks to the proper cylinders in the correct order.   It will be noted that all the primary terminals of the coils are connected and have one common terminal to which the positive side of the battery is attached.

217. Spark Plugs.—The spark plug is the device used in the jump-spark system of ignition that provides the gap for the high-voltage current to jump and ignite the charge.   Numerous types and styles of spark plugs are manufactured for the many different types of engines that use the jump-spark system of ignition.   In general, however, spark plugs may be classified according to their construction as the one-piece type (Fig. 225) and the separable type (Fig. 226).   The two types are made in various sizes and lengths as will be discussed later.   The one-piece plug has the advantage of being perfectly tight, with little danger of leakage around the insulator. The separable type, however, has the advantage of being readily taken apart and cleaned easily. Also it is often desirable to replace an insulator which may become cracked or damaged. In the case of the one-piece plug, this is impossible, and a new plug must be purchased.

Referring to Fig. 226, the ordinary spark plug is made up of the following parts: the steel outer shell, which is threaded and screws into the cylinder block; the insulator, usually made of porcelain but sometimes of mica; the gland or bushing, which holds the insulator in place in the shell; the copper-asbestos gaskets, between the insulator and the shell, to prevent loss of compression and leakage; the electrodes or points; and the terminal.

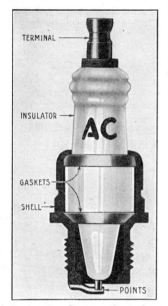

The insulator is the most important part of the spark plug and must be made of high-grade material to withstand the high voltage and prevent leakage of the current or a short

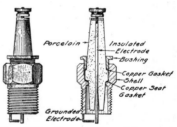

Fig. 225.—Construction of a one-piece spark plug.

Fig. 226.—Construction of the separable type of spark plug.

circuit. It must also withstand high temperatures and pressures. Porcelain is used almost exclusively as the insulating material, but mica withstands heat better than porcelain and, therefore, is a better insulator for spark plugs used under excessively high temperatures. The principal disadvantages of mica are: (1) it is difficult to clean the carbon without injury to the mica, and (2) mica spark plugs are more expensive to manufacture.

The spark-plug points must be made of material that will withstand high temperatures and will not be burned away rapidly by the arcing and heat to which they are subjected. They are usually made of some special nickel, tungsten, or chromium alloy with a very small percentage of iron. The gap at the electrode is very important in the effective and efficient operation of the spark plug in an ignition system. In general, this gap should not be less than 0.020 or more than 0.035 in. This is a bare $\frac{1}{32}$ in. or the thickness of a worn dime.

218. Sizes of Spark Plugs.—There are four common sizes of spark plugs (Fig. 227): the ½-in., with U. S. Standard pipe thread; the ⅞-in. S.A.E. plug; and two sizes of metric plugs. The ½-in. plug uses the same thread as that used on a ½-in. pipe, that is, a pipe whose inside diameter is ½ in. Since a pipe thread is tapered and rather coarse as compared to a bolt thread, this plug, when screwed into the cylinder, gradually becomes tighter and is merely screwed in tight to prevent leakage.

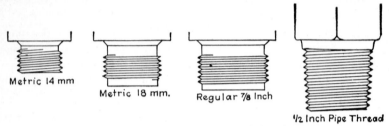

Metric 14 mm   Metric 18 mm.   Regular ⅞ Inch   ½ Inch Pipe Thread

Fig. 227.—Spark-plug sizes and threads.

The ⅞-in. S.A.E. plug has a thread similar to that used on a ⅞-in. bolt with 18 threads per inch. Since it does not taper, it cannot be made tight by screwing in farther and farther as with the pipe thread. This plug, therefore, is equipped with a shoulder and a copper-asbestos gasket (Fig. 228) and is screwed down until it fits snugly in the cylinder block against this shoulder and gasket. It is very important that the gasket be used. Otherwise oil and gas leakage will invariably occur.

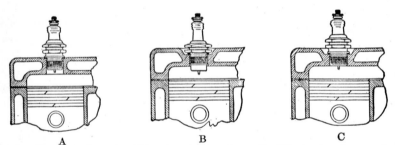

A   B   C

The plug illustrated above is too short. It does not bring the firing points far enough into the combustion chamber.

This plug is too long. It extends too far into the firing chamber. The result will be that the lower end of the plug will become overheated and ignite the mixture too early.

This drawing shows a plug of the correct length. The bottom of the spark plug shell is just flush with the inside wall of the cylinder head.

Fig. 228.—Proper and improper methods of fitting spark plugs.

The large metric plug is very similar to the ⅞-in. S.A.E. plug, except that the threaded part, which screws into the cylinder, is smaller, being 18 mm. or about 0.71 in. in diameter. The small metric plug is 14 mm. in diameter.

Any of these sizes of plugs may be of the one-piece type or the separable type.

A properly installed and fitted spark plug is shown in Fig. 228*C*; that is, the points should not project too far beyond the inner surface of the cylinder head (Fig. 228*B*). Likewise, the shank of the plug should not be so short that the points do not extend through the head, but set back in a pocket, as shown in Fig. 228*A*.

219. Testing and Cleaning Spark Plugs.—Spark plugs will usually give very good service if given occasional or periodic attention and cleaning. The gap will gradually increase with use, owing to the burning off of the points; therefore the plug should be removed every few weeks and the gap adjusted. Excessive fouling and carbon deposits result in improper functioning because such deposits produce leakage of the electric current and eventually a short circuit develops. If the fuel mixture is kept correctly adjusted and the engine does not receive an excessive amount of lubricating oil, or the oil does not work past the piston rings, the average spark plug will give very little trouble from fouling. If trouble of this kind develops frequently, it should be remedied by finding its source.

In cleaning a spark plug, it is important that the porcelain or insulator projecting into the cylinder be well cleaned, as well as the carbon thoroughly scraped out on the inside of the shell. On the other hand, in cleaning the insulator, care should be taken not to roughen it. It is best to clean spark-plug insulators by wiping rather than by scraping.

The outside of the spark plug should also be wiped off frequently because the fine dust that often collects on the exposed porcelain absorbs moisture when the air is damp, and eventually the current jumps on the outside of the plug from the terminal to the bushing and, therefore, does not produce a spark at the gap. An occasional wiping of the exposed porcelain surface will prevent this trouble.

220. Ignition Timing.—One of the very important factors in the correct operation of any internal-combustion engine is the time of occurrence of the ignition. This is particularly true in multiple-cylinder and variable-speed engines. That is, the combustible mixture of fuel and air must be ignited so that the piston will receive the force of the explosion just as it reaches head dead center at the end of the compression stroke and is ready to start on the power or working stroke. If the spark comes too early, the explosion will take place before the piston reaches the end of the compression stroke, and the result will be a knocking effect and appreciable loss of power. If the spark occurs too late, the explosion will be delayed and the maximum effect will not be received by the piston, resulting in considerable power loss and overheating.

Obviously, the time that the explosion is to take place is determined by the time of occurrence of the spark. If the spark comes too early, the explosion will be too early, and, if the spark is too late, the explosion will be late. Furthermore, it should be kept in mind that the occurrence of the spark and the explosion of the mixture are two distinct and separate actions and are not simultaneous as it might appear. This, perhaps, can be best understood by bearing in mind that the fuel mixture when ignited by the spark is "set on fire," so to speak, at one point only. Before an explosion can take place, the ignition must spread throughout the entire mixture. A very short, but nevertheless definite period of time, is required for the propagation of the ignition through the mixture. The rate of ignition is dependent upon several factors, such as the fuel used, the compression pressure in pounds per square inch, the load on the engine, the speed, and the cylinder temperature.

In order to provide for this lapse of time between the ignition of the mixture and its complete combustion by the time that the piston has reached the end of the compression stroke, the spark is advanced; that is, it is made to occur before the piston reaches the end of the stroke by an amount necessary to give the best results. Spark advance is usually designated in degrees of crankshaft travel, that is, an advance of 30 deg. means that the spark occurs when the crankshaft still lacks 30 deg. of being on dead center. A late spark is known as a retarded spark and usually occurs on, or very near, compression dead center. A retarded spark is for cranking the engine only, to prevent its kicking back. As soon as the engine is started and reaches its normal speed, the spark should be advanced the necessary amount to give smooth operation.

As previously stated, the amount of spark advance varies with a number of conditions, the most important of which are engine crankshaft speed, kind of fuel, quality of fuel mixture, compression pressure, and cylinder temperature. It is quite evident that with a given fuel and compression pressure, the greater the speed of the engine, the greater the spark advance must be up to a certain point. That is, if the fuel mixture requires a certain fraction of time to be ignited and exploded, the piston will travel farther in a high-speed engine during this small interval of time than it will in a low-speed engine. Again a lean fuel mixture, or a fuel mixture at a low or medium pressure, will not burn so rapidly as a dense mixture or one under higher compression. Consequently, low-compression engines running at light loads require more spark advance than those operating at higher compression pressures and heavy loads.

A rule that applies very well to most types of gasoline- and kerosene-burning engines is to advance the spark 3 to 5 deg. for each 100 r.p.m. That is, an engine running at 500 r.p.m. would have a spark advance of from 15 to 25 deg. This rule, however, is not a hard and fast one and does not necessarily apply to high-speed engines, such as those running at 1,500 to 2,000 r.p.m. or more. In general, single-cylinder low-speed stationary types of engines, such as farm engines, have a spark advance of from 15 to 30 deg. Variable-speed multiple-cylinder engines, such as those used in automobiles and tractors, usually have a spark advance varying from 30 to 50 deg. Highest speed engines often have a spark advance of 60 to 75 deg.

221. Spark-timing Mechanisms.—In the make-and-break system of ignition, the time that the spark occurs is determined by the time that the igniter is tripped. The tripping mechanism (Fig. 211) is usually operated by the cam gear. Consequently, if the exhaust valve is in time, the spark will be nearly, if not correctly, timed. If it is either too early or too late, it can be adjusted within a range of perhaps 10 to 60 deg. by means of some adjustment on the tripping mechanism or arm itself, as shown. In fact, this adjustment is important on any engine using the make-and-break system of ignition for the reason that more or less wear takes place in the tripping mechanism and this throws the spark partly out of time, requiring periodic readjustment.

In timing the ignition on an engine equipped with the make-and-break system of ignition, the usual procedure is: first, check up the timing of the exhaust valve and cam gear; then, if further adjustment of the spark is necessary, it can be made on the igniter trip arm in some manner, such as by slotted bolt holes or by a slide and set-screw adjustment (Fig. 229).

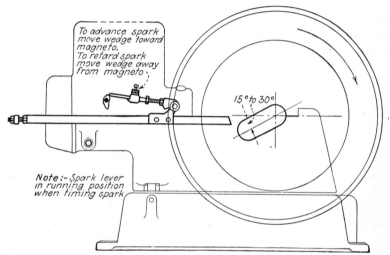

FIG. 229.—Spark-timing adjustment for single-cylinder, make-and-break ignition system.

In the jump-spark system of ignition, the time at which the spark occurs is determined by the time at which the timer makes contact or the breaker points separate. In single-cylinder engines the timer or breaker mechanism is located on or at one end of the camshaft. That is, the rotating member of the timer, or the breaker-point cam will likely be attached to one end of the camshaft. Therefore, if the cam gear and shaft are in time with the crankshaft, the ignition will be in time. Provision is usually made for a small amount of adjustment by means of some slotted holes, or by loosening and turning the breaker cam on its shaft.

References

CONSOLIVER and BURLING: "Automotive Electricity."
Cornell Univ., N. Y. State Coll. Agr., Ext. Bull. 133.

Cornell Univ., N. Y. State Coll. Agr., Ext. Bull. 147.

DUELL: "The Motor Vehicle Manual."

"Dyke's Automobile and Gasoline Engine Encyclopedia."

ELLIOTT and CONSOLIVER: "The Gasoline Automobile."

FRASER and JONES: "Motor Vehicles and Their Engines."

HELDT: "The Gasoline Automobile," Vol. III.

STONE: "Electricity and Its Application to Automotive Vehicles."

TENNEY: "Modern High Speed Ignition."

CHAPTER XIX

MECHANICAL GENERATION OF ELECTRICITY—MAGNETOS

Mechanical devices for generating an electric current are termed either *electric generators* or *magnetos*. Such devices require a magnetic field, which, as we have seen, can be produced either by a piece of steel permanently magnetized, or by means of an electromagnet. Magnetos are distinguished from other electric generators in that they utilize permanent steel magnets for providing the magnetic field, whereas machines ordinarily used for lighting and power purposes have their magnetic field supplied by electromagnetic action.

Fig. 230.—Principal parts of a simple low-tension alternating-current magneto.

222. Magneto Parts and Construction.—A simple magneto is made up of the following parts as shown in Fig. 230: (1) a nonmagnetic metal base to which other parts are fastened directly or indirectly, (2) permanent steel magnets, (3) soft-iron pole pieces, (4) armature, (5) winding or windings, and (6) one or more brushes by which the current is taken off.

Magneto bases were formerly made of bronze, but at the present time aluminum is used almost entirely because it is just as strong and costs less. The base must be made of some nonmagnetic material rather than of iron, because, if the latter were used, the magnetism would flow through it, eliminating or doing away with the field between the poles and nullifying the generating action of the magneto. Again some metal must be used for the base because the electric circuit in nearly all magnetos is completed through the engine block and back through the base to the magneto winding.

The permanent steel magnets are constructed of a high grade of hard steel and are usually made in sections instead of in one solid piece, because small pieces of steel are easier to magnetize effectively than large pieces. These magnets are usually horseshoe shape, as shown. The pole pieces of ordinary soft iron are for the purpose of carrying the lines of force of the magnetic field to the armature so that there will be very little air space for them to travel through. In other words, the pole pieces are cut out to fit the armature and the latter, when mounted in the bearings, clears these pole pieces by the smallest fraction of an inch. An appreciable air gap between the magnet poles and the armature faces would materially weaken the action of the magnetic field and, consequently, reduce the efficiency of the magneto itself.

The armature is the rotating part of the magneto, and may or may not have the winding wound about it. The winding is of insulated copper wire, similar to that used in ignition coils. Some types of magnetos, as explained later, have two or even more windings.

223. Generation of an Electric Current by a Simple Magneto.— The action of any mechanical electric generator is based upon the principle of electromagnetic induction. That is, if a magnetic field is set up about a conductor of electricity and then suddenly weakened or broken down, an electric current will be induced and will flow in the conductor. For example, referring to Fig. 231*A*, with the armature in the position as shown, the magnetic lines of force have a straight path through the soft-iron portion of the armature; that is, the lines of force flow directly through the armature from the north pole of the magnet to the south pole. If the armature is now rotated a fraction of a turn to position shown in *B*, it is observed that the lines of force, since they prefer to continue flowing through the soft iron, are stretched. Or it might be said that the number of lines of force passing through the armature from north to south has decreased. Rotating the armature still farther, as shown in *C*, it is observed that the lines of force are flowing through the armature in the opposite direction, that is, the direction of flow is reversed but they still flow from north to south. It is evident, then, that at some point between positions *B* and *C*, the number of lines of force passing through the armature must have decreased to zero. This change in the number of lines of force about the armature winding sets up an induced electric current in it. If the circuit is complete, the current flows through the ignition system. If the armature is now rotated to the position shown in *D*, the lines of force again enter it, passing through the central core as shown in *A*. They again have a straight path from north to south. Then, as the armature rotates farther, the number

of lines of force passing through it decreases as before. When the armature has reached the position as shown in F, the lines of force have again
reversed their direction of flow with respect to the armature core. At the
instant that this change takes place a second current is induced in the
armature winding. If the armature is now rotated to a position shown
in G, one revolution is completed, and, as we have seen, two currents or
impulses have been induced. The greatest voltage and, therefore, best
spark are produced when the armature approaches the position as shown
by C and F (Fig. 231).

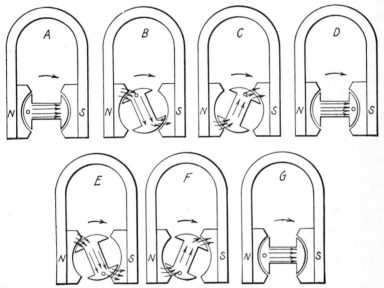

Fig. 231.—Path of the magnetic lines of force through the armature of a shuttle-type
magneto.

Again, the two currents induced during the one revolution of the
armature will be in opposite directions. In other words, in this simple
form of magneto, an alternating current is generated. If the winding of
such a magneto can be connected to the other necessary parts of the
ignition system, such as coil, igniter, or spark plugs, and the current
taken off, a spark will be produced.

It will be observed that a magneto of the type described does not
necessarily generate a steady or continuous flow of current but that it
flows in impulses. This does not mean that it is impractical or useless
for ignition purposes. In other words, if the magneto can be so timed
to the engine that the armature is in the position at which the highest
voltage is set up just when the spark is needed in the cylinder, then the
current will flow and ignition will occur. Although all magnetos are not

constructed exactly similar to the one just described, it will be seen that the theory involved in the generation of the current in all types is essentially the same and is based on ordinary principles of electromagnetic induction.

224. Classification of Magnetos.—Magnetos for ignition purposes are classified in four ways as follows:

1. Alternating current and direct current.
2. Shuttle- or armature-wound and inductor type.
3. Low tension and high tension.
4. Rotary and oscillating.

More specifically, it might be said that any magneto may be described in four ways. For example, the magneto illustrated in Fig. 232 is a shuttle-wound low-tension direct-current rotary magneto. The mag-

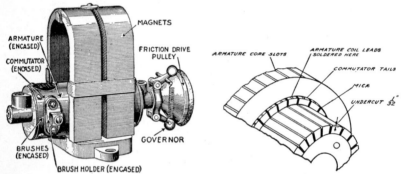

Fig. 232.—A direct-current magneto.   Fig. 233.—Commutator construction.

neto illustrated in Fig. 248 is an alternating-current shuttle-wound high-tension rotary-type magneto, and so on.

225. Alternating- and Direct-current Magnetos.—Practically all magnetos are now of the alternating-current type; that is, the current generated by the magneto flows through the circuit first in one direction, then in the other. Most of them generate two current impulses per armature revolution as previously described, but some generate four current impulses per revolution as will be described later.

Direct-current magnetos have been used to a limited extent for gas-engine ignition, particularly for stationary types of engines. Therefore, a brief explanation of their construction and operation will be given. A direct-current magneto or generator of any type is one that is so constructed that the current generated or the current impulses always flow in the same direction through the external circuit. In other words, a direct-current magneto might be said to generate an alternating current but is so equipped and constructed that the current impulses are sent out through the external circuit in the same direction at all times.

A simple direct-current magneto (Fig. 232) is made up of the same parts as the simple alternating-current magneto with the addition of what is known as a commutator (Fig. 233). That is, any direct-current electric generating device must be equipped with a commutator and brushes that bear upon it and take the current off. The commutator consists essentially of a number of copper blocks or segments, arranged in the form of a ring and insulated from each other. The ring is placed on the armature and turns with it but is thoroughly insulated from the armature metal.

To understand better the operation of a simple direct-current generator, suppose that the armature has one winding and its two ends are fastened to the two segments

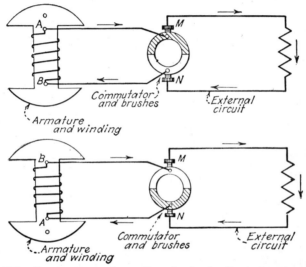

Fig. 234.—Construction and operation of a simple, direct-current magneto.

of a commutator as shown in Fig. 234. Now suppose that the current generated as the armature passes through this position flows through the circuit as shown by the arrows. Now, if the armature is turned one-half a revolution, a second current impulse is generated. The commutator and the winding, likewise, have turned 180 deg. The brushes, however, although remaining stationary, have changed contact with the commutator segments. The current that is now generated, with the armature in this position, flows in an opposite direction through the armature winding and takes the path through the commutator and brushes and out through the external circuit and back to the magneto as shown. It is readily seen that the direction of the current through the external circuit is the same in each position of the armature. In other words, although the current generated in the armature winding alternates in direction, the segmented commutator and two brushes convert it to direct current with respect to the external circuit.

The armatures of direct-current magnetos and similar electric generators bear a number of separate windings. The commutator is also made up of several segments, usually six or more. Consequently, the current generated by the magneto will flow uniformly and steadily, rather than by impulses, and will resemble very closely the current generated by a cell or battery. Since, however, internal-combustion engines do not require a steady flow of current at the spark plug or igniter, it is not essential that magnetos be of the direct-current type. For this reason they have become practically obsolete for ignition purposes.

Another characteristic of direct-current magnetos is that they must be driven at a rather high rate of speed to obtain the correct voltage. The speed is usually obtained by driving the armature by friction through a small leather-faced pulley, which is

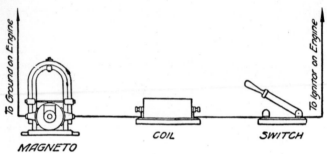

Fig. 235.—Wiring for a direct-current magneto when used for make-and-break ignition.

held against the flywheel of the engine. To prevent excessive speed, which might tear the windings loose from the armature, owing to centrifugal force, this friction pulley is equipped with weights and springs resembling a centrifugal-type governor, so that when the armature reaches a certain speed, the weights expand and draw the pulley away from the engine flywheel.

Direct-current magnetos, when operating at normal speed, generate an electric current of very low voltage, similar to the ordinary ignition battery. Therefore, in addition to the magneto, a coil of the proper type is required, and the system is wired

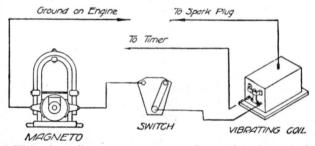

Fig. 236.—Wiring for a direct-current magneto when used for jump-spark ignition.

in exactly the same manner as a battery ignition system. Figure 235 illustrates the make-and-break system of ignition with a direct-current magneto. Figure 236 illustrates the jump-spark system with a direct-current magneto as the source of electricity.

226. Shuttle- or Armature-wound Magnetos.—Many of the magnetos now in use have an H-shaped armature (Fig. 237) with the winding wound about the bar of the H. Therefore, the winding rotates with the armature, and the magneto is said to be of the shuttle- or armature-wound type. Since the winding rotates with the armature it is necessary to take the current off by means of an insulated copper collector ring and a brush. The wire leading to the igniter or spark plugs is attached to a terminal that is a part of the brush and brush holder.

227. Inductor-type Magnetos.—Some magnetos do not have the winding rotating with the armature but have a stationary winding. Such magnetos are of the inductor type. The operation of one magneto of this type is as follows: The armature (Fig. 238) consists of two soft-iron bars

FIG. 237.—Principal parts of a shuttle-type, high-tension magneto armature.

or rotors, spaced about 1 in. apart on the armature shaft and at right angles to each other as shown. The magneto winding is placed on the round piece of iron connecting these two rotors. This part of the armature between the rotors therefore serves as a core for the winding. However, the armature shaft is free to turn in the winding, or it might be said that the core of the winding rotates. The current is generated in the following manner: With the armature in a position as shown in Fig.

FIG. 238.—Armature and coils for K-W magneto.

239*A* the lines of force pass from the north pole, up through the rotor *A*, to the center of the armature shaft. They cannot pass straight across to the south pole, because, as will be seen, the pole pieces are cut off and do not extend upward as far as they do in the shuttle type of magneto. The lines of force, therefore, pass through that part of the armature that serves as the core of the winding to the opposite rotor *B*, then down through its lower end to the south pole. If the armature is now rotated one-quarter turn to the position shown in Fig. 239*B*, rotor *B* is now touching the north pole, and the lines of force come up through it, pass through the armature shaft to rotor *A*, and down through it to the south pole. It is thus observed that, as the armature turns from the first position to the second, the lines of force must die out of that part of the armature serving as the core of the winding, and change their direction

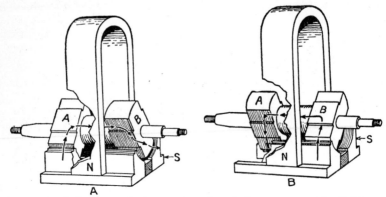

Fig. 239.—Sketch showing path of magnetism through the armature of the K-W magneto.

In other words, at a certain point in the rotation of the armature there is no magnetic field about the winding, but it is made to die out and at that time a current is induced in the winding and flows through the ignition circuit. This particular type of magneto generates four current

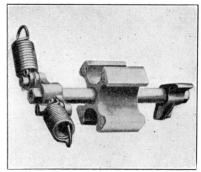

Fig. 240.   Fig. 241.

Fig. 240.—Coils and pole pieces of the Webster oscillating magneto.
Fig. 241.—Armature for the Webster oscillating magneto.

impulses per revolution of the armature but is similar in other respects to other magnetos. It is likewise an alternating-current magneto.

Another inductor-type magneto, the Webster oscillator, is used extensively for stationary gas-engine ignition. The winding is in two parts, one-half being located on each of the two center projections of the two E-shaped soft-iron pole pieces (Fig. 240). The armature (Fig. 241) is shaped like a Maltese cross. When the armature is in the position, as shown in Fig. 242*A*, the lines of magnetic force flow from north to south through the two center projections of the pole pieces and the horizontal part of the armature. In other words, the lines of force are now flowing through that part of the pole piece that serves as the core of the winding. If the armature is now rotated to the position shown in Fig. 242*B*, it is observed that the lines of force have changed their path. They now travel from north to south through the upper and lower projections of the pole pieces and through both parts of the armature; that is, the shifting of the armature resulted in the lines of force changing their path of travel and dying out of the pole-piece projections that serve as the core for the winding. Therefore, since the magnetic field about the winding is thus made to die away or break down, an electric current is generated.

228. Oscillating Magnetos.—The Webster magneto just described is also of the oscillating type; that is, the armature, instead of rotating, is merely turned through a fraction of a revolution and suddenly released and jerked back to its former position by the heavy springs (Fig. 241).

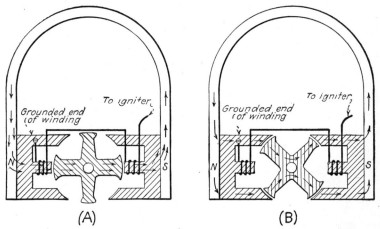

(A)   (B)

Fɪɢ. 242.—Construction and operation of the Webster oscillating magneto.

These springs give the armature a very rapid movement through a fraction of a revolution that is equivalent to rotating the armature at the rate of several hundred r.p.m. The bracket supporting the magneto is a part of the igniter, so that the magneto can be located very close to the igniter and both can be tripped by the same trip arm. This is essential in view of the fact that both must be tripped at the same instant.

Figure 211 illustrates the tripping mechanism that is used with the Webster oscillating magneto. The bracket *R* is clamped to the exhaust-valve push rod. The trip arm *H* is hinged to the bracket by a pin and is supported by the roller *M*. This arm is equipped with the slide or wedge *K*, which slowly lifts it upward as the push rod and bracket are carried backward by the valve cam. As the trip arm moves backward it engages with the finger *G*, which is on the end of the armature shaft of the magneto. This rotates the shaft a fraction of a revolution, and then, as the wedge rides up on the roller, the arm releases the finger *G*, and the armature snaps back to its original position. At the same time, the lower projection of the finger *G* strikes the setscrew *E* of the movable electrode arm *F*, thereby separating the igniter points.

It is observed that there are a number of adjustments possible in connection with this mechanism. The first adjustment is the slide or wedge, which can be moved backward or forward on the trip arm. If it is moved forward, the trip arm will be released quicker and an earlier spark produced. If it is moved backward, the arm will be released later and the spark will be delayed. The setscrew *E*, on the end of the movable electrode arm *F*, must be adjusted so that the push finger *G* hits it at the correct instant. If the setscrew is turned inward toward the push finger, it will be hit sooner and the igniter points will open earlier in the rotation of the magneto armature. If the screw is turned backward, the push finger will hit it later and the points will be opened later in the rotation of the magneto armature. Since the current generated by the magneto is strongest only at a certain point in the armature movement, it is important that this screw be so set that the push finger hits it just at this instant and separates the igniter points. Then the igniter points will be opened at the correct time to produce the strongest possible spark. It will be noted that the mechanism is equipped with a lever *J*, by which the roller may be raised or lowered and the spark thereby advanced or retarded.

229. Low-tension Magnetos.—A low-tension magneto is one that generates a current of low voltage. Such magnetos are generally used only on stationary farm-type engines with the make-and-break system of

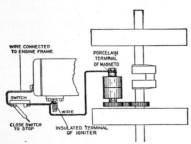

ignition. Figure 243 illustrates an engine equipped with the make-and-break system of ignition and a shuttle-type rotary-armature magneto. It is likewise known as an *alternating-current magneto* and, as previously discussed, generates two current impulses per revolution. The armature is driven by a gear and is so timed that, just as the igniter is tripped, the armature is in the position in which the strongest current is generated. Therefore, a spark will

Fig. 243.—Wiring for a low-tension alternating-current magneto for make-and-break ignition.

occur at the igniter points. It is unnecessary to use a separate coil in connection with low-tension magnetos for make-and-break ignition, for the reason that the winding has so many turns of wire that a sufficiently high voltage is generated to assure a good spark.

Figure 244 shows the circuit taken by the current for an engine equipped with either a rotating or an oscillating alternating-current low-tension magneto. The current, which is generated in the winding of the armature, travels to the end fastened to the collector ring. A small carbon brush bearing against the ring receives the current, and it is led to the terminal; thence, through the conductor or wire to the stationary point of the igniter; thence, through the igniter points to the engine block; thence to the base of the magneto and back to the grounded end of the winding on the armature. The only external wire involved is the one connecting the magneto terminal and the igniter.

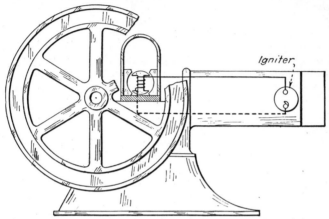

Fig. 244.—Ignition circuit for an engine equipped with make-and-break system with low-tension, alternating-current magneto.

Large-size engines equipped with low-tension rotary alternating-current magnetos usually cannot be started by means of the magneto because the engine cannot be cranked sufficiently fast to create a strong spark. In such cases it is necessary to use a battery for starting. Figure 245 gives a diagram of the wiring of such an engine when equipped with both battery and low-tension alternating-current magneto.

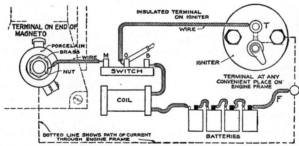

Fig. 245.—Wiring for make-and-break ignition using both battery and low-tension alternating-current magneto.

230. Timing Low-tension Alternating-current Magnetos.—Since a low-tension alternating-current magneto generates two current impulses per revolution, it must be driven at the correct speed ratio with respect to the crankshaft, namely, at a speed of 1:1, 1:2, 1:4, and so on. For a single-cylinder four-stroke-cycle engine, such magnetos are usually driven at crankshaft speed. This would mean that three current impulses are generated that are unused. However, since the igniter points are not tripped, except when a spark is needed, the current merely flows through the circuit without producing a spark, and no harm results.

The correct timing of the ignition for an engine equipped with the make-and-break system and a low-tension rotary alternating-current magneto is very important. The procedure is as follows: First, check the timing of the igniter trip, as previously described under Spark-timing Mechanisms, so that the igniter trips at the correct time in the piston travel. Second, turn the engine until the igniter is just tripped and then stop. Third, turn the magneto armature until it is in the position at which it generates the strongest current impulse. This is done simply by rotating the armature slowly with the fingers until the strongest pull or resistance to rotation is felt. Fourth, mesh the gears and fasten the magneto in place.

THE HIGH-TENSION MAGNETO

231. Construction and Operation.—A high-tension magneto is one that is used for jump-spark ignition and generates a current of suffi-

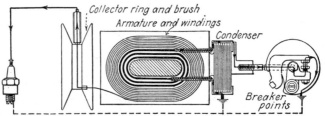

Fig. 246.—Sketch showing electrical circuits for a high-tension magneto.

ciently high voltage to jump the gap at the spark plug. In fact, such a magneto is a complete jump-spark system in one compact unit with the exception of the spark plugs.

To generate this exceedingly high voltage, the high-tension magneto must be equipped with a number of parts that are not found in the low-tension types. The important parts of such a magneto are as follows: (1) nonmagnetic base, (2) permanent steel magnets, (3) armature, (4) soft-iron pole pieces, (5) primary winding, (6) secondary winding, (7) breaker points, (8) condenser, (9) collector ring and brush, and (10) safety gap. For convenience and clearness, so that the construction and operation of the high-tension magneto will be readily understood, the electrical circuits and parts involved are shown by Fig. 246.

The operation of the magneto is as follows: As the armature rotates and passes through the position at which the lines of force change their path through it, a current is induced in the primary winding of coarse insulated wire. Just as this current is induced in the primary winding, the breaker points, which are in the primary circuit, are separated by a cam, interrupting the current and breaking down the strong electromagnetic field set up by it. This results in the instantaneous induction of a very high voltage in the secondary circuit, which consists of many thousands of turns of very fine wire. Therefore, if the breaker points separate at the correct time in the travel of the engine piston, and the secondary winding can be connected to the spark plug, a spark will be produced and will ignite the mixture.

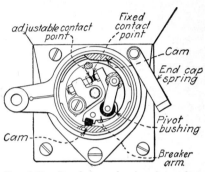

FIG. 247.—Breaker mechanism for a shuttle-type, high-tension magneto.

232. Breaker Points.—The breaker points (Fig. 247) are similar to those used in the nonvibrating coil system of ignition with battery. They consist of a stationary insulated point and a movable noninsulated point and arm that comes in contact with the cam. These points are located at one end of the armature. If the magneto is of the inductor type, that is, with stationary winding, the points will be stationary, and the cam that separates them will rotate with the armature. On the other hand, if the magneto is of the shuttle type, the winding and points rotate and the cams are stationary.

The contact points themselves are made of a high-grade alloy steel so that they will not pit up or get rough too quickly. One point is usually adjustable so that the space between them, when separated, can be adjusted. This space should be 0.015 to 0.020 in.

233. Condenser.—The high-tension magneto must be equipped with a condenser similar to that used in a jump-spark coil, for the purpose of eliminating the arcing at the breaker points and intensifying the current in the secondary circuit. The condenser is connected in the primary circuit as shown in Fig. 246.

234. Primary Circuit.—Tracing the primary circuit of a high-tension magneto (Fig. 246), the current generated in the primary winding travels to one end fastened to the insulated stationary breaker point, through the breaker points to the metal of the magneto, and back to the grounded end of the winding on the armature. It is thus observed that the primary winding does not pass outside of the magneto itself.

235. Secondary Circuit.—Tracing the secondary circuit for an ordinary shuttle-type high-tension magneto (Fig. 246), the current induced in the winding travels to the collector ring, from the collector ring to the brush, then to the spark plug, and back through the engine block to the base of the magneto, and, thence, to the grounded end of the secondary winding. If the magneto is of the inductor type, the collector ring is unnecessary, and, for a single-cylinder engine, the wire carrying the current from the magneto to the spark plug is fastened directly to one end of the secondary winding.

236. Safety Spark Gap.—If the wire that carries the current from the magneto to the spark plug should become broken or disconnected at any point, it is obvious that the high-voltage current induced in the secondary winding could not take its regular path and, therefore, might dissipate itself by breaking through the insulation in the winding on the armature. This would create a permanent short circuit and, therefore, make the magneto useless until it could be rewound. To eliminate the possibility of such injury in case of a broken secondary circuit, all high-tension magnetos are equipped with what is known as a safety spark gap.

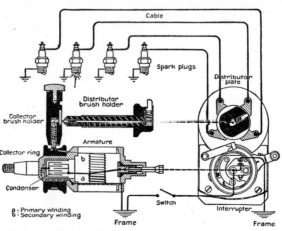

Fig. 248.—Electrical circuits for a shuttle-type, high-tension magneto for a four-cylinder engine.

This gap consists of a point fastened to that side of the secondary winding that leads to the spark plug or the collector ring, and another point exactly opposite the first and grounded to some metal part of the magneto. It must be about ¼ to ⅜ in. wide, which is somewhat greater than the spark-plug gap. Otherwise the current would have a tendency to jump at this point at all times, rather than at the spark plug.

237. Distributor.—A high-tension magneto for a multiple-cylinder engine, in addition to the parts named, requires a distributor and brush (Fig. 248) to distribute the secondary current to the different spark plugs in the correct order, just as the nonvibrating coil system of ignition with battery requires a distributor to distribute the secondary circuit to the spark plugs in the case of an engine with two or more cylinders.

The circuits involved in the operation of a high-tension magneto for a multiple-cylinder engine are shown by Fig. 248. It is observed that the primary circuit is not different from that in a magneto for a single-cylinder engine; that is, it consists only of the primary winding and the breaker points. The secondary circuit includes the distributor and takes the following path: From the end of the winding, the current goes to the

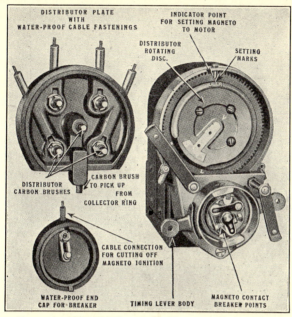

Fig. 249.—Construction of the Eisemann G-4 magneto.

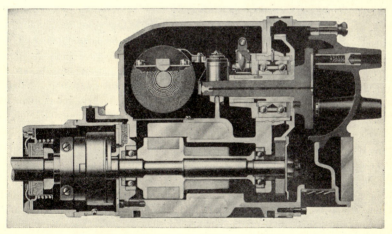

Fig. 250.—Sectional view of Eisemann CT-4 magneto.

collector ring if the magneto is of the shuttle type; if of the inductor type, it goes directly to the distributor brush or rotor. From this brush it is received by the distributor segments and terminals. Wires attached to the distributor terminals conduct the current to the spark plugs in the order in which they are fastened to the terminals. From these plugs the current returns through the engine to the base of the magneto, and thence to the grounded end of the secondary circuit. In other respects, the magneto is exactly the same as that used for a single-cylinder engine.

HIGH-TENSION MAGNETO TYPES

The different makes of high-tension magnetos vary somewhat in detailed construction and operation, but the same general principles are followed in all makes. The greatest improvement made in recent years is the complete enclosure of the working parts, thus providing greater protection from dust, dirt, and moisture.

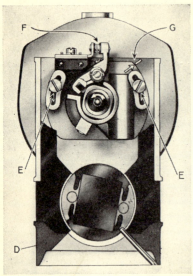

FIG. 251.—End view of Eisemann CT-4 magneto.

238. Bosch Magneto.—The Bosch magneto is of the shuttle-wound type, its construction and circuits being shown by Fig. 248.

239. Eisemann Magnetos.—The Eisemann Model G-4 (Fig. 249) has the conventional shuttle-wound construction with the distributor placed directly over the collector ring and carrying the collector-ring brush.

The Eisemann CT-4 magneto (Figs. 250 and 251) is an inductor type. The single horseshoe magnet lies near the base in a horizontal position, and the breaker points are located on the distributor rotor assembly. The cam is rotated by the distributor gear. Obviously this cam must have as many lobes as there are cylinders.

240. Fairbanks-Morse Type RV Magneto.—The Fairbanks-Morse Type RV magneto (Fig. 252) is an inductor type with revolving magnets, that is, the armature carries the permanent steel magnets, and, as rotation occurs, the polarity changes with respect to the pole pieces and coil core every half turn, thus generating two current impulses per revolution. The stationary coil, interrupter points, and distributor are located just above the armature and are accessible through removal of the top cover, which carries the distributor brushes. The breaker cam is rotated by the distributor gear.

241. Wico Type AP Magneto.—The Wico Type AP magneto (Figs. 253 and 254) is an inductor type having the coil located above the armature and having the breaker mechanism and distributor assembly mounted on the end of the armature as shown.

242. International Magnetos.—McCormick-Deering tractors are equipped with either a Model E-4A (Fig. 255) or a Model F (Fig. 256)

Fig. 252.—View of windings, breaker points, and distributor for Fairbanks-Morse Type RV magneto.

Fig. 253.—Wico Type AP tractor magneto.

International magneto. The former is of the conventional shuttle-wound construction. The Model F is an inductor type with the stationary coil mounted above the armature under the arch of the magnets. The breaker points are operated by a cam on the end of the armature.

243. Oscillating High-tension Magnetos.—Figures 257 and 258 show the Wico high-tension magneto, which is used extensively on single-cylinder farm engines. Referring to the sketch (Fig. 259), this magneto

is constructed and operates as follows: The permanent steel magnet consists of several small straight bars clamped together and bridging the ends of two laminated soft-iron core members, which might also be called pole pieces. The opposite ends of these two core members are bridged by a laminated soft-iron yoke, which is moved up and down on a guide and is really the armature of the magneto. Two windings, a primary and a secondary, are placed on these core members, one-half of each winding being placed on each core member and so connected that the current generated in each of the two halves will flow in the same direction. One end of the primary winding is grounded to the magneto frame. The other end is connected to the fixed contact (insulated from frame) of the breaker

Fig. 254.—End view showing coil and breaker points for Wico Type AP tractor magneto.

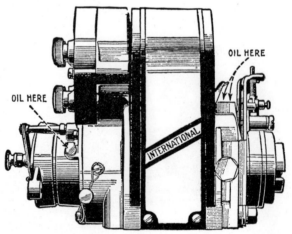

Fig. 255.—International Type E-4A high-tension magneto.

points or interrupter. Since the moving contact is attached to the armature and is thereby grounded, there is a complete electrical circuit when the points are closed.

Like any other high-tension magneto, the secondary winding consists of many turns of very fine wire. One of the ends is grounded to the magneto frame. The other end is connected by a heavily insulated lead wire to the spark plug. Thus the secondary circuit is completed through the engine.

In generating a high-tension current, the armature is held firmly in contact with the ends of the poles by magnetic attraction. The lines of magnetic force now have a complete soft-iron path from north to south, as indicated by arrows in Fig. 259. The breaker points are also in contact. Now, if the armature is suddenly pulled downward away from the pole pieces, the magnetic path will be broken and the lines of force will be reduced in the two core members; and the farther the armature is removed

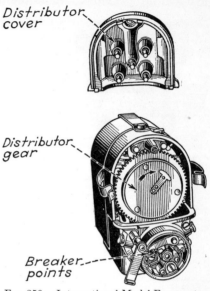

Fig. 256.—International Model F magneto.

from the poles the greater will be the reduction in the lines of magnetic force. This decrease in lines of magnetic force induces a cur-

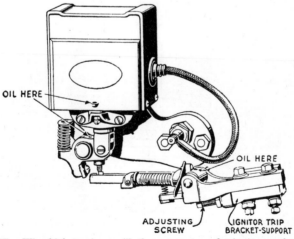

Fig. 257.—Wico high-tension oscillating magneto and tripping mechanism.

rent in the primary circuit. As soon as a current flows in the primary circuit, it acts as an electromagnet producing lines of magnetic force in the core in the same direction as those of the permanent magnet. Thus, it will be seen that the total magnetic lines of force will not be reduced as

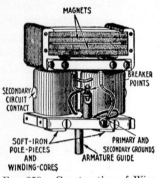

MAGNETS

SECONDARY
CIRCUIT
CONTACT

BREAKER
POINTS

SOFT-IRON
POLE-PIECES
AND
WINDING-CORES

PRIMARY AND
SECONDARY GROUNDS
ARMATURE GUIDE

FIG. 258.—Construction of Wico high-tension magneto.

much as they would without the primary winding. As soon as the armature has moved far enough away from the poles to allow a considerable reduction of the lines of magnetic force, and at such a point that the sustaining action of the primary is still adequate, the breaker points are opened (thus breaking the primary circuit), which removes all the sustaining action and allows the lines of magnetic force to decrease instantly to a very small value. This rapid changing of the lines of magnetic force induces a current in the secondary winding of sufficiently high voltage to jump the gap in the spark plug. A condenser is connected across the breaker points to produce a quick break of the

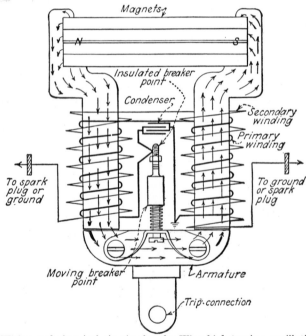

Magnets

N   S

Insulated breaker
point

Condenser

Secondary
winding

Primary
winding

To spark
plug or
ground

To ground
or spark
plug

Moving breaker
point

Armature

Trip connection

FIG. 259.—Wiring and electrical circuits for the Wico high-tension, oscillating magneto.

primary current when the breaker points are opened and, incidentally, prevents arcing at the points.

The obvious advantages of the Wico magneto are its great simplicity—the absence of numerous moving parts. The armature and the movable breaker point are all that move. The tripping mechanism either is operated from the exhaust-valve push rod, as shown in Fig. 257, or is actuated by an eccentric, as shown in Fig. 260. In either case, the quick snapping of the armature away from the poles is accomplished by the expansion of a spring, known as the drive spring, which has been compressed before the time for producing the spark. Excessive downward movement of the armature is prevented, and its positive return to the pole pieces is assured by another spring known as the return spring. In either case, means are provided in the tripping mechanism for advancing and retarding the spark and also for adjustment of the timing with the engine.

244. Armature and Distributor Speeds.—All high-tension magnetos must be driven by gears at a certain speed ratio with respect to the

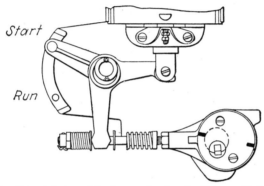

FIG. 260.—Eccentric-type tripping mechanism for Wico oscillating magneto.

crankshaft speed of the engine. Most of them are of the shuttle type and generate two current impulses per revolution. Therefore the breaker points are made to open twice per revolution. The K-W magneto (Fig. 239) generates four current impulses per revolution, but this offers no advantage except for engines having eight or twelve cylinders. In such cases they can be operated at one-half the speed of other magnetos supplying only two sparks per revolution.

Referring to Table XIII, it will be noted that, for all one-, two-, and four-cylinder four-stroke-cycle engines equipped with high-tension magnetos, the armature speed is the same as the crankshaft speed. A single-cylinder engine would receive a sufficient number of sparks if the armature turned at one-fourth crankshaft speed and two cams were used. The same would be true of a two-cylinder engine with a 360-deg. firing interval if the armature rotated at one-half crankshaft speed and two cams were used. It is desirable, however, to use these higher speeds in order to insure the generation of a good strong spark at all times. For example, if a single-cylinder engine were equipped with a magneto running at one-fourth crankshaft speed and the engine were slowed down to 300 r.p.m., the magneto speed would be only 75 r.p.m. The spark would likely be so weak as to make the engine misfire, skip, or even stop. By using only one cam and thereby opening the breaker points but once per armature revolution, only half as many sparks are generated. This still gives twice as many as are necessary for a single-cylinder engine; that is, there will be a spark every crankshaft revolution. However, these extra sparks will come at the end of the exhaust stroke. Therefore, no harm results.

It will be noted from Table XIII that the distributor gear speed is one-half crankshaft speed in all cases. This is explained by the fact that in any multiple-cylinder four-stroke-cycle engine, all cylinders fire once in two crankshaft revolutions or one distributor-gear revolution.

TABLE XIII.—MAGNETO CHARACTERISTICS ACCORDING TO NUMBER OF CYLINDERS

Number of cylinders	Engine, r.p.m.	Explosions per minute	Armature, r.p.m.	Number of breaker cams	Number of sparks generated	Distributor speed
1.........................	1,000	500	1,000	1	1,000	
2 (360-deg. firing interval)	1,000	1,000	1,000	1	1,000	500
2 (180- and 540-deg. firing interval)...............	1,000	1,000	1,000	2	2,000	500
4.........................	1,000	2,000	1,000	2	2,000	500
6.........................	1,000	3,000	1,500	2	3,000	500

245. Impulse-starter Couplings.—With the exception of a few large machines, tractors are not equipped with self-cranking devices such as are used on automobiles. Hand cranking is the usual means of starting. Since most tractor engines are of such a size that in cranking they can be turned only a fraction of a revolution at a time or rotated very slowly at best, a special provision of some kind is necessary to permit the magneto to supply a strong spark at this cranking speed. The device now used for this purpose is known as an impulse-starter coupling since it always serves as a part of the magneto-drive coupling.

The operation of an impulse starter is best understood by reference to Fig. 261. The flanged metal disk M is fixed to the magneto shaft and bears a projection B. A similar flanged disk N is fixed on the drive shaft and bears a projection D. If these two disks are now brought together as in L and a stiff spring C is inserted on the inside between the two projections, the disk N, through its projection and the spring, will drive disk M and, therefore, the magneto. Now suppose a stationary catch H is released and allowed to engage in a notch in disk M. The armature will stop rotating, but disk N, being fixed to the drive shaft, will continue to turn. This will cause the spring to be compressed between the projections B and D and to exert a high pressure on the disk M. When disk N has advanced about one-fifth of a turn, a trip cam F on its outer rim strikes and lifts the catch H. This releases the disk

M, and, owing to the excessive pressure of the spring, it jumps ahead at a high rate of speed until it catches up with the driving disk N, and the spring is again extended. If the release takes place at the end of the compression stroke of a certain piston and the magneto breaker points are opened during this sudden movement of the armature, a good spark will be produced that will ignite the charge. When the engine reaches

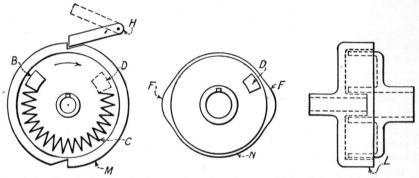

FIG. 261.—Sketch showing fundamental parts and construction of simple impulse-starter coupling.

a speed of about 150 r.p.m. the cams F strike the catch H a sufficiently hard blow to permanently disengage it and cause it to be hooked up by a simple device not shown in the figure. The magneto then rotates in the normal manner.

246. Automatic Impulse Starters.—The impulse starters just described are manually controlled; that is, the catch H must be engaged by hand

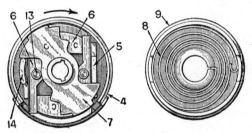

FIG. 262.—Automatic impulse starter.

each time the engine is cranked. Some magnetos are equipped with impulse starters that engage automatically whenever the speed drops below 150 r.p.m. Such a device is shown by Fig. 262. The action is dependent upon centrifugal force, which causes certain weighted members to move outward sufficiently to disengage the coupling when the proper speed is attained.

247. Impulse-starter Troubles.—Impulse-starter troubles sometimes develop for two reasons, namely, lack of lubrication and worn parts.

From the preceding descriptions it is observed that all impulse starters are made up of certain small moving parts whose free movement is essential to the proper functioning of the device. This is particularly true with respect to the automatic types. If lubrication is neglected, or if dust and dirt gum up the parts, poor action is likely.

Most of the wear takes place in the catch block, catch notch, and catch release. The catch block, when released, is under considerable pressure. Consequently, the end of the catch and the notch holding it are subject to a certain amount of wear. Eventually the catch may fail to engage or may slip out of its own accord. Also the starter may function normally when the engine is cranked but release the moment firing starts. The engine fails to pick up and must be cranked again. All impulse starters have a distinct click when working properly. Therefore, if this click is not heard or is weak, the trouble is likely due to the causes mentioned.

Since the spark made by an impulse starter is for starting the engine, it must always occur just as the piston reaches the end of the compression stroke. In other

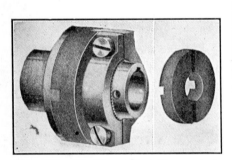
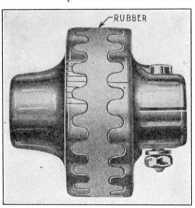

FIG. 263.—Flexible, adjustable magneto coupling.　FIG. 264.—Special type of magneto coupling.

words, an impulse spark is always a retarded spark regardless of the position of the spark advance and retard lever, provided, of course, that the magneto is correctly timed. Even though the spark lever is advanced, since the starter remains engaged and holds the armature stationary until the piston reaches compression dead center, a kickback is impossible. On the other hand, if the starter catch releases too easily or fails to engage at all, as previously mentioned, a kickback is liable to occur, especially if spinning the engine by hand is resorted to.

248. Magneto-drive Couplings.—The connection between the drive shaft on the engine and the magneto armature shaft must be positive, flexible, and adjustable. It must not permit slippage, as the magneto would get out of time. A rigid coupling would likely subject the magneto bearings to strain and excessive wear if the drive and magneto shafts were not in exact alignment. An adjustable coupling is necessary in timing to permit the armature to be moved any small fraction of a turn and therefore brought into exact time with the engine.

Figure 263 shows one type of flexible, adjustable coupling. The fiber disk is grooved on each side in opposite directions and fits between the two coupling members on the drive and magneto shafts. The latter have lugs that slip into the grooves. This arrangement permits a limited sliding action in any direction if the two shafts are out of line. The timing adjustment is in the drive-shaft member, which consists of two pieces: a collar keyed to the shaft and a clamp locked to it. Both the collar and the clamp are splined. By loosening and shifting the clamp one or more splines, any adjustment can be secured.

Figure 264 shows another coupling, which consists of three parts: a toothed flange, bearing 19 teeth and bolted to the drive shaft; a 20-toothed flange, keyed to the magneto shaft; and a double-toothed rubber disk, which fits between the two flanges. The disk absorbs shocks and strains due to misalignment.

To change the magneto timing remove the flange bolt on the drive shaft and slide back one flange. Disengage the rubber disk, rotate in the desired direction as many

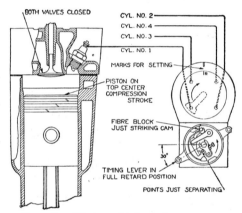

FIG. 265.—Timing a high-tension magneto.

teeth as necessary, and replace. Then slide back the flange and replace bolt. Owing to the one-tooth variance on the two flanges it is possible to secure as little as one degree of adjustment.

Another coupling uses the grooved-disk arrangement as previously described to give flexibility, but has two flanges each bearing a circular row of holes. One flange has one more hole than the other so that only one pair of holes can line up at any time. A bolt through these two holes connects the coupling. If an adjustment is necessary the bolt is removed, one flange is moved the necessary amount, and the bolt is replaced in the two new holes that are lined up.

249. Spark Advance and Retard.—The advancing and retarding of the spark on engines equipped with high-tension magnetos are accomplished by changing the time at which the breaker points are opened. That is, if the magneto is of the shuttle-wound type, the cams are stationary and are carried on a ring as shown in Fig. 265. If this ring is shifted or turned slightly, the breaker points will hit the cams later, and therefore the spark will come later. If the magneto is of the inductor type, the points themselves are shifted, so that the rotating cam hits the movable

point sooner or later.   A stop screw and slot of some kind usually serve
to limit the range of spark advance and retard.

250. Timing High-tension Magnetos.—The timing of a high-tension
magneto requires a clear understanding of the magneto construction and
operation.   High-tension-magneto timing may be divided into two parts,
namely, (1) the timing of the magneto within itself and (2) the timing of
the magneto with the engine.   Timing a high-tension magneto within
itself involves (*a*) the timing of the breaker points with the armature
rotation and (*b*) the timing of the distributor with the armature and
breaker points.

If the breaker points are not separated at the correct instant in the
rotation of the armature, there will be a weaker current to interrupt,
and a strong current will not be induced in the secondary circuit.   Con-

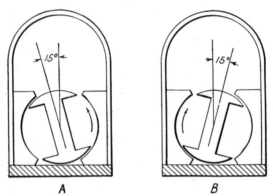

A   B

Fɪɢ. 266.—Armature positions when maximum current is generated.

sequently, the breaker points must be set so that the cam opens them
when the armature is in the position at which it is generating its strongest
current.   This position is shown by Fig. 266.   If the points open before
or after the armature reaches this position, the spark produced at the
spark plug will obviously be weaker and less effective.

The timing of the breaker mechanism with the armature is comparatively simple.   In all shuttle-type magnetos, a keyed hub on the breaker
disk fits into a keyway in the hollow end of the armature shaft.   Likewise,
in replacing the breaker cam on an inductor-type magneto, a similar key
and keyway are provided to position correctly the cam with respect to the
armature.

If the magneto is equipped with a distributor for multiple-cylinder
ignition, the distributor gear must be timed with the armature gear,
depending upon the direction of rotation, so that the distributor rotor
will be in full contact with one of the segments when the breaker points
separate.   Manufacturers usually mark certain teeth (see Fig. 267) so

that, in case the magneto is taken apart and reassembled, the marked teeth are meshed together again. The distributor segments are rather long so that, with the distributor running slower than the armature, the brush will be in contact with a certain segment, even though the breaker points are shifted and open at different times as in retarding and advancing the spark.

Two distributor-gear teeth are usually marked, because all high-tension rotary-type magnetos can be made to rotate in either a clockwise or a counterclockwise direction as viewed from the driving end of the armature. The direction of rotation depends upon the engine construction. An arrow on the end plate usually indicates this direction for any particular magneto.

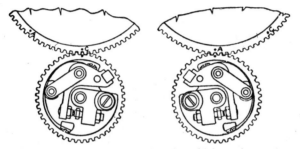

Fig. 267.—Distributor-gear markings for a high-tension magneto.

251. Timing High-tension Magneto with Engine.—When a magneto is disconnected and removed from an engine, it is essential that it be correctly timed when replaced, so that the spark will come at the right instant in the piston travel. The following procedure for timing a high-tension magneto is applicable to any engine (Fig. 265), regardless of the number of cylinders:

1. Place piston 1 on head dead center at the end of the compression stroke.
2. Select or determine the distributor terminal to be connected to spark plug 1. Unless the wires are already cut to an exact length and fitted, any terminal may be chosen for spark plug 1.
3. Place distributor rotor on terminal brush or segment 1. To do this it is usually necessary to remove the distributor cover.
4. See that breaker mechanism is in the retard position.
5. Slightly rotate or move armature in driven direction until breaker points are just opening. In doing this, the amount of rotation should not be sufficient to disturb the position of the distributor rotor.
6. Without moving the armature, place magneto in position on the engine and connect the drive coupling.
7. Check the timing by turning the engine over very slowly. See that the breaker points open just as the piston reaches head dead center at the end of the compression stroke. If the magneto is equipped with an impulse starter it should snap just as piston 1 reaches compression dead center. If it snaps slightly before or after, carefully

disconnect the coupling, rotate the armature a trifle forward or backward as the case may be, and again connect and recheck.

8. Connect the wires to the distributor and spark plugs in the proper manner according to the firing order of the engine.

252. High-tension-magneto Care and Adjustment.—The high-tension magnetos that are now being manufactured require very little attention and should seldom give any trouble. They are well enclosed and almost completely dust- and moistureproof. This does not mean that they do not require even reasonable care. Occasional lubrication, according to the manufacturer's instructions, and prevention of excessive dirt accumulation around the magneto are important.

The parts of a high-tension magneto requiring lubrication are the armature bearings and the distributor rotor. Since practically all magnetos have ball bearings for the armature, very little oil is needed. From 2 to 5 drops of a very light grade of machine oil placed in the small oil openings at either end about once a month, or after 100 hr. of operation, is all that is necessary. A heavy oil or an excessive amount should be avoided by all means, as gumming, short circuits, and other troubles will develop.

In a few cases the bearings are packed in a light grease and so inclosed that it cannot escape. These require no further lubrication.

If it is desirable to clean the distributor rotor, brush contacts, breaker points, and the collector ring at any time, only a clean rag and gasoline should be used.

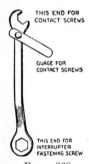

253. Breaker-point Adjustment.—The proper functioning of a high-tension magneto is largely dependent upon the breaker points. An occasional inspection should be made and the following noted:

THIS END FOR CONTACT SCREWS

GUAGE FOR CONTACT SCREWS

THIS END FOR INTERRUPTER FASTENING SCREW

Fig. 268.— Magneto-adjusting wrench.

1. The contact points should be smooth, flat, and free from dirt, oil, or moisture and make a good full contact. If they are rough, use only a very fine file, and file off as little as possible.

2. The points should have the required opening, namely, about 0.015 or $\frac{1}{64}$ in. This can be checked with the gage supplied with every magneto as shown by Fig. 268. One point is adjustable and is fixed by a lock nut (Fig. 269). By loosening the lock nut, the point can be turned in or out the required amount.

3. The movable point should hinge freely on its pivot. Corrosion caused by moisture may develop sluggish action, which usually results in the magneto failing to spark. The movable arm should be removed and the hole and pivot pin carefully smoothed up.

254. Tractor-ignition Troubles.—With the great improvement and refinement in high-tension magneto construction and its almost universal adoption for tractor ignition, troubles from this particular source are comparatively limited. Obviously they must be confined to the magneto, spark plugs, and cables.

Ignition troubles may be classified as:

1. Starting troubles.
   a. Engine fails to start.
   b. Engine fires a few times and stops.
2. Running troubles.
   a. Engine misses on one or two cylinders.
   b. Engine stops suddenly.

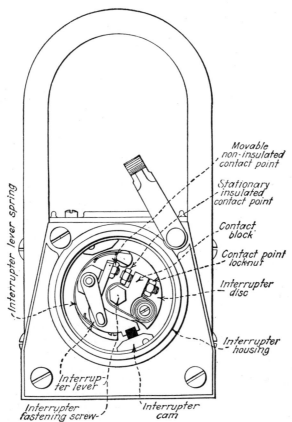

Fig. 269.—Breaker-point construction and adjustment for shuttle-type, high-tension magneto.

Starting troubles due to ignition failure may result from the following:

1. Spark lever is in retard position so that magneto is grounded. Practically all tractors have a ground-contact device on the breaker housing (Fig. 270), which cuts off the ignition and stops the engine when the spark lever is in full retard position.
2. Switch closed (on tractors equipped with an ignition switch).
3. Impulse coupling not engaged or fails to catch.

4. Impulse coupling releases too quickly before engine picks up speed.

5. Moisture or oil on spark-plug points, porcelain, or distributor terminals.

6. Moisture, oil, or dirt inside magneto on collector ring and brush, distributor contacts, or breaker points.

7. Breaker points rough, out of adjustment, or sticking.

8. Magneto out of time, or cables incorrectly connected to plugs.

9. Magnets weak.

If an engine is running but misses on one or more cylinders, the trouble, if due to faulty ignition, will likely be found in the spark plugs or cables. It can be due, however, to moisture, dirt, or oil on one or more of the distributor contacts.

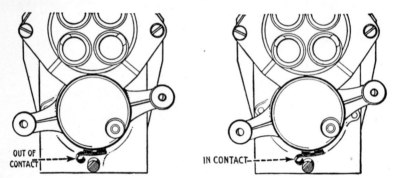

Fig. 270.—Short-circuiting device for high-tension magneto to stop engine.

If the engine suddenly stops owing to ignition failure, the trouble in nearly every case is in the magneto. It should be checked for:

1. Stuck points.

2. Broken insulation.

3. Burned-out winding.

4. Short circuit.

5. Drive coupling slipped out of place.

255. Tracing Tractor-ignition Trouble.—The first step in diagnosing tractor-ignition trouble is to determine whether it is in the spark plugs, cables, or magneto. The action of the engine, as just described, will indicate the location to a certain extent.

The best procedure is to disconnect the cables from the spark plugs and hold each one about $\frac{1}{8}$ in. from the engine block while the engine is rotated slowly. If a spark is observed when the impulse coupling snaps, the magneto is functioning properly and the trouble is in the plugs. If one or more cables do not show a spark, look for breaks and examine their respective distributor terminals and contacts. If, after a number of trials and a complete check of the retard lever, switch, and impulse coupling, no spark can be secured from any of the distributor terminals, it is quite evident that the trouble is in the magneto. In this case examine and clean the distributor, collector ring, and breaker points. Disassemble the magneto with great care and only to the extent necessary.

In disassembling shuttle-type magnetos be sure to remove the ground-return brush in the base and the collector-ring brush and holder before removing the armature.

256. Testing High-tension Magneto Armatures.—A simple method of testing a high-tension magneto armature for short or broken circuits is best understood by reference to Fig. 271.

Using any 6-volt battery such as an automobile storage battery, a "hot-shot" battery, or four dry cells in series, one battery wire *B* is grounded to the armature, and the other wire *A* is rubbed on the breaker-point retaining screw *D*, which is the insulated end of the primary winding. A third wire *C* has one end grounded on the armature while the other end is held about ⅛ in. from the collector ring. If, when the end of wire *A* is rubbed on screw *D*, a good spark appears at the gap at the collector ring, the windings are in good condition. If no spark is produced it indicates that there is a short circuit in one or the other of the windings.

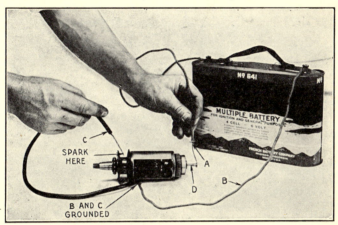

Fig. 271.—Testing a high-tension magneto armature for short or broken circuits.

This test can be made without removing the armature and breaker points from the magneto by placing a piece of paper between the breaker points and manipulating the wires as before.

References

Consoliver and Burling: "Automotive Electricity."
Cornell Univ., N. Y. State Coll. Agr., Ext. Bull. 133.
Cornell Univ., N. Y. State Coll. Agr., Ext. Bull. 147.
Fraser and Jones: "Motor Vehicles and Their Engines."
Heldt: "The Gasoline Automobile," Vol. III.
Magneto Characteristics, *Trans. Amer. Soc. Agr. Eng.*, Vol. 10, No. 2.
Oscillating Magnetos, *Trans. Amer. Soc. Agr. Eng.*, Vol. 5.
Stone: "Electricity and Its Application to Automotive Vehicles."

CHAPTER XX

ELECTRIC GENERATORS—STARTERS—LIGHTING

The rather common practice of operating tractors at night during the rush season or for economic reasons requires certain additional items of electrical equipment to provide light. Likewise the convenience of the automobile-type electric starter is proving advantageous for tractors under certain conditions. Therefore information relative to such equipment is of interest and value to the tractor user or operator.

257. Lighting.—If lights only are desired for a tractor, the current may be obtained directly from a storage battery or from a generator.

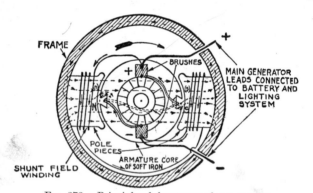

Fig. 272.—Principle of shunt-wound generator.

The use of a battery permits burning the lights without operating the engine, whereas, if a generator is used alone, light is available only when the engine is running. A combination generator and battery arrangement is most satisfactory and eliminates the necessity of periodically removing the battery for recharging. The usual lighting voltage is 6 unless the tractor engine is large and is equipped with a 12-volt or larger battery for starting.

258. Generators.—Generators for tractor lighting and starting are direct current and shunt wound (Fig. 272). A cutout relay (Fig. 273) automatically opens and closes the circuit between the generator and the battery, according to the engine speed and corresponding generated voltage, thereby preventing the battery from discharging through the generator when the engine is inoperative or idling.

This relay consists essentially of a small iron core, the fine or voltage winding, the coarse or current winding, and the points. As noted, when the generator is not operating the points are held open by a small spring. As the generator starts operating, a voltage is built up in the shunt-field winding. The fine or voltage winding of the relay, being a part of the field circuit, thereby carries a weak current that magnetizes the relay core and pulls the movable point into contact with the stationary point. The main or charging circuit is thus completed, and the stronger current now flowing through the heavy winding on the relay core holds the points firmly in contact. If the generator speed drops sufficiently to permit the generator voltage to fall below the battery voltage, there is a reverse flow of current, the core is demagnetized, and the points open again.

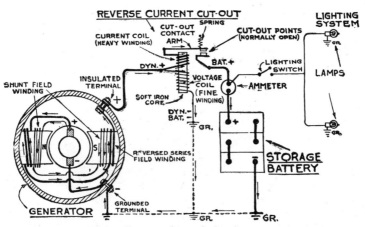

Fig. 273.—Wiring diagram of a typical reverse current cutout.

Generators are rated according to voltage as 6, 12, 24, and so on, according to the voltage of the battery to be charged. However, the actual generated voltage at normal speeds must slightly exceed the battery voltage in order to force the current into the battery. This explains the noticeably brighter lights when the generator is running.

259. Generator Control.—In any simple shunt-wound generator, the tendency is for the voltage to vary more or less directly with the armature speed; nevertheless, a more or less constant voltage is essential, if possible, over a considerable speed range. A slightly low voltage gives poor lights or a low battery-charging rate, whereas a slightly high voltage may burn out the lights or charge the battery at too high a rate.

Automotive-type generators, therefore, utilize a number of different methods of controlling the voltage and corresponding current output as follows:

1. Adjustable third brush.
2. Current-controlled magnetically operated vibrating relay.
3. Voltage-controlled magnetically operated vibrating relay.
4. Combination of voltage- and current-controlled vibrating relays.

260. Third-brush Control.—A simple shunt-wound generator requires but two brushes as shown in Fig. 272. Figure 274 shows how a third brush is utilized to carry the field circuit from the commutator to the field windings. The use of the third brush in this manner permits locating it on the commutator in such a position that there will be an entirely different reaction between the magnetic field and armature windings as the armature speed varies. In other words, owing to a phenomenon known as field distortion, the field strength and resulting

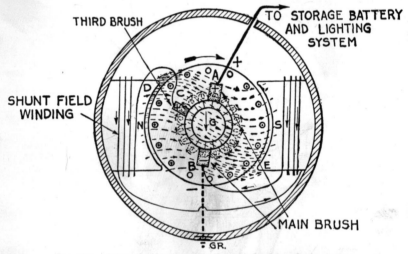

Fig. 274.—Diagram showing principle of third-brush regulation.

voltage does not increase with increased armature speed, beyond a certain point. On the other hand, by adjusting the brush position through a limited range with respect to the commutator and armature rotation, the charging rate of the generator can be varied considerably. For example, moving the third brush opposite the direction of the armature rotation as far as possible will reduce the maximum current output to 3.5 amp. Moving the brush in the direction of armature rotation as far as possible will produce the maximum current output of 20 to 25 amp.

The principal objection to a plain third-brush-regulated generator is the danger of burning out the windings, lights, and so on, if the battery-charging circuit is unexpectedly broken. What happens in such a case is that the generator, having been suddenly relieved of the back pressure or voltage of the battery, builds up an excessive field strength that in turn produces a high voltage in the main circuit. Whenever it becomes necessary to operate a third-brush-regulated generator that is not connected to the battery, either the brushes should be lifted off the commutator, or the external circuit should be short-circuited with a low-resistance conductor.

261. Current-controlled Vibrating Regulator.—A current-controlled generator is one that uses a vibrating relay, operated by the charging current, to vary the field resistance and thereby control the amount of current generated. Referring to Fig. 275, it will be observed that the generated current passes through a coil wrapped around an iron core. Therefore, when this current reaches a given value, the magnetized core separates the points, thus breaking the normal field circuit and cutting in a field-resistance unit. Thus the field current, and therefore

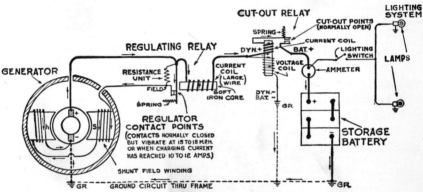

Fig. 275.—Circuit diagram of typical vibrating-type regulator to obtain constant current regulation of the generator.

the generated current, is reduced. Ordinarily the contact arm vibrates rapidly and keeps the generator output practically constant.

Current-controlled vibrating-relay-regulated generators are little used for the reason that they have the same objection as the third-brush-controlled generator, namely, that if the main charging circuit is broken or encounters high resistance, a high voltage builds up in the field circuit, which, in turn, produces an excessive voltage and current flow in the generator windings, lights, and so on, and often causes damage.

262. Voltage-controlled Vibrating Regulator.—Referring to Fig. 276, a voltage-controlled generator employs a set of magnetically operated contact points, similar to the current-controlled device just described. However, it will be noted that the coil that magnetizes the core and separates the points is not in the charging circuit, but consists of fine wire that forms a separate high-resistance circuit in parallel with the charging circuit. Consequently, when the generated voltage attains the desired value in this coil, the points are separated and a high-resistance unit is cut into the field circuit. This obviously reduces the field strength and the resulting generated voltage and current. When the voltage drops slightly, the points close, the voltage rises again, the points open,

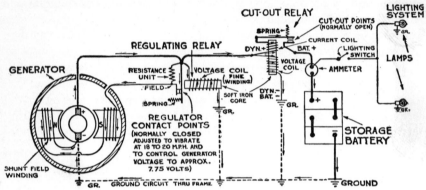

FIG. 276.—Circuit diagram of typical vibrating regulator to obtain constant voltage regulation of the generator.

and so on. In other words, beyond a certain speed, the points will vibrate rapidly and thereby maintain a constant voltage and relatively constant charging rate.

From a study of this device it will be observed that if the charging circuit is broken or the battery is disconnected, no harm will result for the

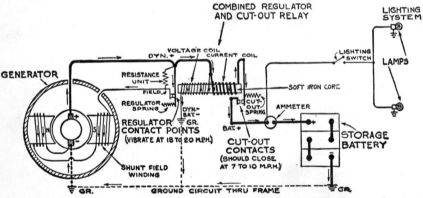

FIG. 277.—Circuit diagram of typical combination vibrating-type regulator and cutout to obtain constant current and voltage regulation of the generator.

reason that the relay will continue to operate and maintain a constant and proper field voltage.

263. Combined Current- and Voltage-controlled Vibrating Regulator.—In a few cases, generators are equipped with a combined current and voltage regulator of the vibrating type, as shown by Fig. 277.   This offers no special advantages other than combining and providing all the advantages of both types.

264. Starter Motors.—Starter motors are similar to generators, with the exception that they have a series-wound instead of a shunt-wound

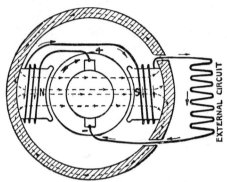

Fig. 278.—Diagram of a series-wound motor.

field circuit, as shown by Fig. 278.   Also, both the field and armature windings are of very coarse low-resistance wire, which will allow a high-amperage current to flow.   A starter motor must exert a strong turning effort or torque.   To provide this, a heavy current must flow through the circuits and create a very strong magnetic field and consequent pronounced reaction between the armature and the field.

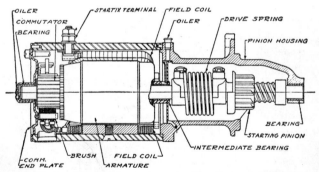

Fig. 279.—Sectional view of a starting motor using Bendix-type starter-drive mechanism.

The principal variation in different starters is in the number of poles and brushes.   For small engines, two-pole starters with two brushes are usually used.   Larger engines use four-pole starters with four brushes. Otherwise the construction and operation are the same.

265. Starter Drives.—The starter drive is the device that connects
and transmits the power to the engine flywheel. A small spur pinion
on the armature shaft meshing with teeth on the flywheel is the usual
means of transmitting the power. However, the drive must embody
some convenient means of disengaging as well as engaging the pinion
which will not give trouble or injure the starter motor when the engine
begins to fire or if it kicks back.

Two common types of starter drives are the Bendix (Fig. 279) and
the manual type (Fig. 280). In the former, the construction is such that
engagement of the pinion takes place as soon as the starter switch is closed
and the armature begins to rotate. A threaded sleeve on the armature

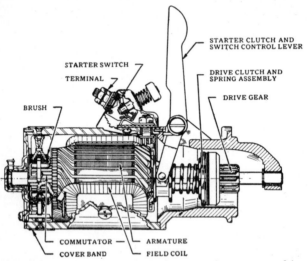

FIG. 280.—Sectional view of a starter motor using manual-type starter-drive mechanism.

shaft carries this pinion and permits it to move along the shaft and into
and out of engagement. If the armature starts turning, the thread
carries the pinion into engagement with the flywheel gear. As soon as
the engine fires and turns the pinion faster than the armature shaft is
turning it, the pinion is carried out of engagement by the threads. The
device is equipped with a heavy spring to absorb the shock as the pinion
comes into mesh with the flywheel or when the engine kicks back.

The manually operated starter drive is so called because the pressure
on the starter foot lever first slides the pinion into mesh with the flywheel
gear. Pushing the foot lever still farther then closes the switch, and the
armature rotates and cranks the engine. To prevent damage to the
starter motor when the engine begins to fire and turn the pinion at a
higher speed than the armature is turning it, an overrunning clutch is
built into the pinion and connects it to the armature shaft in such a

manner that the pinion is automatically disconnected from the armature even though the foot lever is not released.

References

Consoliver and Burling: "Automotive Electricity."
"Dyke's Automobile and Gasoline Engine Encyclopedia."
Elliott and Consoliver: "The Gasoline Automobile."
Stone: "Electricity and Its Application to Automotive Vehicles."

CHAPTER XXI

LUBRICATION AND LUBRICATING SYSTEMS

Lubrication plays an important part in the design and operation of any type of internal-combustion engine. The life and service given by any engine are dependent largely upon the consideration and care given to its lubrication, both in the design of the system and during its use and operation. The fundamental purpose of lubrication of any mechanical device is to eliminate friction and the resulting wear and loss of power. Other important functions of lubrication of an internal-combustion engine are (1) to absorb and dissipate heat, (2) to serve as a piston seal, and (3) to act as a cushion to deaden the noise of moving parts.

In the internal-combustion engine, lubrication is even more difficult than in other machines, for the reason that a certain amount of heat is present, particularly in and around the cylinder and piston, as a result of the combustion of the fuel. The lubrication of these heated parts is a somewhat more difficult problem than it would be if the heat were not present.

266. Friction.—Whenever two materials of any kind move against each other, a certain amount of a force known as *friction* tends to oppose the movement. For example, when a liquid or a gas is made to flow through a passage such as a pipe or a conduit, the flow is more or less retarded by the friction between the liquid or gas and the inner surface of the pipe. Such friction is undesirable. On the other hand, the transmission of power by belts and pulleys is entirely dependent upon the friction between the belt and the pulley surfaces. The greater the frictional contact, the more effective the arrangement and the lower the power loss.

The friction due to metal parts of machines moving in or against each other is usually highly undesirable, not only because of the power required to overcome it, but also owing to rapid wear. Likewise, it often results in the generation of a certain amount of heat, which may damage the parts or produce a fire hazard. This friction cannot be completely eliminated, but it can be reduced to such an extent by the use of a suitable lubricant that the operation of the machine is greatly improved and its life lengthened.

LUBRICATING OILS AND GREASES

There are a large number and variety of lubricants available for the many different lubrication requirements. One particular kind, quality, or type of lubricant has a comparatively limited use.

264

Practically all lubricants are derived from one or the other of three sources: namely, animal, vegetable, or mineral matter.

267. Animal Lubricants.—The oils and greases obtained by rendering the fat of animals, such as swine, cattle, sheep, fish, and so on, are utilized to a very limited extent for the lubrication of mechanical devices. Some of the common animal lubricants are lard oil, tallow, sperm oil, and fish oil. The principal disadvantage is that they cannot stand heat and readily combine with oxygen, becoming waxy or gummy. Also, under certain conditions, fatty acids that are harmful to the machine parts are liberated.

268. Vegetable Lubricants.—The oils obtained from vegetable materials, such as seeds, fruits, and plants, are likewise limited in their utility for general lubrication purposes for the reason that they also have a tendency to oxidize readily and become gummy. Some common vegetable oils are castor oil, cottonseed oil, olive oil, and linseed oil. Only castor oil has ever proved of any value for engine lubrication. It is used to a limited extent in some types of airplane and marine engines because it does not congeal except at low temperatures, it does not mix readily with engine fuels, and it retains its lubrication properties at high temperatures.

269. Mineral Lubricants.—Crude petroleum, obtained from the earth and, therefore, usually classed as a mineral, serves as the greatest source of lubricating materials as well as fuels for internal-combustion engines. Petroleum lubricants are not only more plentiful and therefore less expensive than animal and vegetable oils, but also they retain their lubricating properties when subjected to abnormal temperatures and other conditions.

Crude petroleum varies in quality, character, and chemical make-up, according to its geographical source. For example, some crude oils contain a high percentage of lighter products, such as gasoline and kerosene, and are therefore more valuable as a source of fuel. Other crudes, usually heavier in gravity, are low in these lighter products but contain a higher percentage of products that, when refined, produce lubricating oils and greases. The so-called paraffin-base crudes are usually lower in specific gravity and yield a high percentage of fuel products when distilled. The asphalt-base crudes are usually higher in specific gravity, darker in color, and yield a high per cent of lubrication products.

Considerable argument has been put forth in the past concerning the relative merits of lubricants produced from different crude oils. Owing to the great development and advance made in the refining process, however, it is doubtful if any conclusive and consistent proof can be shown in favor of any particular crude-oil base as a base for a lubricant for a particular engine or purpose.

270. Manufacture of Lubricants.—Crude oil consists of a mixture of many hydrocarbons having different boiling points. Therefore, the first process in the conversion of any crude oil into its various products consists of heating it and distilling off the lighter fractions, such as gasoline, kerosene, gas oil, and so on (see Fig. 83). When a temperature of 350 to 400°C. is reached, the heavier fractions, suitable for lubricants, distill over. If it is desired to secure a large quantity of lubricating oil from the crude, steam is introduced into the stills and they may be operated under a partial vacuum, in order to prevent these heavy fractions from cracking or breaking up into lighter hydrocarbons. The cracking process has already been referred to and explained in Chap. VIII. The lubricating-oil distillates may be redistilled with steam under a vacuum and thereby separated into heavy and light oils. Following distillation, the oils are subjected to the refining process, which consists of treatment with sulphuric acid or filtering through fuller's earth to remove certain undesirable elements and impurities and to lighten the color. This is followed by washing with water, neutralizing with an alkali, a second washing with water, and a final action, involving hot air, to remove all traces of moisture.

When a crude oil is distilled until the fuel products and the light and heavy lubricating oils have been removed there usually remains a certain quantity of a heavy, thick, dark oil that, when subjected to a purifying treatment, is utilized for steam-engine-cylinder lubrication and for the lubrication of automobile, truck, and tractor transmissions, and other similar mechanisms.

271. Classification of Engine Oils.—Oils for lubricating internal-combustion engines are classified according to their body or viscosity, as light, medium, heavy, and extra heavy. Many oil companies use certain distinguishing symbols to designate these various grades.

Light oils are recommended for engines having close-fitting pistons and bearings, or engines operating at light loads and high speeds. Likewise a lighter grade of oil is usually recommended when an engine is operated in cold weather or under similar conditions of temperature.

Medium oils are used under average conditions in automobile engines in cold or temperate weather, provided the engine is not badly worn. A medium-body oil will give satisfactory results with a plain splash system of lubrication or with a force-feed system.

Oils having a heavy body are used in engines operated in very warm weather, or in truck and tractor engines that operate under a medium to heavy load. Heavy oils are also recommended for air-cooled engines and for engines having badly worn pistons, cylinders, and piston rings, as a heavy oil, under such conditions, tends to hold the compression better.

Extra heavy oils are used in some tractor engines, for the reason that these engines when in operation are subjected to a uniformly heavy load, which means a higher operating temperature.   As a general rule, the load carried by an engine and, therefore, its operating temperature determines, to a large extent, the grade of oil to be used.   The hotter the engine, the heavier the oil recommended, up to a certain limit.

272. Semisolid and Solid Lubricants.—In addition to the fluid lubricants already discussed, certain other materials in a semisolid or a solid form are used to a very limited extent for internal-combustion-engine lubrication.   Cup grease or hard oil is the most common semisolid lubricant.   It is made by thoroughly mixing an ordinary mineral fluid oil with some animal fat, usually beef tallow, that has been saponified. The proportions vary from 80 to 90 per cent mineral oil and 10 to 20 per cent fat.   Cup greases are prepared in the following grades:

> No. 1—very soft.
> No. 2—soft.
> No. 3—medium.
> No. 4—hard.
> No. 5—very hard.

The softer the grease the greater the mineral-oil content.

Flake graphite, mica, or soapstone is sometimes mixed with cup grease for special purposes, particularly, where the bearing surfaces are rough or high pressure exists.

Graphite, talc, and mica are solid materials having a limited use as lubricants.   Of these, graphite in a flaky or amorphous form is the most valuable.   These materials are used primarily where high pressure exists and a fluid or semisolid oil would not give good results.

273. Importance of Correct Lubrication.—The market is literally flooded with hundreds of brands and grades of oils of varying quality. Consequently, the owners and operators of gas engines, automobiles, trucks, tractors, airplanes, and so on, realize the great importance of choosing the best possible oil for the engine.   The price should not be more than a minor determining factor.   During the early development and use of these machines, the selection of the correct lubricant was, in many cases, left with the operator.   As time went on, the engine manufacturer realized the important part played by lubrication and lubricating oil in the satisfaction and service rendered by his product and, consequently, assumed the responsibility for determining and recommending the proper grade of oil to be used.   As a rule, thorough tests are made either by the engine manufacturer or by the oil refiner, usually the former, before any definite lubrication recommendations are offered.   Such being the case, the user and operator of any type of internal-combustion engine should adhere closely to such advice.   However, some knowledge of the fundamental factors involved in the choice of an engine oil is valuable and important.

ESSENTIALS OF A GOOD LUBRICATING OIL

As a general rule, the majority of the standard or better-known brands of engine oil now on the market are reliable products and will give good results provided the correct grade is used.   There are no hard and fast rules or specifications by which one may choose the proper oil to use in a given case.   Likewise, there are no simple chemical or physical tests that the average individual may apply to an oil to determine its character.   Such things as the color, feel, and general appearance mean nothing.

Thomsen[1] states that in order to satisfy certain fundamental lubricating requirements, a lubricant:

1. Must possess sufficient viscosity and lubricating power—oiliness—to suit the mechanical conditions and conditions of speed, pressure, and temperature. Too little oiliness means excessive wear and friction; too high a viscosity means loss of power in overcoming unnecessary fluid friction.

2. Must suit the lubrication system.

3. Must be of such a nature that it will not produce deposits during use when exposed to the influence of the air, gas, water, or impurities with which the oil may come into more or less intimate contact while performing its duty.

274. Tests of Lubricating Oils.—There are a large number of physical and chemical tests to which an oil may be subjected for the purpose of determining its quality and adaptability to a given purpose.   Many of these, however, are of minor importance except to the oil refiner, the chemist, or the lubricating engineer.   In a bulletin, published by the American Society for Testing Materials and entitled, "The Significance of Tests of Petroleum Products," the following statements are made:

The rapid growth of the petroleum industry has been accompanied by the development of a variety of physical and chemical methods for the testing of petroleum products.

Physical tests are more widely used than chemical tests.   This is natural, in view of the fact that the utility of petroleum products depends to a large extent upon their physical characteristics.   Some of these physical tests are of little value except as they serve the refiner in controlling manufacturing processes, while others are useful both to consumer and to manufacturer as an index of the value or fitness of products for particular uses.

Such chemical tests as now exist serve principally to protect against impurities or undesirable constituents.   This is because petroleum is an extremely complex raw material, varying greatly in composition in the various producing fields,

[1] THOMSEN, "Practice of Lubrication."

with each new field bringing its own peculiar problems. Little is known about the chemistry of petroleum at the present time, but it may reasonably be expected that a considerable advance along this line will be made in the future.

Some of the more important tests, together with their definition or meaning, the recommended procedure followed in making them, and their significance and value, are as follows:

275. Gravity.—In testing oils for gravity, the same method is followed as that discussed in Chap. VIII on the Gravity of Fuels. The A.S.T.M. bulletin previously referred to states: "The property of gravity is of importance in the control of refinery operations. It is of little significance as an index of the quality or usefulness of a finished product and its use in specifications is to be avoided."

Regarding other tests of lubricating oils of more or less importance, the above-mentioned bulletin states as follows:

276. Color.—The color of petroleum products is described and defined in terms of color by reflected light or by transmitted light. Color requirements of lubricating oils are frequently overemphasized as color does not necessarily indicate quality. A fallacy, which is prevalent among consumers, is that pale color indicates low viscosity.

277. Cloud and Pour Points.—Petroleum oils become more or less plastic solids when sufficiently cooled, due either to partial separation of wax or to congealing of the hydrocarbons composing the oil. With some oils the separation of wax becomes visible at temperatures slightly above the solidification point, and when that temperature is observed, under prescribed conditions, it is known as the "cloud point." With oils in which wax separation does not take place prior to solidification, or in which the separation is not visible, the cloud point cannot be determined. That temperature at which the oil just flows under prescribed conditions is known as the "pour point," irrespective of whether the cloud point is observable. Oils vary widely in these characteristics depending upon the source of the crude oil from which they are made, upon the grade or kind, and upon the method of manufacture.

The cloud point is of value when the oil is to be used in wick-feed service, or when a haze or cloud in the oil above a given temperature would be objectionable for any reason. However, the test may give misleading results if the oil is not dry, due to the separation of water, and the test should always be interpreted with this fact in mind. In general, the cloud point is of more limited value and narrower in range of application than the pour point.

The pour point gives an indication of the temperature below which it might be dangerous to use the oil in gravity lubricating systems. No single test can be devised which can be taken as a positive and direct measure of the performance of an oil under all conditions of service, and the pour test should be regarded as giving only an indication of what may be expected. Consequently, cloud and pour points should be interpreted in the light of actual performance under the particular conditions of use.

278. Viscosity.—The viscosity of a fluid is the measure of its resistance to flow. In most commercial work an expression involving the time in seconds required for a measured volume of oil to flow, under specified conditions, through a carefully standardized tube, is known as the viscosity of the oil. In practice, at least three different instruments must be employed in commercial testing: one for light distillates, one for lubricating oils, and one for heavy fuel and road oils. As viscosity changes rapidly with temperature, a numerical value of viscosity has no significance unless both the temperature and the instrument are specified.

279. Viscosimeters for Lubricating Oils.—The Saybolt Universal Viscosimeter (Fig. 281) is now used almost universally in the United States for determination of the viscosity of lubricating oils. On the lighter oils the measurements are

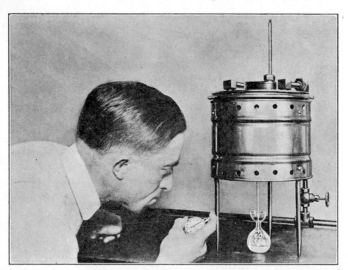

Fig. 281.—Testing oil for viscosity.

made at 100 or 130°F., and on the more viscous oils at 210°F. The instrument gives reliable results at any temperature between about 31 and 210°F., provided the oil is not near its solid point or its flash point. Above 210°F., radiation losses become so large that the results are not satisfactory, and recourse must be had to instruments such as the Ostwald type of viscosimeter, described by Bingham, which can be completely immersed in a high-temperature bath.

The Saybolt Furol Viscosimeter is applicable to, and generally recommended for, the determination of the viscosity of heavy fuel and road oils. The temperature usually employed is 122°F.

280. Significance of Viscosity.—The significance of viscosity depends upon the purpose for which the oil is to be used.

For lubricating oils, viscosity is the most important single property. In a bearing operating properly, with a fluid film separating the surfaces, the viscosity of the oil at the operating temperature is the property which determines the bearing friction, heat generation, and the rate of flow under given conditions of load, speed, and bearing design. The oil should be viscous enough to maintain a fluid film between the bearing surfaces in spite of the pressure tending to squeeze it out. While a reasonable factor of safety is essential, excessive viscosity means unnecessary friction and heat generation. Since the rate of change of viscosity with temperature varies with different oils, viscosity tests should, in general, be made at that standard temperature which approximates most closely the temperature of use.

281. S.A.E. Viscosity Numbers.—In order to provide some standard means of designation of motor-oil grades, the Society of Automotive Engineers, cooperating with the oil refiners and automotive manufacturers, has worked out an oil-grading system based upon viscosity numbers as indicated in Table XIV. As noted, the measured viscosity in seconds by the Saybolt Universal Viscosimeter (Fig. 281) is taken at 130 and 210°F. to provide a knowledge of the relative change in fluidity between average operating crankcase temperatures. The oil manufacturer designates the various grades by means of numbers instead of in the usual manner, the number being stamped on the container. The user, knowing the viscosity number recommended by the engine manufacturer, selects the oil accordingly. Obviously, the adoption of this system will go far in eliminating the necessity for using descriptive terms, such as light, medium, heavy, etc., which at best have proved somewhat misleading.

Other lubricating oil tests, of little significance except to the refiner, are flash and fire tests, carbon residue, and neutralization number.

ENGINE LUBRICATION

The lubrication of an engine may be considered under two distinct heads: namely, (1) the choice and use of the correct kind and grade of lubricant and (2) the choice, design, construction, and operation of the system with which the engine is equipped. A good lubricating system must be efficient in operation, reliable, troubleproof, and simple. Yet, even though it possesses all these important features, if a poor-quality lubricant or one of incorrect grade is used, unsatisfactory service is likely to result.

As previously stated, in selecting the oil to be used, satisfactory results are more likely to be obtained if the advice and recommendations of the engine manufacturer are followed. Then, having selected a suitable oil, one should familiarize himself with the construction and operation of the lubrication system of the engine itself and see that it functions properly at all times.

282. Parts Requiring Lubrication.—The most important parts of a gas engine requiring lubrication are as follows:

1. Cylinder walls and piston.
2. Piston pin.
3. Crankshaft and connecting-rod bearings.
4. The camshaft bearings.
5. The valves and valve-operating mechanism.
6. Other moving parts, such as cooling fan, water pump, ignition mechanism, and so on.

TABLE XIV.—CRANKCASE LUBRICATING-OIL VISCOSITY NUMBERS
(S.A.E. Recommended Practice)

S.A.E. viscosity number	Viscosity range,[1] Saybolt Universal, sec.			
	At 130°F.		At 210°F.	
	Min.	Max.	Min.	Max.
10	90	Less than 120		
20	120	Less than 185		
30	185	Less than 255		
40	255	. . . .	. . .	Less than 75
50	. . .	. . . .	75	Less than 105
60	. . .	. . . .	105	Less than 125
70	. . .	. . . .	125	Less than 150
10W	5,000 to 10,000 (at 0°F.)			
20W	10,000 to 40,000 (at 0°F.)			

In the case of prediluted oils, S.A.E. viscosity numbers, by which the oils are classified, shall be determined by the viscosity of the undiluted oils.

Wherever the S.A.E. viscosity numbers are used on prediluted oils, the container labels should show in some suitable manner that the S.A.E. number applies to the undiluted oil.

[1] Oils with viscosities falling between the ranges specified shall be classified in the next lower grades.

283. Systems of Engine Lubrication.—The following outline gives, in a general way, the methods and systems commonly used in lubricating the types of engines mentioned:

1. Stationary Farm Engines.
    a. Bearings by compression grease cups, and cylinder and piston by gravity sight-feed oiler.   Other parts by hand.
    b. Simple splash system.
2. Stationary Heavy-duty Engines.
    a. External mechanical force feed.
    b. Circulating force feed and splash.
3. Multiple-cylinder Medium- and High-speed Engines.
    a. Simple circulating-splash system.
    b. Internal force-feed and splash system.
    c. Internal force feed alone.
    d. External force feed.

284. Farm Gas-engine Lubrication.—The lubrication of the common types of farm gas engines is comparatively simple, for the reason that the engine is simple in construction and operates at a rather low speed.

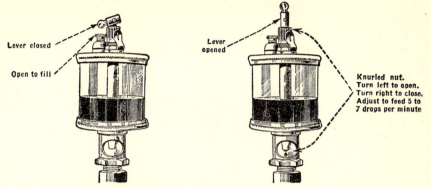

Fig. 282.—Construction and operation of a gravity sight-feed oiler.

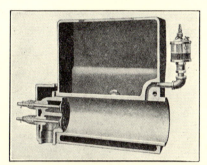

Fig. 283.

Fig. 284.

Fig. 283.—Cylinder lubrication using a gravity oiler.
Fig. 284.—Farm-engine lubrication by means of grease cups.

Most of the engines of this type are lubricated by method 1*a* in the above outline. A gravity sight-feed oiler (Fig. 282) supplies oil for the cylinder and piston, as shown by Fig. 283. A hole in the piston and connecting rod also permits this oil to lubricate the piston pin and bearing.

The oil drops from the main chamber of the oiler to a small drip chamber underneath and then into the tube or passage leading to the cylinder. The amount of oil fed is controlled by means of a knurled nut, which is so adjusted that the cylinder receives a certain number of drops per minute. This varies from 5 to 20 depending upon the size of the engine. Frequent adjustment is necessary owing to temperature changes or variation in the viscosity of the oil. Oil will drop slower in

Fig. 285.—Construction and lubrication of a plain crankshaft bearing on a farm engine.

cold weather or when the engine is cold.   Likewise, a heavy oil will have a tendency to feed slower than a lighter oil.   The oil feed can be cut off or turned on instantly by means of the small trip lever just above the adjusting nut.   In addition to the parts already mentioned, this type of oiler is equipped with a check valve just underneath the drip chamber. This permits the oil to pass downward to the cylinder but prevents the

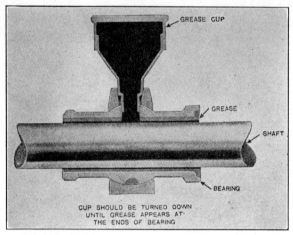

FIG. 286.—Compression grease-cup construction and operation.

back pressure from the explosion from blowing the oil upward through the oiler and thus interfering with the uniform dropping and flowing of the oil.

The other important parts of the engine requiring lubrication, such as the connecting-rod and main crankshaft bearings, are usually lubricated by means of compression grease cups, as shown by Fig. 284.   Figures 285 and 286 show in detail how these bearings are constructed and how the grease is supplied to them.   Figure 287 illustrates a compression grease cup of the automatic-feed type, which is somewhat more convenient than the plain type of cup.

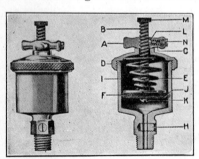

FIG. 287.—A special type of compression grease cup.

Figure 288 shows an engine with the crankcase enclosed.   To facilitate greasing the connecting-rod bearing, the crankshaft is drilled or hollowed out from one end into the crank journal.   A grease cup attached to this end of the crankshaft forces grease through the passage to the bearing.

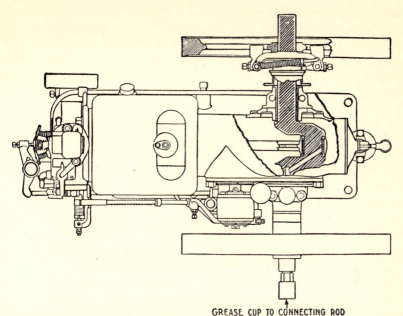

GREASE CUP TO CONNECTING ROD

FIG. 288.—Crankshaft drilled to supply grease to crank bearing.

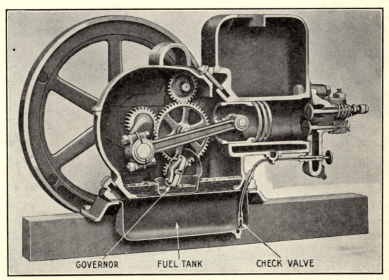

GOVERNOR   FUEL TANK   CHECK VALVE

FIG. 289.—A farm-type engine with splash lubrication.

Other parts, such as the push rod, governor, valve stems, and ignition mechanism, are usually oiled at frequent intervals by means of an ordinary squirt can when the engine is in operation.

Some of the smaller, later models of farm engines are oiled almost entirely by a plain splash system, which necessitates almost complete enclosure of the working parts with an oiltight dustproof housing. Figure 289 illustrates an engine of this type in which the governor weights splash the oil to the working parts.

The principal advantages of this system are simplicity, convenience, decreased oil consumption, and more positive lubrication. On the other hand, care must be taken to see that leaks in the housing and around the bearings do not develop. Engines with enclosed crankcases are usually equipped with breathers to relieve the pressure in the crankcase due to the pumping action of the piston.

LUBRICATION OF MULTIPLE-CYLINDER AND TRACTOR ENGINES

The proper lubrication of a multiple-cylinder engine, such as that used in an automobile, tractor, or truck, is a somewhat more difficult problem, owing to the greater number of parts and the higher operating speed. Likewise lubrication is the most important factor in the satisfactory operation and service of the farm tractor. Because a tractor operates under conditions that are conducive to rapid wear, such as heavy loads, high temperatures, exposure to dust and dirt, and travel over rough ground, any slight defect in the lubrication system or temporary neglect in this respect will likely result in unexpected trouble. Faithful attention to all lubrication details by every tractor owner and operator is of utmost importance if the greatest possible service is to be secured from the machine.

The lubrication of a farm tractor may be considered under three heads: namely, (1) engine, (2) transmission, and (3) chassis. Each is lubricated in a different manner and will be discussed separately.

There is considerable variation in the construction and operation of the engine-lubrication systems found on different tractors. However, the systems used may be placed under one or the other of the four systems indicated in Par. 283 as adapted to multiple-cylinder engines.

285. Circulating Splash.—In the circulating-splash system the lubrication of all the principal engine parts is dependent directly upon the splashing and spraying of the oil by the connecting rods dipping into it. The one important requirement for its successful operation is the maintenance of a uniform oil level under the rods. This is done by maintaining a continuous flow of oil from the sump or reservoir into a splash pan that has a depression or trough under each rod.

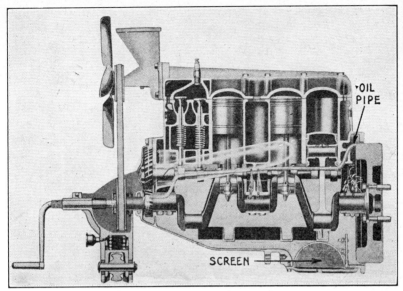

Fig. 290.—Circulating-splash oiling system using flywheel to circulate the oil. (*Courtesy of Ford Motor Company.*)

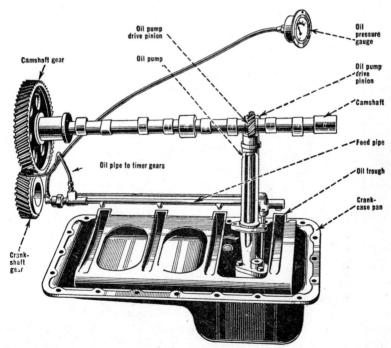

Fig. 291.—Circulating-splash oiling system using pump to circulate the oil.

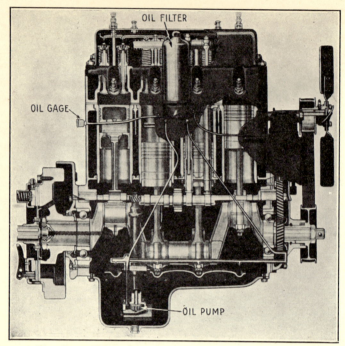

Fig. 292.—Circulating-splash oiling system with oil filter.

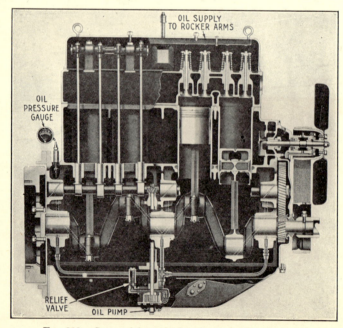

Fig. 293.—Internal force-feed and splash oiling system.

Figure 290 shows the Fordson tractor engine, which utilizes the flywheel to circulate the oil. The flywheel housing serves as a reservoir, and a tube leading from a point directly over the flywheel to the front near the timing gears keeps the troughs full. The oil, as it returns to the reservoir, must pass through a screen which removes any sediment, dirt, or other foreign matter. The proper oil level is determined by two test cocks, the lower one indicating the low level and the upper one the high level. In this tractor the engine oil also lubricates the clutch and steering gear.

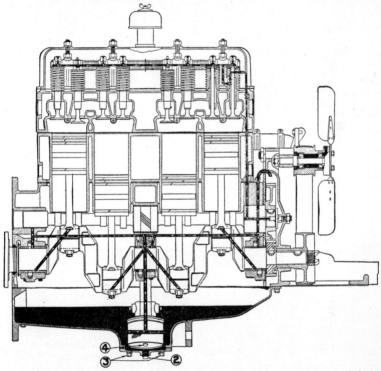

Fig. 294.—Vertical, four-cylinder tractor engine equipped with full internal force-feed lubrication system.

Figure 291 shows the lubrication system used in some McCormick-Deering tractors. A gear pump, driven from the camshaft, pumps the oil from the reservoir up into the splash pan. A dial-type indicator shows whether the pump is working or not. A fine screen around the pump inlet prevents sediment and dirt from being circulated. Figure 292 shows this system with a special externally connected filter. Two test cocks on the side of the sump are provided for determining the oil level.

286. Internal Force Feed and Splash.—In this system (Fig. 293) the oil is forced directly to the main crankshaft, connecting-rod, and

Fɪɢ. 295.—Horizontal, two-cylinder tractor engine equipped with full internal force-feed lubrication system.

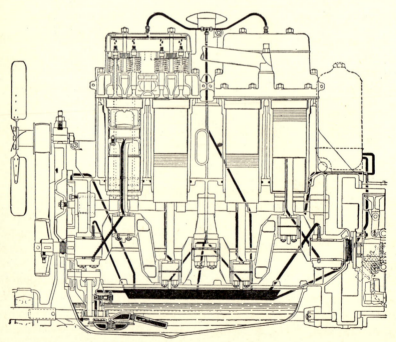

Fɪɢ. 296.—The full internal force-feed system of lubrication.

camshaft bearings. Drilled passages in the crankshaft carry the oil from the main bearings to the connecting-rod bearings as shown. The oil oozing out of these bearings creates a spray that lubricates the cylinder walls, pistons, and piston pins. The connecting rods do not dip in the oil, and a splash pan is unnecessary. A pressure indicator, connected

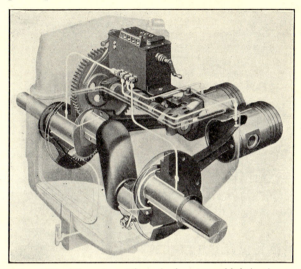

Fig. 297.—The external force-feed system of lubrication.

as illustrated, shows whether the pump is working and the pressure being maintained. The valve rocker arms are oiled from a trough attached to the underside of the cylinder-head cover. In other engines (Figs. 294 and 296) using the internal force-feed and splash system, the valve mechanism is oiled by pressure from the crankcase as shown.

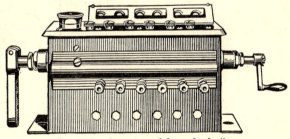

Fig. 298.—An external force-feed oiler.

287. Full Internal Force Feed.—This system goes a step farther and forces the oil, not only to the main crankshaft, connecting-rod, and camshaft bearings, as previously described, but also to the piston-pin bearings through tubes or passages that lead from the connecting-rod bearing, up the connecting rod to the piston pin, as shown in Figs. 295

and 296. The cylinders and pistons receive their oil from the piston pins and from the mist created by the oil issuing from the various bearings. The valve mechanism likewise is oiled, usually by a pressure feed as indicated.

288. External Force Feed.—Some heavy-duty engines are lubricated by means of an external force-feed oiler (Fig. 297).

The oiler (Fig. 298) consists essentially of a small rectangular reservoir mounted on the engine and containing a number of small pumping units immersed in the oil.

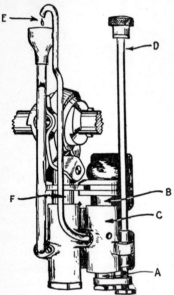

These pumps force the oil under high pressure through small, metal tubes to the cylinders, pistons, bearings, and other working parts that need a continuous supply. The oiler may be driven by a small pulley and belt or by an oscillating arm and ratchet device. A sight-feed arrangement on the top enables one to observe the number of drops per minute that are pumped. If the amount is incorrect it can be changed by means of the adjusting screw D (Fig. 299).

The construction and operation of a pumping unit are as follows: The unit itself is made up of two plungers B and F (Fig. 299). The oil from the reservoir is first received by cylinder C and delivered by plunger B through the sight-feed arrangement at E. It is then received by plunger F and forced through a connecting tube to the bearing.

Fig. 299.—Pumping unit and mechanism for an external force-feed oiler.

Although there are a number of makes of these oilers and they differ somewhat in construction and operation, the general principles are very similar. They are made in sizes varying from one or two pumping units for single-cylinder steam and gas engines up to 20 or more pumping units and leads for special multiple-cylinder engines.

With this system, fresh oil is supplied constantly to the working parts. However, this often results in the consumption of a greater quantity of oil than the engine would use if equipped with a splash or other system of lubrication. Another advantage is that these oilers will handle heavy oils or oils thickened by low temperatures very satisfactorily.

289. Oil-circulating Pumps.—Two types of pumps are used for circulating the oil in engines equipped with either the circulating-splash, the internal force-feed and splash, or the full internal force-feed system. The most used type is the gear pump (Fig. 300). It consists of two small spur gears held in a horizontal position and enclosed in a close-fitting, oil-tight housing. A vertical shaft from the camshaft of the engine drives

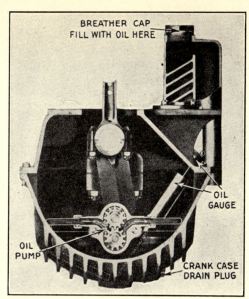

Fig. 300.—A gear-type oil pump.

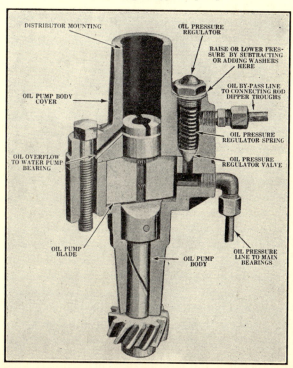

Fig. 301.—A vane-type oil pump.

one pump gear and it, in turn, drives the second gear.   The oil enters on that side of the housing on which the gear teeth are turning away from each other, or going out of mesh, and is carried between the teeth and the inner surface of the housing around to the opposite side and discharged.

A second type of oil pump, known as a revolving-vane pump, is shown in Fig. 301.   This type is limited in use.

290. Oil Pressures—Relief Valves.—The force or pressure applied to the oil by the pump depends largely upon the lubrication system.   A plain circulating-splash system requires very little pressure, usually from 2 to 5 lb. per square inch.   On the other hand, a full force-feed system requires that the oil be pumped a greater distance through numerous tubes and passages and in sufficient quantity at all times to insure the

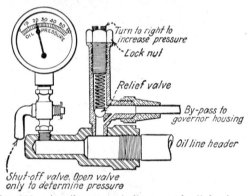

Fig. 302.—An oil-pressure indicator and relief valve.

proper lubrication of the various parts.   Therefore, in different engines, a higher pressure ranging from 10 to 40 lb. is necessary.

In order to maintain the correct pressure and control the quantity of oil circulated, a relief valve (Fig. 302) is connected to the oil-distribution system at some point, usually near the pump-discharge line.   This valve consists essentially of a ball held in place over an opening by an adjustable spring.   The valve operates in such a way that it will permit a certain amount of the oil to by-pass back to the oil reservoir as it leaves the pump.   That is, if the spring tension is decreased, more oil will by-pass, and less will be forced through the system.   If the tension is increased, less oil will go by the valve and back to the reservoir and more will be forced through the system.

291. Oil Gages and Indicators.—Any engine having an enclosed crank-case and using a circulating-splash or an internal force-feed system of lubrication must have two oil indicators as follows:

1. An oil-level indicator.
2. An oil-pressure or circulation indicator.

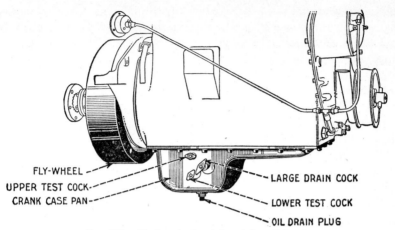

FLY-WHEEL
UPPER TEST COCK
CRANK CASE PAN
LARGE DRAIN COCK
LOWER TEST COCK
OIL DRAIN PLUG

FIG. 303.—Test cocks for determining oil level.

FIG. 304.—The bayonet-type oil-level gage.

The oil-level indicator shows the quantity of oil in the crankcase sump or reservoir and whether it is low, half full, or full, and so on. The usual types are (1) test cocks (Fig. 303) (one for low level and one for high level), and (2) bayonet gage (Fig. 304). In any case the oil level should not be allowed to drop very much below the upper or full point. The lower the level, the less the quantity of oil being circulated and the greater the rate of circulation and absorption of sediment and heat.

The maintenance of the correct pressure in any internal force-feed system is very important. For most engines a pressure of 25 to 40 lb. is recommended. As the bearings and cylinders wear, there will be a slight drop in pressure. Should this become appreciable, it can usually be restored by adjusting the relief valve (Fig. 302). If the drop develops suddenly rather than gradually, the engine should be stopped immediately and the cause determined. The trouble may be:

1. Too thin oil.
2. Lack of oil.
3. Oil too cold or too heavy to flow.
4. Broken pump parts or oil lines.
5. Clogged oil screen or oil line.

An oil-pressure gage of some kind, connected to the distributing lines, is usually placed at some convenient external point to show the operator that the oil is being circulated and to indicate the pressure being maintained. Figure 295 illustrates the common type of gage used. Another type consists merely of a slender vertical plunger which extends upward through the engine housing. As soon as the engine starts and the oil starts circulating, this plunger rises vertically and is held in this position by the oil pressure. If, for any reason, the oil is not circulating or the pressure is low, the indicator fails to rise.

292. Oil Filters.—Some tractors are equipped with an oil filter (Fig. 305), through which all the oil must pass before reaching the bearings and parts to be oiled. Referring to the figure, the oil comes through the inlet passage and fills the space between the outer housing and the felt filtering element. Since it is under pressure, it readily passes through the filtering material, which takes up the sediment and other solid matter. The oil then enters the central passage and passes downward and out to the bearings. These filters should be cleaned periodically by removing the element and washing it in gasoline according to directions.

In cleaning the oil filter shown in Fig. 306, the element is not removed. Instead the drain plug is unscrewed and the engine started and allowed to run until about 2 quarts of oil drain from the filter; then the plug is replaced. The removal of the drain plug allows the oil-flow reversing valve to drop by spring pressure, thus closing the oil passage leading from the pump to the outside of the filtering element. This forces the oil in the opposite direction through the differential safety valve and felt element, carrying with it all the dirty oil that has collected on the felt. When the drain plug is replaced it forces the reversing valve back in place so that the oil will flow through the filtering element normally.

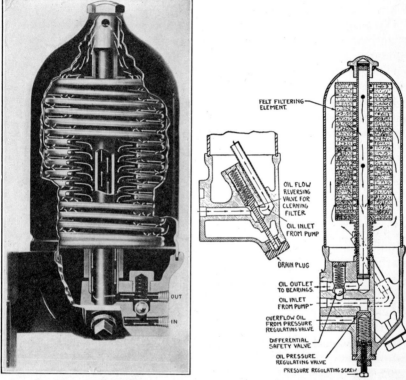

FELT FILTERING ELEMENT

OIL FLOW REVERSING VALVE FOR CLEANING FILTER

OIL INLET FROM PUMP

DRAIN PLUG

OIL OUTLET TO BEARINGS

OIL INLET FROM PUMP

OVERFLOW OIL FROM PRESSURE REGULATING VALVE

DIFFERENTIAL SAFETY VALVE

OIL PRESSURE REGULATING VALVE

PRESSURE REGULATING SCREW

OUT

IN

FIG. 305.—Oil filter with removable filtering element.   FIG. 306.—Oil filter which is drained and cleaned by reversing the oil flow.

293. Engine Lubrication Troubles.—As already stated, the primary requirements for the proper lubrication of an engine under all operating conditions are (1) the design and construction of a dependable system of oil circulation and distribution, and (2) the selection of a lubricant of the correct grade and quality. However, the satisfactory operation of the engine from the standpoint of lubrication is also dependent upon certain other conditions. Some of these are:

1. Periodic and regular changing of the oil.
2. Careful observation and regulation of the oil-reservoir supply and the quantity distributed or fed to the parts requiring lubrication.

3. Prevention of pollution of the oil from any or all of the following sources:

 a. Solid matter of a foreign nature, such as dust, dirt, seeds, particles of vegetable matter, iron and steel particles, and so on, which may get into the reservoir through unprotected openings or by carelessness in handling the oil or working around the engine.

 b. Water from leaks in the cooling system or through unprotected openings.

 c. Liquid fuel which may be taken into the cylinders in an unvaporized condition and, remaining unburned, may get by the piston and rings, diluting the oil.

 d. Particles of carbon due to excess carbon deposits in the combustion space and the piston-ring grooves.

The above discussion applies particularly to enclosed crankcase engines in which there is a continuous circulation of the same oil.

294. Changing Oil.—Regular and periodic changing of the crankcase oil is very important, not only because it may become polluted as outlined above, but because it becomes thinner and less oily, so to speak, owing to exposure to the high engine temperature and the resultant partial decomposition. No set rules can be laid down and applied to any and all types of engines concerning the best procedure to follow. However, the following general suggestions are more or less uniformly applicable:

1. Change the oil in new or recently overhauled vehicle engines after from 100 to 500 miles of running, depending on whether it is an automobile, truck, motorcycle, etc.

2. In used or so-called run-in vehicle engines, change the oil every 500 to 1,000 miles.

3. In new or recently overhauled tractors, change the oil after 10 to 25 hr. of operation.

4. In used or run-in tractors, change the oil after 60 to 70 hr. of operation.

5. Always drain the engine when hot, as the oil is thinner and will drain out quicker and more completely.

6. When changing the oil, flush the crankcase, using only a special flushing oil or a similar thin grade of oil. Do not use kerosene for flushing, because it lodges in depressions and pockets and thins the fresh lubricant. In flushing an engine, put in the required amount of flushing oil and operate engine from 30 sec. to 1 min. before draining it out.

7. Clean or change oil-filter element and pump screen, if removable, at frequent intervals.

295. Oil Consumption.—One of the most vital problems connected with the satisfactory lubrication of any internal-combustion engine is that of oil consumption. A limited or reasonable oil consumption commensurate with satisfactory engine performance under all operating conditions is the goal of the designer and the demand of the user.

Oil is consumed or lost in several ways as follows:

1. By working past the piston into the combustion chamber and being burned.

2. By escaping from the crankcase as a mist or vapor.

3. By leakage.

The important factors affecting the oil consumption of a given engine are:

1. Engine speed.
2. Engine design and changes due to wear.
3. Oil characteristics.

Engine speed is the most important factor affecting oil consumption, as shown by the curves in Fig. 307.[1] Numerous tests as well as actual experience provide conclusive proof of this fact. The reasons for

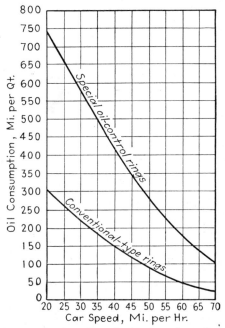

FIG. 307.—Curves showing effect of engine speed on oil consumption.

increased oil consumption at increased engine speeds are (1) the higher oil temperature decreases the oil viscosity; and (2) more oil is thrown on the cylinder walls owing to the greater pumping pressure and the increased centrifugal force of the crankshaft.

The loss of oil that works past the pistons is commonly termed *oil pumping*. A limited loss in this manner can be expected unless the engine is new or nearly so and is in almost perfect mechanical condition. As the period of service lengthens, the oil loss by pumping increases because the piston rings and cylinder become worn, and carbon accumulates in the ring grooves. The use of special oil-control piston rings and the drilling of small holes in the lower ring grooves appear to be the best

[1] *S.A.E. Jour.*, Vol. 29, No. 3. – Hougen

means of reducing oil pumping.   Oil-control rings should not be used in more than the two lowest grooves, and, in general, a single oil-control ring in the lowest groove is sufficient.

Oil viscosity has an important bearing upon oil consumption.   However, the practice of using an oil having a viscosity higher than is ordinarily recommended for a given engine in order to reduce pumping losses, is to be discouraged for the reason that such an oil may prove too heavy to provide proper lubrication, particularly at low temperatures.

Very often two different brands of oil of the same viscosity specifications will vary in their loss characteristics.   This variation is probably due to a difference in volatility.   Regarding the volatility of lubricating oils and its effect on losses, Wilson[1] states as follows:

Volatility of an oil is a quality which depends very largely upon the process of refining the oil.   Even some of the higher priced oils contain an appreciable amount of volatile material.

Since the advent of crankcase ventilation, the matter of volatile material in the oil has become more important because of its effect upon oil consumption. And not all the loss through the ventilating outlet is in the form of vapor, for the draft of air carries along some oil in the form of fog or mist.   The amount so carried away depends upon the driving speed.   The amount of volatile material, which is lost at high speed, is not only the volatile material contained in the new oil, but also the amount resulting from any breaking up of the oil due to the high temperature reached by that which comes in contact with hot surfaces or gases.   Volatility is of little consequence until the oil temperature exceeds 170°F. A motor oil should be nonvolatile at cylinder-wall temperature but volatile at combustion-chamber temperature.

296. Pollution of the Oil.—Foreign matter may find its way into the engine crankcase in a number of ways.   Metal particles are being constantly worn or broken off from bearings and other parts.   Dust and dirt particles will be drawn in through the breather opening if the latter is not provided with an effective screen or filter.   A breather is a necessary part of an enclosed-crankcase engine for the reason that the pistons create a pumping action and, therefore, an uneven pressure, which might otherwise force the oil out through the crankcase joints or upward past the pistons and into the combustion chamber.   The breather also usually serves as the oil-filler opening.

Care should likewise be taken that dirt and small pieces of metal do not drop into the crankcase or engine housing when repairs and adjustments are being made.

The principal objection to this solid foreign matter is that it is apt to clog the oil lines or scratch the cylinder and piston surfaces.

[1] *Univ. Wis., Eng. Exp. Sta., Bull.* 78.

Water that may find its way into the oil is liable to cause rusting and corrosion of the working parts. It also forms an emulsion with the oil and renders the latter less effective. Water accumulating in the crankcase is usually indicated by the rising oil level and its appearance in the bottom of the crankcase when the drain plug is first removed.

Dilution of the lubricating oil by the liquid fuel may take place owing to (1) a poor carburetor setting that permits too rich a mixture to enter the engine, (2) excessive choking in an attempt to make the engine start, particularly in cold weather, and (3) loose, badly worn piston rings that permit the fuel to pass into the crankcase and condense.

297. Carbon Deposits.—A certain amount of carbon and carbon residue eventually forms and becomes deposited on the closed end of the piston, in the ring grooves, and in the cylinder head, as a result of the burning and decomposition of small quantities of lubricating oil in the normal operation of the engine. This carbon is undesirable for several reasons: (1) it often gets very hot and causes preignition of the fuel mixture; (2) it may cause the piston rings to wear rapidly or stick in the grooves; (3) particles work their way into the lubricating oil and contaminate it; (4) particles may get under the valves or cause them to stick or fail to function; and (5) it may accumulate around the spark plug or igniter and result in the failure of the sparking device to ignite the fuel charge.

The use of the correct grade, quality, and amount of lubricating oil and the replacement of worn piston rings and badly worn or damaged pistons and cylinders are the most important requirements in the prevention of excessive carbon deposits.

The frequent removal of carbon from the cylinder head, piston-ring grooves, and other parts is advisable for best results. The best procedure is to take the engine apart and clean the parts thoroughly by scraping. Carbon removal by burning, by means of oxygen and a special torch and apparatus, or by the use of kerosene or special carbon-removing compounds is seldom more than partially effective.

TRACTOR TRANSMISSION LUBRICATION

The parts involved in the transmission of the power from the engine to the rear wheels include the clutch, change-speed gears, differential, and final-drive mechanism. The lubrication of the gearing and bearings involved is most important because of high pressures developed.

298. Clutch Lubrication.—With the exception of the multiple metal-disk clutch, found in the Fordson tractor, all tractor clutches are said to be of the dry type. The Fordson clutch operates in a bath of oil, since it is located in the flywheel housing, which holds the engine oil. The only lubrication needed by a dry-type clutch is in the throwout collar and bearing (Fig. 317).   This is usually taken care of by means of a grease cup or an ordinary oil can.

299. Transmission Lubrication.—Lubrication is necessary in a tractor transmission to eliminate friction and wear in the gear teeth, chains and sprockets, and shaft bearings.   Since all tractor transmissions are enclosed in oiltight, dustproof, cast-iron housings, lubrication is accomplished by maintaining an oil level such that all parts either will operate directly in the lubricant or will have it thrown onto them. Figure 338 shows a typical tractor transmission and its lubrication.

Plugs are provided for putting in fresh oil, draining the used oil, and also for testing or indicating the proper level.

300. Transmission Lubricant.—The instructions of the manufacturer should be followed in the selection of the lubricant to be used in a tractor transmission.   Practically all machines, however, use the regular transmission lubricant recommended for automobiles, trucks, and tractors.

Changing the oil in the transmission is not so important as changing the engine oil.   It is not subjected to such high temperatures and, therefore, retains its body more or less indefinitely.   Likewise, if proper precautions are taken, very little solid matter should find its way into the oil.   Unless the tractor is used almost continuously for tractive work, it is unnecessary to change the transmission lubricant more than once

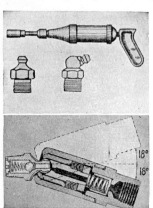

FIG. 308.—Pressure grease gun and fittings showing cross section of nozzle and fitting.

or twice a year.   In exceptional cases, however, where the tractor is used extensively under adverse conditions, it is generally advisable to change it once every month or two. Best results will be secured by draining the oil when the tractor is warm or, if necessary, by warming up the transmission housing thoroughly with a blow torch or similar heating arrangement.

301. Chassis Lubrication.—There are a number of points on any tractor that cannot be reached or conveniently lubricated by the engine or transmission oil but require at least periodic lubrication.   These are found largely on the so-called chassis parts and include axle bearings, front wheels, steering mechanism including steering rods, knuckles, arms, pins and spindles, and control-lever bearings.   The water pump, clutch-shifting collar, fan, and similar accessories are usually lubricated in the same way.

The usual procedure in chassis lubrication is to apply a semifluid lubricant or a soft grease to the bearing by some compression device. In the earlier tractors, compression grease cups (Fig. 286) were used entirely.   Eventually, however, the cups would get lost, the passages would become plugged up, or the operator would neglect to keep the cups filled and screwed down. Therefore, the compression grease cup is being discarded to a large extent and being replaced by a high-pressure grease-gun system that is more convenient and positive because of the high pressure obtainable.

The high-compression greasing system requires a fitting of some kind at each point to be lubricated. A grease gun (Figs. 308 and 309), when applied to the fitting, forces the lubricant under great pressure to the bearing.

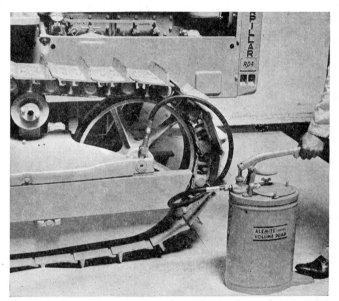

Fig. 309.—Method of lubricating track roller and other bearings of track mechanism.

Front-wheel bearings are frequently lubricated by filling the hub caps with grease as in automobiles.

As a rule, chassis parts should be lubricated at least once daily.

References

American Lubricating Greases, *Va. Poly. Inst., Eng. Ext. Div., Bull.* 35.
A Study of Engine Oil Filters, *S.A.E. Jour.*, Vol. 25, p. 625.
CROSS: "Handbook of Petroleum, Asphalt and Natural Gas."
HELDT: "The Gasoline Automobile," Vols. I and IV.
LOCKHART: "American Lubricants."
Lubricants and Fuels for Tractor and Motor Vehicle Engines, *Agr. Eng.*, Vol. 11, No. 7.
Lubricating Oils for Kerosene Burning Tractors, *Trans. Amer. Soc. Agr. Eng.*, Vol. 11.
Oil and Gasoline Information for Motorists, *Univ. Wis., Eng. Exp. Sta., Bull.* 78.
"Symposium on Motor Lubricants," *Amer. Soc. Testing Materials, Bull.*
"The Significance of Tests of Petroleum Products," *Amer. Soc. Testing Materials, Bull.*
THOMSEN: "Practice of Lubrication."

CHAPTER XXII

CLUTCHES

Some means of disconnecting the power unit from the transmission gears and drive wheels of a tractor are necessary because (1) the internal-combustion engine as used in a tractor must be cranked manually or by a special starting mechanism; (2) this type of engine must attain a certain speed before it will have any power; (3) shifting of the transmission gears must be permitted for the purpose of securing different traveling speeds; and (4) stopping the belt pulley must be permitted without having to stop the engine. All these can be taken care of by placing a clutch between the engine and the transmission gears and belt pulley.

302. Clutch Location.—The location of the clutch depends upon the general layout of the tractor. From Fig. 310*A*, it will be seen that if the engine is placed crosswise of the frame, the clutch is built in the belt pulley on one end of the crankshaft. Figure 310*B* shows that if the engine is placed lengthwise, the clutch is just behind the flywheel on the rear end of the crankshaft and the belt pulley is gear driven.

303. Clutch Requirements.—A satisfactory tractor clutch must fulfill the following requirements: (1) It should not slip, grab, or drag; and, (2) it should be convenient, accessible, and easy to operate, adjust, and repair.

TYPES OF CLUTCHES

Tractor clutches may be classified as follows:

1. Contracting band.
2. Cone.
3. Expanding shoe.
4. Multiple disk.
    a. Dry disk (lined).
    b. Oil type (unlined).
5. Single plate.
    a. Hand-lever operated.
    b. Foot operated.
6. Twin disk.

Of the types listed above the first two were used to a limited extent in early machines but are now obsolete. Likewise, the multiple-disk clutch has been practically superseded by the single-plate and twin-disk types. Disk-type clutches seem more nearly to meet the requirements of a satisfactory clutch as previously mentioned.

304. Contracting Band.—The contracting-band clutch (Fig. 311) is a very simple device, consisting essentially of a lined steel band that fits around a flange projection on the flywheel. One end of the band is

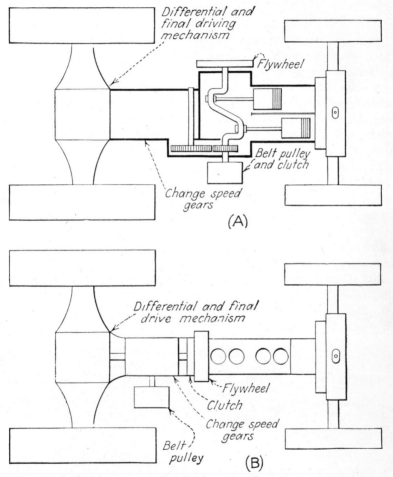

Fig. 310.—Conventional tractor transmission layouts: (*A*) engine crosswise of the frame; (*B*) engine lengthwise of the frame.

fastened to a pivoted arm, whose inner end is moved outward from the shaft center by a sliding cone. This causes the band to contract about the flange and turn with the flywheel. Since the band, arms, cone, and complete clutch assembly are fastened rigidly to a shaft carrying the belt pulley and some of the transmission gears, the engagement of the clutch connects the engine with the pulley and the transmission. A threaded bolt with nut, which supports one end of the band as shown,

Fig. 311.—Contracting-band clutch.

provides a simple means of adjustment to overcome wear and slippage.

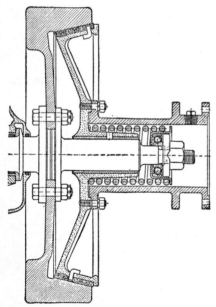

Fig. 312.—Cone clutch.

The ordinary asbestos-fabric brake-band lining is used for this clutch.

305. Cone Clutch.—The principal member of the cone-clutch assembly (Fig. 312) is a metal disk with a conical peripheral surface that engages with a similarly shaped recess in the flywheel. The cone is faced with ordinary brake lining. A heavy spring, placed behind the cone, exerts sufficient pressure to insure its positive engagement. A sleeve fastened to the cone extends back and is bolted to the transmission shaft. Therefore, engagement of the clutch connects the gears with the engine. Disengagement is produced by sliding the cone and sleeve backward on the flywheel shaft extension against the spring pressure.

306. Expanding-shoe Clutch.—The expanding-shoe clutch (Fig. 313) consists of two or more iron shoes, which pivot on pins attached to a cast-iron spider that has as many arms as there are shoes. The spider-and-shoe assembly is fastened rigidly to the transmission shaft. The shoes

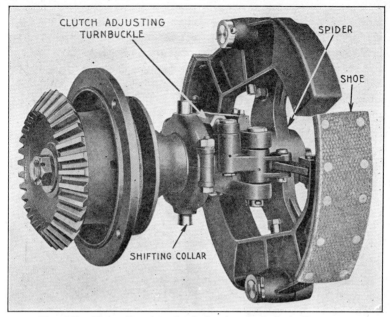

Fig. 313.—Expanding-shoe clutch showing method of adjustment.

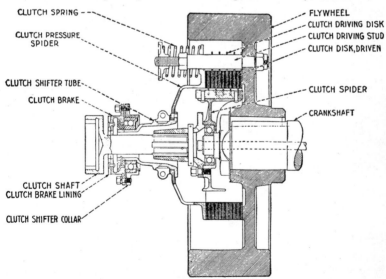

Fig. 314.—Multiple-disk clutch.

are connected to a sliding clutch collar by means of a bell crank and lever mechanism so that in engaging the clutch this collar pushes the shoe faces against the inner rim surface of the flywheel, thus connecting the engine with the transmission.

The adjustment of this clutch involves shortening or lengthening the connection between the sliding collar and the shoe. The two-shoe clutch (Fig. 313), has a turnbuckle that permits adjusting both shoes at the same time and therefore insures equal adjustment.

307. Multiple-disk Clutch.—A multiple-disk clutch (Fig. 314) consists of a number of thin metal plates, at least five or more, arranged alternately as driving and driven disks. One set is attached to the flywheel and the other to the clutch shaft and transmission. If the plates are firmly pressed together the clutch is said to be engaged and power is transmitted. This pressure is secured by means of a housing and a set of heavy springs as shown. The clutch throwout collar is attached to the rear part of the housing so that the depression of the operating lever

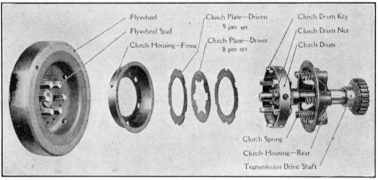

Fig. 315.—Multiple-disk clutch which runs in oil. (*Courtesy of Ford Motor Company.*)

slides the housing backward and compresses the springs. This enlarges the plate space and permits one set to rotate free and independent of the other set.

In the dry-type multiple-disk clutch (Fig. 314) the driven plates are faced on each side with friction material. The only part requiring lubrication is the throwout collar. The plates should be kept clean and dry. If slippage develops due to wear, it can be overcome by increasing the tension of the three springs by tightening up on the nuts.

The clutch shown in Fig. 315 is a multiple-disk type that runs in oil. The thin steel plates are perfectly smooth and unlined. The eight driving plates are notched to fit the six square flywheel studs and therefore always turn with the engine. The nine driven plates are notched to fit in the clutch drum, which is fastened to the transmission shaft. When the two halves of the housing are bolted together, the springs tend to force the entire housing backward on the shaft. Since the drum is fixed, this action pinches the plates firmly together between it and the front half of the housing, and the entire assembly turns as a unit. When the clutch foot lever is depressed, it slides the entire housing forward and compresses the springs. This widens the plate space between the front half of the housing and the drum and permits the two sets of plates to slip past each other and turn independently. No adjustment is provided. Owing to the fact that the entire mechanism operates in oil, very little wear occurs. If slippage develops, it will be due to weak springs or badly worn plates. In such cases replacement of these parts is necessary. If the clutch sticks or does not release easily, it is due to bent plates or gum from the lubricating oil. The only remedy is to take the clutch apart, clean the plates, and replace those which are bent or worn.

308. Single-plate Clutch.—The single-plate clutch (Fig. 316) is a disk-type clutch in which a single, thick, iron plate, faced with friction material on both sides, serves as the driven member. It engages directly

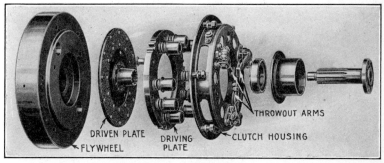

Fig. 316.—International single-plate clutch.

with the flywheel on one side and with an unlined iron driving plate on the other. The pressure produced by a number of springs, located between the driving plate and the housing, which is bolted to the flywheel, holds the friction surfaces firmly in contact. Three arms are hinged to the housing and have their outer ends connected to eyebolts that screw into the driving plate. When the inner ends of the arms are pushed toward the flywheel by the throwout collar, the driving plate is pulled away from the driven member and the clutch is thereby disengaged. This clutch does not have any means of adjustment.

Unlike the one just described, the single-plate clutch (Fig. 317) does not have a separate housing member, but the thick iron driving plate serves as such. A number of heavy coil springs exert a strong pressure that holds the plates firmly in engagement against the flywheel. The releasing action consists of sliding the driving plate backward on the clutch shaft and the pins that bear the springs. This compresses the latter and releases the pressure on the driven disk.

The single-plate clutch, shown in Fig. 318, differs in certain respects from those previously described. Instead of using spring pressure to produce positive plate contact, it utilizes short bell-crank levers that press the plates together. Instead of a foot pedal, a hand lever connected to the throwout collar actuates the bell cranks and thereby releases or engages the clutch. Unlike a foot-operated spring-compressed clutch, this clutch when disengaged will remain so without holding the lever.

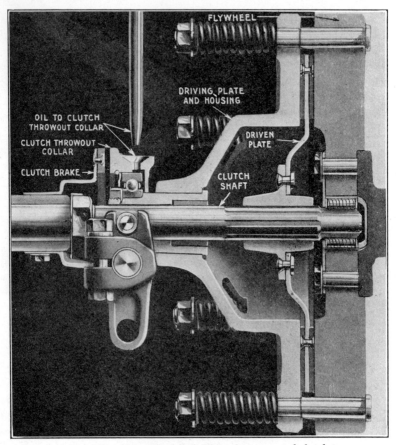

Fig. 317.—Caterpillar single-plate, foot-operated clutch.

309. Twin-disk Clutch.—The twin-disk or three-plate type of clutch resembles, in certain respects, the single-plate clutch but differs in that the flywheel rim does not serve as a friction surface. Instead, as shown in Fig. 319, there are three separate plates: a lined driving plate pinned to, and rotating with, the flywheel and two unlined driven plates attached to the clutch shaft. The driving pressure in all clutches of this type is obtained by bell cranks and a sliding cone-shaped collar rather than by heavy springs. Some light springs placed between the driven disks are used merely to insure the prompt release of the disks when disengagement is desired. The clutch fingers or bell cranks are attached to the threaded adjusting collar. Turning this collar to the right brings the fingers closer to the movable clamping disk and thus compensates for any wear of the lining. The spring-controlled lock-pin fits into any one of a series of closely spaced holes and provides for any amount of adjustment desired.

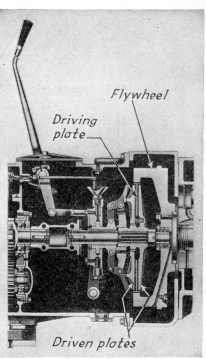

FIG. 318.—Caterpillar, hand-lever-operated, single-plate clutch.

310. Belt-pulley Single-plate Clutch.—Figures 320 and 321 are sectional and disassembled views, respectively, of a single-plate-type clutch as built into the belt pulley of a two-cylinder horizontal engine placed crosswise of the tractor frame. A heavy cast-iron plate C is fixed on the end of the crankshaft, whereas the belt pulley with its attached spur gear is free to turn on and independently of the crankshaft. An outer cast-iron plate A is attached to the belt pulley by means of three bolts that extend through to the gear side of the latter and are connected to the bell-crank arms, which in turn are attached to a sliding collar and hand lever. This outer plate fits loosely on the three bolts and has a limited lateral movement. Two heavy asbestos-composition rings B and D serve as friction facings. One is located between the outer plate A and the fixed plate C, and the other is between plate C and the belt pulley. As shown by Fig. 320, the engaging action consists of drawing the plate members and friction disks together against the pulley so that the entire assembly rotates as a unit. Three light springs, placed between the pulley and the outer plate A, provide instant release and prevent dragging when the operating lever is in the disengaged position. The adjustment of this clutch involves screwing up an equal amount on the three nuts that hold the outer plate A in place. It should be so adjusted that a fairly hard pull of the hand lever is required to snap the clutch into engagement.

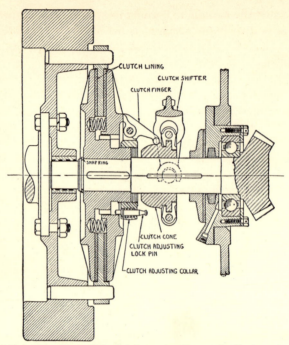

Fig. 319.—Twin-disk-clutch construction and adjustment.

Fig. 320.—Belt-pulley-type, single-plate clutch.

Eventually, when the two liners become too thin, further adjustment will be impractical and their replacement is necessary.

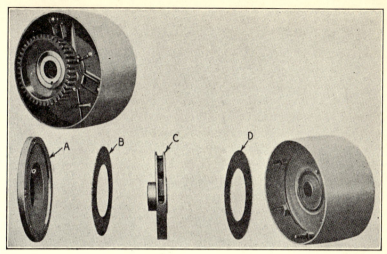

Fig. 321.—Belt-pulley, single-plate clutch showing driving plate *C*, driven plate *A*, and friction rings *B* and *D*.

311. Clutch Brakes.—When a clutch is disengaged for the purpose of shifting the transmission gears, the driven member may continue to rotate owing either to the slight but normal friction between the parts or to a slight dragging caused by incomplete disengagement of the driving and driven members. Since some of the gears are on the clutch shaft, any attempt to slide them into mesh with the other transmission gears results in a clashing action, which either prevents their engagement or, if they are forced into mesh, may produce wear or even break the teeth.

In order to stop the clutch shaft the instant disengagement occurs, thereby permitting immediate shifting of the gears without danger of breakage, most clutches are provided with a brake of some sort that engages the driven member when the clutch lever is moved to the disengaged position. For example, Fig. 317 shows the location of

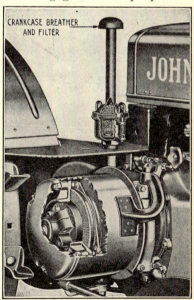

CRANKCASE BREATHER AND FILTER

Fig. 322.—Belt-pulley, plate-type clutch showing clutch and pulley brake.

such a brake for a single-plate clutch.　For belt-pulley-type clutches the brake is attached as shown by Fig. 322, so that it engages the face of the pulley when the clutch is released and instantly stops it, thus permitting the gears to be shifted without clashing.　It also stops the pulley quickly in doing belt work.　This type of brake is usually adjustable for wear.

312. Clutch Linings.—Clutch linings for dry-type clutches are either woven or molded.　In the manufacture of the woven facing, long-fiber asbestos, together with a small percentage of cotton, is twisted around fine brass wires to form threads.　These threads are then woven into a straight band of the proper width, impregnated with a suitable compound to act as a binder, and then formed into rings.

The molded type of facing is made of short-fiber asbestos pressed into a hard sheet.　The facing rings are then cut by a die into rings, which are impregnated with a suitable binder and cured.　After curing, the rings are ground down to size, drilled, and countersunk.　The molded facing is the predominating type.

313. Clutch Troubles and Adjustments.—The modern tractor clutch, if given ordinary care and attention, will seldom fail to perform satisfactorily.　Any abnormal or unusual action is nearly always due to one of the following causes:

1. Failure to keep clutch fully engaged, due to riding the clutch pedal or permitting hand lever to remain only partly in the engaged position.
2. Failure to keep clutch adjusted to compensate for normal wear.
3. Permitting oil and grease to get on friction surfaces.
4. Failure to lubricate the throwout collar and bearings as directed.

Slippage is the most common clutch trouble and develops from one or more of the first three causes just named.　It is immediately indicated and detected by the tendency of the engine to speed up, when the clutch is engaged and a load is applied, without apparently exerting any appreciable tractive or belt power.　Naturally, if for any reason the clutch surfaces are not firmly pressed together, there will be some slippage and wear.　If this condition exists for any length of time, it is apparent that undue wear, heat, and other effects will develop, which will produce slippage even with the lever fully engaged.　The precaution, therefore, is always to keep a clutch either completely disengaged or completely engaged.　If it is desired to let the engine run with the tractor standing still, the best practice is to put the gears in neutral and engage the clutch. Leaving the gears engaged and holding the clutch disengaged causes heating and wearing of the throwout collar and may wear the clutch facings.

A dragging clutch is one which does not completely disengage.　It is indicated by failure of the tractor or pulley to stop, and clashing of the gears when shifting.　A dragging clutch is usually caused by incorrect adjustment of the clutch or clutch lever, or both.

The methods of adjusting the various types of clutches have already been mentioned. Before making any adjustment it is suggested that the clutch construction and operation be thoroughly understood and the manufacturer's directions followed carefully.

References

ELLIOTT and CONSOLIVER: "The Gasoline Automobile."
HELDT: "Motor Vehicles and Tractors."

CHAPTER XXIII

TRACTOR TRANSMISSIONS—DIFFERENTIALS—FINAL DRIVES —ACCESSORIES

The mechanism involved in transmitting the engine power to the drive wheels or tracks of a tractor ordinarily includes three distinct parts, namely: (1) the change-speed gears, (2) the differential, and (3) the final-drive mechanism. There are some exceptions to this as will be pointed out later. Unlike the automobile, there is considerable variation in the construction, arrangement, and operation of these three units in

Fig. 323.—Steam-tractor transmission gearing.

the different makes of tractors. In general, a tractor transmission must serve the following purposes:

1. It must provide a means of self-propulsion at different road speeds with the same engine speed.

2. It must provide the proper speed reduction between the engine crankshaft and the traction members.

3. It must provide for an equalization of the power transmission to the traction members on both gradual and short, quick turns.

4. It must provide a means of reversing the direction of travel.

314. Change-speed Gears.—The steam tractor has certain distinct advantages over the gas tractor with respect to transmitting the power

306

for traction because of the great speed and corresponding power variation of the engine and the ease with which it can be reversed. In other words, both the rate and the direction of travel are readily controlled by the speed and direction of rotation of the engine and only a single train of gears (Fig. 323) is needed. On the other hand, it is desirable to operate the gas-tractor engine at a fairly uniform speed and to reverse it is imprac-

Fig. 324.—Early type of gas-tractor transmission.

tical. Therefore, the necessary change in rate and direction of travel must be taken care of in the transmission itself.

The early gas tractors were provided with rather heavy cast-iron exposed gears, similar to those on steam engines. Provision was made for two forward speeds and reverse as noted in Fig. 324. Obviously, these transmissions were heavy, noisy, and subject to rapid wear.

With the introduction of the lighter tractors, certain improvements were incorporated in them such as (1) machined and hardened steel gears, which provided the same strength with less weight, (2) partial or complete enclosure of gears in an oil- and dust-tight housing, (3) antifriction instead of plain bearings, and (4) three to six forward speeds. Today these features are found in nearly all tractor transmissions whether large or small.

As in automobiles, the different gear changes in tractors are designated as either low, intermediate, or high, or as first, second, third, and so on. However, the fundamental reasons for having three or more speeds are not necessarily the same. In the case of an automobile, there is not nearly so great a speed-reduction ratio between the engine and the rear axle when it is in high gear as there is in a tractor. Therefore, if an attempt were made to start an automobile directly in high gear, either the engine would stall, or the car would be subjected to too great a jerk or strain. Consequently, the machine is gradually brought to a certain speed by means of the low and intermediate gears and finally shifted into high without producing undue strain on the engine, gears, or other parts involved. In other words, the primary function of the low and intermediate gears in an automobile is to permit a gradual rather than a sudden speed acceleration. Incidentally, these low-speed gears also provide the gear reduction necessary under unfavorable road conditions, such as mud, sand, and hills. Table XV shows the travel speeds of representative wheel and track-type tractors.

TABLE XV.—GEAR CHANGES AND RATES OF TRAVEL OF TRACTORS

Tractor	Rate of travel, m.p.h.					
	First gear	Second gear	Third gear	Fourth gear	Fifth gear	Sixth gear
Allis-Chalmers WC............	2.5	3.5	4.8	9.0		
Case Model CC...............	2.6	3.7	5.1			
Caterpillar 22.................	2.0	2.6	3.6			
Caterpillar D4................	1.7	2.4	3.0	3.7	5.4	
Caterpillar D8................	2.0	2.8	3.3	3.8	4.6	6.2
Farmall 20....................	2.3	2.8	3.3	3.8		
John Deere A.................	2.3	3.0	4.8	6.3		
McCormick-Deering T-40.......	1.8	2.3	2.8	3.3	4.0	
Minneapolis-Moline Z..........	2.2	2.6	3.1	4.8	14.3	
Oliver Row-Crop 70...........	2.4	3.3	4.3	5.9		

In tractors, even in high gear, the engine-to-rear-axle speed ratio is rather high, and the machine travels at a very slow rate as compared to an automobile in high gear. Therefore, the machine will start off directly in high gear without first going through low and intermediate. In tractors the speed changes are provided primarily for handling different drawbar loads, or to permit a certain field machine to be drawn along at the correct speed. In general, the low gear is for extremely heavy loads or for very slow field speeds. The second and third gears are probably the ones that are used most for plowing, harrowing, and similar work.

A number of tractors have four forward speeds and some have as many as five or six. For wheel tractors having four or five speeds, the high speed is usually too high to permit any appreciable pulling effort and is used primarily for faster traveling over highways or in going to and from the field.

The principle involved in securing these different speeds is to bring into mesh, with each other, certain gears having different numbers of teeth. For example, in a tractor the ratio of the engine speed to the rear-axle speed may be 25:1 in high gear and 30:1 in intermediate. This means that in the one case certain smaller gears, known as driving gears in the change-gear set, mesh with other driven gears having a greater number of teeth. In the second case, these same small driving gears are meshed with other driven gears having a still greater number of teeth. Consequently, the speed is reduced proportionately.

As previously stated, the change-speed gears, as well as all other transmission parts of tractors, are now made of hardened steel and are completely enclosed and run in a bath of oil. The change-gear set is usually so located in the transmission train that the power first passes through it from the clutch, and thence through the differential to the final-drive mechanism and wheels. The different speed changes are made by sliding certain gears into or out of mesh with other gears having different numbers of teeth. In order to slide the gears and at the same time make them turn with the shaft, the latter is splined, that is, it has a number of ribs running lengthwise and cut integral with it. The gears have corresponding grooves in the hub to receive these splines. The splines and grooves are carefully machined and just enough clearance is allowed to permit the gear to slide freely when lubricated. The spline shaft and gears are clearly shown in most of the transmission illustrations in this chapter. Such transmissions are known as the selective sliding-gear type.

In reversing the direction of travel of a tractor when the engine cannot be reversed, it is necessary to interpose an additional gear, known as a reverse idler gear, in the change-gear set. By transmitting the power around through this gear, the direction of rotation of the countershaft, differential, and final-drive mechanism is reversed and the machine moves backward. This reverse gear is in constant mesh with some other gear and therefore turns at all times.

Figure 325 illustrates how the different speed changes are secured and the power is transmitted in a three-speed gear set. Figures 326 and 327 likewise illustrate very clearly other three-speed transmission layouts.

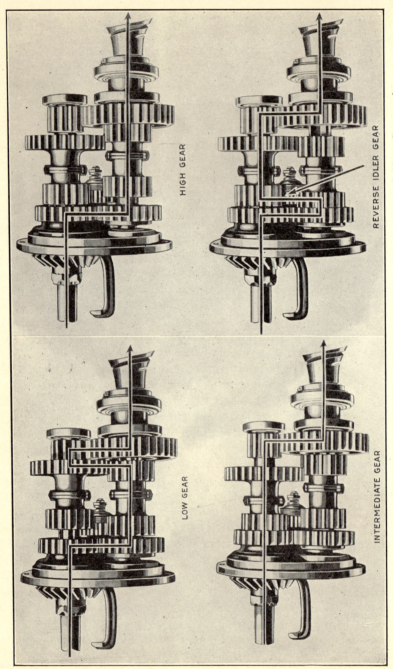

HIGH GEAR

REVERSE IDLER GEAR

LOW GEAR

INTERMEDIATE GEAR

Fig. 325.—Path of power through a three-speed transmission. (*Courtesy of Ford Motor Company.*)

Fig. 326.—Three-speed, all-purpose tractor transmission.

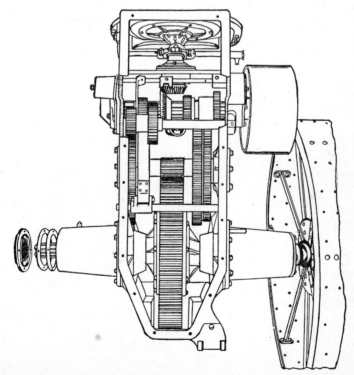

Fig. 327.—Three-speed, general-purpose tractor transmission.

In each case the engine crankshaft is at right angles to the rear axle, and a bevel gear and pinion are necessary at some point in the assembly.

FIG. 328.—Four-speed transmission used in John Deere tractor.

In the one case (Fig. 327), it will be noted that this pinion is directly

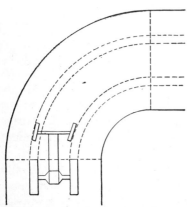

FIG. 329.—Sketch showing effect of turning on rear-wheel travel.

behind the clutch and meshes with a larger bevel gear on the spline shaft. In the other case (Fig. 326), the spline and gear-set shafts are kept parallel to the crankshaft, and the bevel gears come between the change-gear set and the differential gear. Figure 328 illustrates a transmission in which bevel gears are unnecessary, because the engine crankshaft is parallel to the rear axle.

315. Differentials.—A differential is a special arrangement of gears so constructed and located in the transmission system of an automotive machine that it will permit one driving member to rotate slower or faster than the other and at the same time propel its share of the load. For example, referring to Fig. 329, it is quite evident that, in making a turn, the outside wheel of an automobile or tractor must travel farther and therefore turn faster than the inside wheel. If some special device were not provided to permit this unequal travel and at the same time equalize the pull, obviously slippage, excessive strain, and abnormal wear would result.

316. Differential Construction.—The construction and operating principles of the more common bevel-gear-type differential are best understood by referring to Fig. 330. The main drive pinion B meshes with and drives the large gear C. A differential housing E is bolted rigidly to one side of gear C. This housing may be solid or split and carries one or more studs F, upon each of which is mounted a differential pinion G, which is free to rotate on the stud.

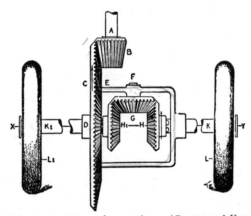

Fɪɢ. 330.—Differential construction and operation. (*Courtesy of Vacuum Oil Company.*)

The two halves of the axle or shaft to be driven, namely, K and K_1, are inserted, one from each side as shown, and bevel gears H and H_1 placed on the ends. These two gears and their respective shaft ends are splined or keyed in such a manner that they must turn together. At the same time, gears H and H_1 mesh with differential pinion G. Main drive gear C and its attached housing E are free to rotate as an independent unit about the shafts K and K_1.

317. Differential Operation.—1. *First condition* with wheels L and L_1 fastened to outer ends of shafts K and K_1, respectively, and entire mechanism raised up so that the wheels are clear and free to rotate: Power is received by the drive pinion B and in turn is transmitted to gear C turning it in the direction indicated. This likewise rotates the housing E in the same direction. Since the wheels L and L_1, shafts K and K_1, and gears H and H_1 are free to move, they rotate in the same direction also; that is, the entire mechanism rotates as one unit. The differential pinion G does not turn on its stud F because the bevel gears H and H_1 are turning at the same rate and, therefore, lock pinion G. Consequently, with this condition existing, that is, with an equal resistance applied to both wheels, the latter will turn at the same rate and the tractor will move straight ahead.

2. *Second condition* with one wheel locked and the other clear and free to rotate: Suppose wheel L is resting on a rough, firm surface and wheel L_1 is raised and is free to rotate. Power is again applied to large gear C from pinion B. This again rotates housing E as indicated and carries pinion G around with it. But since wheel L and its shaft K are subjected to much greater resistance than wheel L_1 and its shaft K_1, L, K, and gear H remain stationary. Therefore pinion G, as it is carried around by the housing E, is also forced to turn on its stud F and, in so doing, causes gear H_1, shaft K_1, and wheel L_1 to rotate in the same direction as the drive gear and housing. Thus the pinion G, in making one revolution with the housing E, also makes one complete revolution on its stud by rolling on the stationary gear H. The axle gear H_1 is thus subjected to two rotative actions—one revolution due to its being in mesh with the differential pinion G, which has been bodily revolved about the axis XY, and the other due to the rotation of the pinion G on its stud as it is rolled once around the stationary axle gear H. The free wheel thus makes two complete forward revolutions while the drive gear and housing are making one revolution. This is the condition existing which causes one wheel of an automobile or tractor to spin while the other remains stationary when there is unequal resistance applied to them as in the case of one being in soft mud and the other on firm footing.

In a similar manner it will be observed that any difference in the rotation of the wheels is compensated for by the rotation of the differential pinion G on its stud F while also revolving with the entire housing about the axis XY. Any rotation of this differential pinion on its stud means that it rolls on one of the axle gears, and the amount of motion in rolling on one gear is transmitted to the other as additional turning and driving effort. Any retarded motion of one wheel results in accelerated motion of the other. The power and driving force are thus transmitted to the wheels in proportion to the distance each must travel.

318. Spur-gear Differential.—A spur-gear-type differential is used in a few cases but is not so common as the bevel type, probably because of difficulty in making it as compact. It does eliminate the spreading action or side thrust which is always present with bevel gearing.

As the name implies, a spur-gear differential uses spur-differential pinions and axle gears instead of bevel pinions and gears. Otherwise it works in exactly the same manner.

Referring to Fig. 331, the differential pinions are arranged in pairs and rotate free on their respective spindles attached to the housing. The two paired pinions mesh with each other and one in each pair meshes with one of the spur axle gears while the other meshes with the opposite axle gear. With equal resistance applied to both axle shafts and wheels, the entire assembly will rotate as a unit, but the pinions will remain stationary on their studs. On the other hand, if one wheel is held stationary, the spur pinions meshing with this axle gear will roll around on it, and therefore will be rotated on their studs. They, in turn, will drive the other set of pinions meshed with the other axle gear and the latter will be rotated. Therefore, the one axle gear and shaft will be rotated by the rotation of the entire housing and also by the pinions turning on their studs.

Fig. 331.—Spur-gear-differential construction.

The main driving gears, regardless of the type of differential, may be of either the spur, bevel, or worm type. In fact, in automobiles and trucks the main differential gear, often called the ring gear, is usually a spiral spur gear. Some trucks also use a worm gear, while most tractors use a spur gear.

319. Differential Location.—In some tractors the differential is located on the rear axle the same as in all automobiles, as shown by Fig. 332. In such cases both drive wheels are fastened to the axle halves, and the latter transmit the power to the wheels in addition to supporting the machine. This is known as live-axle construction.

Referring to Fig. 333, it is noted that the differential is located on the countershaft in front of the rear axle and that the latter is of one piece. In this case the power is transmitted directly to the wheels by means of a heavy internal spur gear attached to each as shown. The wheels themselves turn free and independent of each other on the axle shaft, which serves primarily as a support. This is known as dead-axle construction. The arrangement is but little used at present.

A third-type of construction, as illustrated by Fig. 328, combines the two arrangements just described. The differential is on a short countershaft that carries two spur pinions, one on each side of the assembly. These pinions mesh directly with two larger spur gears on the inner ends of the two-piece axle. The wheels are keyed to the outer ends of the axle shafts so that the latter transmit the power to the wheels.

Thus, each wheel has its own driving gears and axle shaft and can rotate independently.   This again is a live axle.

Figure 335 illustrates this same type of differential location and power transmission, but roller chains and sprockets are used instead of gears.

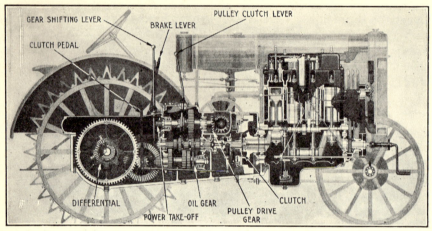

Fig. 332.—Sectional view of McCormick-Deering general-purpose tractor.

320. Final Drives.—The illustrations accompanying this chapter show quite clearly the common types of final drives for tractors, that is, the means by which the power is transmitted finally to the rear axle and

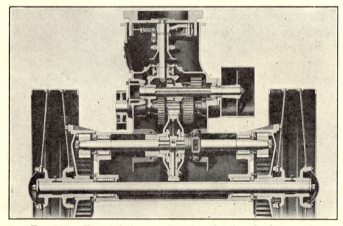

Fig. 333.—Transmission construction showing dead rear axle.

wheels.   With the differential located on the rear axle as just described, the final drive usually consists of a heavy spur pinion on the countershaft meshing with a large spur differential gear as shown by Figs. 327 and 332.

Figure 336 shows the Fordson tractor transmission using a worm-and-worm-wheel final drive.   This permits securing the proper speed reduc-

BOLT HOLES
SLOTTED TO PERMIT
ADJUSTING CHAIN TENSION

Fig. 334.—Chain-and-sprocket final-drive mechanism showing method of adjusting chain tension.

tion without the use of a very large diameter gear.   In other words, the gear and housing are smaller and more compact.   A worm drive is also

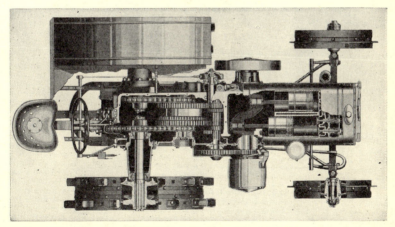

Fig. 335.—Top view of John Deere chain-and-sprocket-drive transmission.

quieter in operation.   A steel worm with a bronze wheel is used.   There are certain objections to a worm drive such as (1) greater friction with its accompanying heating effect, wear, and power loss; and (2) difficulty may be encountered in dragging the tractor, because the pitch of the worm may not be such as to permit the worm wheel to turn it. In other words, with a low-pitch worm, the power transmission is not always

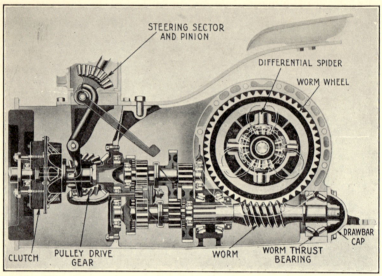

Fig. 336*A*.—Side view of worm-and-worm-wheel final-drive mechanism. (*Courtesy of Ford Motor Company.*)

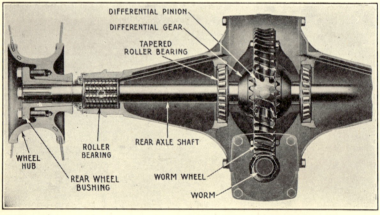

Fig. 336*B*.—Worm-and-worm-wheel final drive showing rear-axle construction. (*Courtesy of Ford Motor Company.*)

reversible. For a worm-and-worm-wheel drive, the gear reduction is equal to the quotient obtained by dividing the number of teeth in the worm wheel by the number of worm threads. A worm drive when subjected to heavy pulls as in trucks and tractors always generates considerable heat due to friction. Therefore, careful attention to lubrication, both as to kind and to correct amount of oil, is very important.

Figures 334 and 335 illustrate final drives by sprocket wheels and roller chain. This type of drive provides a limited amount of flexibility and thereby relieves the engine, clutch, and other parts of excessive shock under certain conditions. A chain final drive may also simplify the transmission construction by eliminating extra shafts, gears, and bearings, which would otherwise be required to give the proper speed reduction and carry the power back from the change-speed gears to the rear axle.

The principal objection to a chain drive is the tendency toward looseness due to wear and slight stretching. As will be noted, these drives are well enclosed and run in a bath of oil, so that once the chain is "broken in" it will wear very little and therefore slacken very slowly. Provision is made in most cases to take up the chains from time to time, as necessary. Of course, a small amount of slack should always be allowed. Otherwise the wear and power loss are apt to be greater than with the chain too loose.

SPECIAL TRANSMISSION TYPES

321. All-purpose Tractor Transmissions.—With the development of the all-purpose tractor and its adaptation to row-crop production, certain variations in tractor-transmission construction and operation have necessarily appeared. The need for sufficient clearance and a certain wheel tread are the principal reasons for these variations. For example, some machines locate the differential in the center of a countershaft, as shown by Figs. 326 and 337. This countershaft, with its housing, extends outward to each side and carries spur gears that drive the main wheels. The latter are mounted on separate stub axles bearing a larger spur gear. Then this final-drive mechanism is enclosed in a large, flat, vertical housing that also serves as a sort of bracket to give the tractor sufficient clearance without the need of unusually high wheels.

Figure 338 illustrates the use of high wheels to secure clearance. Here we have a typical live-axle drive with the wheels mounted directly on the differential shafts.

322. Four-wheel-drive Transmission.—The Massey-Harris all-purpose tractor is one of the few wheel-type machines of recent development that utilizes all four wheels for driving. This feature provides better traction than two drive wheels but involves a somewhat more complicated construction, in order to permit steering the machine with the same wheels that are driving it and to supply a certain degree of flexibility of the traction members necessary in traveling over uneven ground.

Referring to Fig. 339, the power is transmitted from the engine to the change-speed gears located just behind the front axle. From these

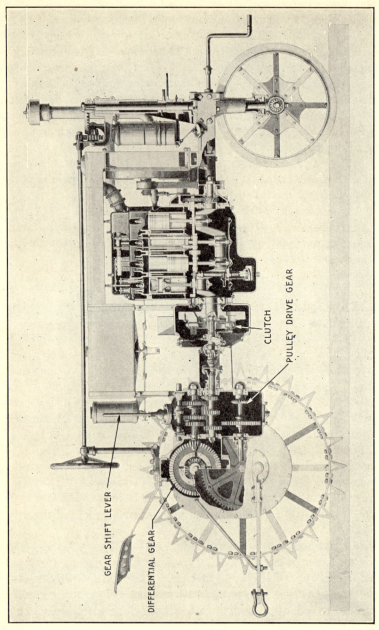

GEAR SHIFT LEVER

DIFFERENTIAL GEAR

CLUTCH

PULLEY DRIVE GEAR

FIG. 337.—Sectional view of McCormick-Deering Farmall tractor.

gears the power is transmitted to the two sets of differential gears and axles. Flexibility is provided for by constructing the housing that encloses the drive shaft leading to the rear axle in such a manner that the differential and axle assembly can rotate a certain amount about this drive shaft as an axis.

Fig. 338.—Sectional view of Oliver Row Crop "70" tractor.

Steering is accomplished through the front wheels by mounting them with heavy cast-iron brackets and spindles and inserting a universal joint at the outer end of each of the two halves of the axle shaft (Fig. 340) to permit power to be transmitted positively, even when the wheels are angled.

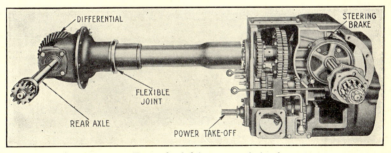

Fig. 339.—Four-wheel-drive transmission layout.

Two hand-operated brakes on the front-axle halves (Fig. 339) permit quick, short turns.

323. Track-type Tractor Transmissions.—The transmissions for the track-type tractors are not unlike those in wheel machines except that the steering mechanism is incorporated in them. That is, the ordinary wheel tractor is guided by means of the front wheels, whereas the conventional track-type tractor, having but two traction members, must be both propelled and guided through the latter.

Fig. 340.—Four-wheel-drive transmission and axles showing universal joint to facilitate steering.

324. Caterpillar Transmission.—Figure 341 shows the change-gear set for the Caterpillar tractor. The power is transmitted to a countershaft (Fig. 342), by means of bevel gears, thence, through two steering

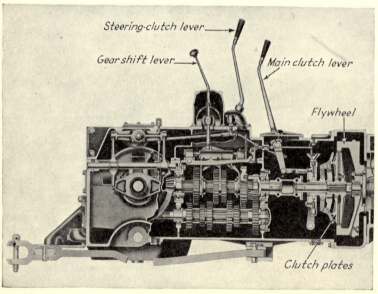

Fig. 341.—Caterpillar tractor transmission showing clutch and change-speed gears.

clutches located on this countershaft on each side of the bevel gear to the spur-type final-drive gears, and thence to the track sprocket. Steering is accomplished through the multiple-dry-disk steering clutches. That is, by means of hand levers, either clutch can be disengaged, which obviously cuts off the power to that particular track and causes the tractor to make a turn. Each clutch is equipped with a foot-operated brake

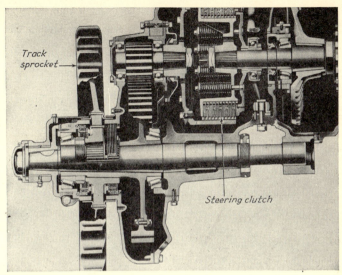

Fig. 342.—Caterpillar tractor transmission showing steering clutch and final-drive mechanism.

(Fig. 343), which acts on the outside of the clutch drum carrying the driven plates. If a quick, short turn is desired, the clutch is not only

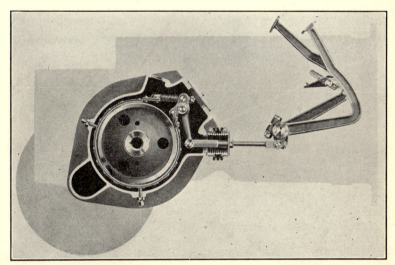

Fig. 343.—Steering-clutch brakes used on Caterpillar tractor.

released but the brake for that particular clutch is actuated and the track movement virtually stopped on the one side. With all the power going to the other track, the machine obviously will turn very short.   By pressing on both brakes at the same time, the machine can be stopped almost

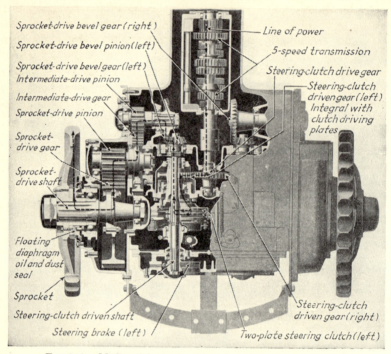

Sprocket-drive bevel gear (right)
Sprocket-drive bevel pinion (left)
Sprocket-drive bevel gear (left)
Intermediate-drive pinion
Intermediate-drive gear
Sprocket-drive pinion
Sprocket-drive gear
Sprocket-drive shaft
Floating diaphragm oil and dust seal
Sprocket
Steering-clutch driven shaft
Steering brake (left)
Line of power
5-speed transmission
Steering-clutch drive gear
Steering-clutch driven gear (left)
Integral with clutch driving plates
Steering-clutch driven gear (right)
Two-plate steering clutch (left)

Fig. 344.—McCormick-Deering track-type tractor transmission.

instantly.   It should be noted that a differential is unnecessary in the Caterpillar transmission.

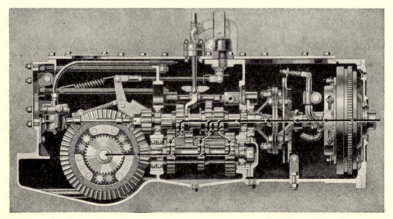

Fig. 345.—Cletrac tractor transmission and differential.

325. McCormick-Deering Trac-tractor Transmission.—The McCormick-Deering Trac-tractor transmission (Fig. 344) is similar to the

Caterpillar transmission in that two clutches are used for steering and a differential is unnecessary. However, the construction is different in that the steering-clutch shafts are parallel to the engine crankshaft, and two sets of bevel gears are required to transmit the power from the steering-clutch shafts to the intermediate drive shafts and thence to the track-sprocket drive gears.

326. Cletrac Transmission.—The Cletrac transmission (Fig. 345) has the same general arrangement as the Caterpillar but uses a differential and brake for steering instead of special clutches. The spur-gear differen-

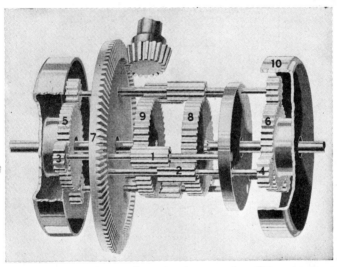

Fig. 346.—Differential gearing and steering drums used on Cletrac tractor. (1 and 2, differential pinions; 3 and 4, external steering-drum pinions; 5 and 6, steering-drum gears; 7, main-drive gear; 8 and 9, main-drive and differential gears; 10 steering drums.)

tial (Fig. 346) is located directly back of the change-speed gears in the center of the countershaft. A brake drum and band are placed on each side of the differential as shown. The drum carries a spur gear that meshes with three spur pinions. The latter, in turn, are attached to the three differential spur pinions that mesh with the differential shaft gear on that particular side. Therefore, if the brake band is tightened on one side or the other, that particular differential gear and shaft will turn slower, the track will slow down, and the machine will turn in that direction. The tractor is steered by two hand levers that control the steering brakes.

TRANSMISSION ACCESSORIES

327. Brakes.—A transmission brake of some sort on a tractor is essential for controlling the machine on steep hills or for holding it perfectly stationary in doing belt work. As shown by Fig. 347, this brake consists of a metal pulleylike drum attached to one of the intermediate transmission shafts, with a flexible band about it. In some cases, the brake is placed inside the transmission, and in others on the outside. It may be either foot or hand operated. Ordinary brake-lining material is used for lining the band, and an adjustment is always provided for taking up wear.

One of the essentials of an all-purpose tractor is the ability to make a short, quick turn with ease, particularly in row-crop operations in order to save time at the ends. These tractors, therefore, are provided with two foot-operated differential or countershaft brakes, which permit holding one driver practically stationary and turning on it as a pivot. One make of tractor has the two brakes connected indirectly to the steer-

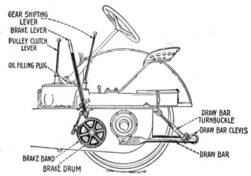

Fig. 347.—Sketch showing transmission brake and adjustable drawbar.

ing mechanism itself by means of a cable, so that turning the steering wheel tightens the one brake or the other and makes the tractor turn quickly and easily in the desired direction.

328. Drawbars.—A tractor drawbar should be strong, convenient, and adjustable both vertically and horizontally. The average height varies from 8 to 20 in., depending upon the size and type of tractor.

329. Belt Pulley.—The belt pulley on wheel-type tractors is usually located on the right-hand side and to the rear of the engine. If the engine is placed crosswise of the tractor frame, the pulley is located on the crankshaft and driven direct (Fig. 328). If the engine is placed lengthwise, the pulley must be driven through a special bevel gear, as shown by Fig. 333. Track-type tractors, as a rule, have the pulley on the rear just above the drawbar.

There is considerable variation in the size and speed of tractor belt pulleys. As a consequence manufacturers of belt-driven farm machines must supply a number of different sizes of pulleys for their machines so that they can be operated at the correct speed by any tractor. Inconvenience and difficulty in this respect could be overcome if all tractor manufacturers would adopt one belt speed as a standard.

The belt speed for a given pulley is equal to the product of its circumference and revolutions per minute. That is, given a pulley 16 in. in diameter operating at 500 r.p.m., the belt speed is equal to

$$\frac{16 \times 3.1416 \times 500}{12} - 2,094.4 \text{ ft. per minute.}$$

That is, belt speed depends on both pulley diameter and speed. The tractor maker, therefore, would not necessarily be restricted to any given pulley speed or size. On the other hand, the use of a single standard belt speed for all tractors would mean that manufacturers of belt-driven machines would need to supply but one size of pulley for a given machine. Tractor belt speeds usually range from 2,300 to 3,500 ft. per minute.

In any case, in doing belt work, the transmission gears are kept in neutral and the pulley controlled, as a rule, by means of the main clutch. A few tractors have a special pulley clutch so that the pulley can be kept stationary without stopping the engine or holding the main clutch disengaged. In such cases the pulley can also be held stationary when the tractor is doing field or road work.

Belt pulleys vary in width and diameter, depending upon the size of the tractor, engine speed, and pulley speed. An important consideration in choosing a tractor for belt work is to see that the pulley is convenient to operate and gives the belt sufficient clearance.

330. Power Take-off.—Most tractors are now provided with a power-take-off attachment for the purpose of utilizing the engine power to drive the mechanism of a field machine, such as a grain binder, corn picker, and so on. In other words, in such cases the engine supplies power for two distinct purposes, namely, (1) for drawing the field machine through the field and (2) for operating the mechanism of the field machine. The adoption and extensive use of the power take-off are a rather recent development in tractor utilization. Heretofore such machines drove their own mechanism through heavy ground wheels.

A power take-off consists essentially of an arrangement of shafts and gears connected to the transmission and terminating at the rear of the tractor so that it can be connected conveniently to any machine. In most cases the power take-off is built into the transmission and is completely enclosed with the exception of the projecting end of the shaft to which the connection is made. Such a device is shown by Fig. 348. A special hand lever permits disengaging the power-take-off drive at any time.

Certain standards and recommendations concerning power take-offs have been adopted by the American Society of Agricultural Engineers as follows:

1. The 1⅜-in. shaft is recommended as desirable for future design where practicable.

2. The end of the power-take-off shaft on the tractor shall be 25 in., plus or minus 5 in. above the ground line.   (Measurements to be taken less lug equipment.)

3. The power-take-off shaft or an extension thereof shall extend far enough to the rear of the tractor to clear the fenders and platform, and the hitch point of the drawbar shall be 10 to 15 in. back from the end of the splined portion of the shaft.

4. The horizontal location of the power-take-off shaft shall be as near the tractor center line as possible.   (Tolerance 5 in. right or left of center.)

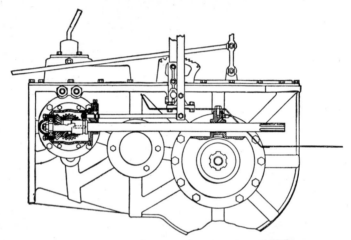

Fig. 348.—Inclosed-type power-take-off mechanism.

5. The tractor manufacturer shall furnish the standard power-take-off shaft.

6. The tractor manufacturer shall adequately shield the power-take-off shaft and tractor universal joint, providing protection for the operator against the telescoping member attached thereto, assuming connection between tractor and implement is according to recommended practice.

7. The manufacturer of the power-driven machine shall furnish the power-drive parts up to the tractor spline shaft, the necessary hitch parts to attach to their recommended drawbar location, and all shields except the one attached to tractor, covering the spline-shaft fitting or universal joint.

8. The tractor power-take-off drive shall be provided with a throwout clutch operating independently of the tractor travel, of a design safe against accidental engagement, and with a control located conveniently to the operator.

9. The telescoping members of the power shaft shall be so arranged that the tractor universal joint and the adjacent tool universal joint cannot be placed in improper relationship.   The tractor universal joint is in correct relationship to the adjacent tool universal joint when the two shaft yokes are in the same plane.   Slip clutches, if placed between these joints, should be designed to obviate improper alignment.

10. In plan view, the tractor drawbar hitch point should be located midway between pivotal centers of the tractor universal joint and the adjacent driven-tool

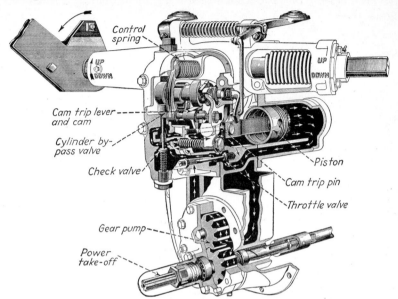

Fig. 349.—Hydraulic-type power lift—tool being lowered.

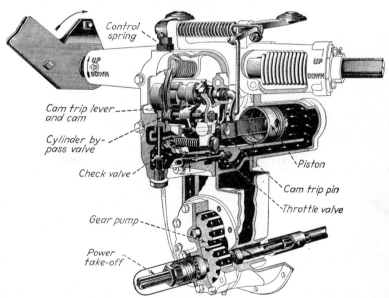

Fig. 350.—Hydraulic-type power lift—tool being raised.

universal joint, where the driven tool is being driven in normal working position. The object of this location is to equalize universal joint angularity when turning.

11. In side elevation the tractor drawbar hitch should be as close as practicable to the drive shaft connecting the tractor universal joint and the adjacent driven-tool universal joint, the object being to reduce the telescoping of the drive shaft to minimum over rough ground.

331. Power Lift.—All-purpose tractors can usually be equipped with a device by means of which the planter, cultivator, or similar equipment attached to the tractor, can be raised or lowered at any time by the tractor power itself.   Such a device is known as a *power lift*.   It not only relieves the operator of this heavy work but also permits raising and lowering the equipment quicker and usually without stopping the tractor.

These devices are attached to the rear of the transmission or rear-axle housing and frequently have the power-take-off shaft and drive combined with them.

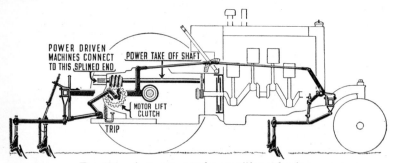

Fig. 351.—A gear-operated power-lift mechanism.

The two common types of power lifts are the hydraulic type (Figs. 349 and 350) and the mechanical or gear-driven type (Fig. 351).   The former utilizes oil pressure to actuate the lifting arm.   A gear pump driven by the power-take-off shaft circulates the oil and creates pressure on a piston or plunger.   The movement of the piston in one direction lifts the tool, and the movement in the opposite direction lowers it.   The lift is tripped at any time by a foot lever.   Whenever the lift is not in use the pump continues to operate but the oil is by-passed and recirculated.

The mechanical lift (Fig. 351) is driven directly from the power-take-off shaft by a worm and gear.   A ratchet-type clutch and convenient foot trip engage and disengage the lift at any time.

332. Transmission Bearings.—Antifriction bearings of one type or another have almost entirely replaced plain bearings in all tractor transmissions.   The illustrations accompanying this chapter show typical installations in which all types, namely, ball, spiral-roller, and tapered roller bearings are used.   Although the cost of manufacture is greater, this is more than offset by such advantages as (1) longer life, (2) reduced power losses, (3) ease of lubrication, and (4) possible adjustment for wear.

References

A Review of Power Take-off Standardization Work, *Agr. Eng.*, Vol. 12, No. 1.

ELLIOTT and CONSOLIVER: "The Gasoline Automobile."

HELDT: "Motor Vehicles and Tractors."

————: "The Gasoline Automobile," Vol. II.

Proposed American Society of Agricultural Engineers Standard Power Take-off. *Agr. Eng.*, Vol. 12, No. 1.

The Power Take-off for Tractors, *Trans. Amer. Soc. Agr. Eng.*, Vol. 19.

CHAPTER XXIV

TRACTOR CHASSIS—FRAMES—TRACTION DEVICES— PNEUMATIC TIRES—STEERING MECHANISMS

As previously mentioned, the chassis of a tractor includes the frame, wheels or other means of traction, and the steering mechanism. Heretofore there has been considerable variation in tractor-chassis construction, but the present trend is toward more or less similarity in this respect.

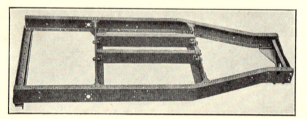

Fig. 352.—Built-up-type structural-steel tractor frame.

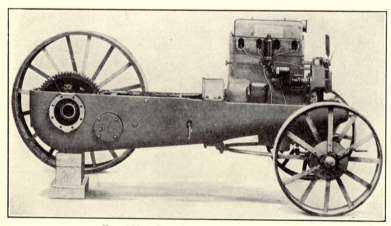

Fig. 353.—One-piece, cast-iron U-frame.

333. Frames.—Figure 352 shows the earlier type of frame construction involving the use of structural-steel members of the correct size and strength, shaped and riveted together to produce a convenient framework to which all other units of the machine were attached. As the trend in design changed to smaller, lighter weight and more enclosed tractors, the built-up steel frame was displaced either by (1) a one-piece

332

pressed-steel or cast-iron U-frame (Fig. 353), or by (2) the so-called frameless construction (Fig. 354). The U-frame provides simple but rigid construction and also serves as part of the engine and transmission housing. Frameless construction involves the use of two distinct castings, one of which serves as an engine crankcase and the other as a transmission housing. These must be made rather heavy and securely

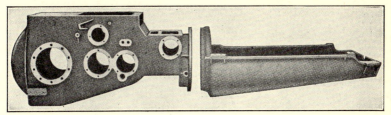

Fig. 354.—Frameless tractor construction.

bolted where they join. Frameless construction is the most common at present.

TRACTION DEVICES

The proper working of a tractor under certain conditions often depends upon its means of securing traction and the use of the correct equipment. All manufacturers of both wheel and crawler types now supply equipment to meet practically any condition encountered.

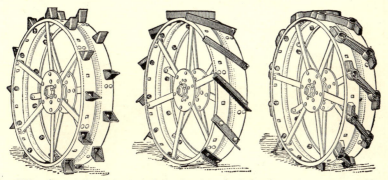

Fig. 355.—Conventional-type steel-rim tractor wheel showing common types of lugs.

334. Wheels.—The ordinary wheel-type tractor is equipped with two rear-drive wheels and two front steering wheels. Some wide-tread, all-purpose tractors have two rear-drive wheels and only one front steering wheel. Although there may be two front wheels, they are so close to each other that they are not unlike a single, wide wheel.

The average tractor wheel (Fig. 355) has a steel rim with rolled edges. A cast-iron hub is attached to the rim by means of flat steel spokes riveted at both ends. Special wheels for industrial purposes may be solid or have round spokes and are equipped with pneumatic tires. Some farm tractors have solid cast-iron front wheels.

FIG. 356.—Skeleton wheel with reversible hub.

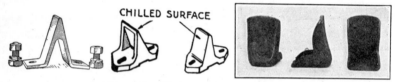

CHILLED SURFACE

FIG. 357.—Various types of spade lugs.

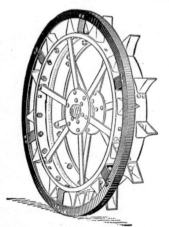

FIG. 358.—Road band for highway protection.

FIG. 359.—Extension rim.

Certain all-purpose tractors can be equipped with special skeleton wheels as shown by Fig. 356. This type of wheel reduces soil packing and the so-called balling up of the wheel occurring with certain soils, especially when moist or wet.

The size of a wheel involves both diameter and rim width. Wheel sizes vary largely according to the size, weight, and type of the machine.

Figures 355, 356, and 357 show some of the common types of lug equipment. For most farm jobs, the *spade lug* gives the best results. It gives better traction in loose soil conditions and is more or less self-cleaning. Figure 357 shows some of the various types of spade lugs. *Angle cleats* tend to chop out the section of earth between the two adjacent cleats on which the wheel is resting and attempting to secure footing, and thereby may actually produce slippage and partially destroy their purpose. *Road lugs* are low and blunt for the purpose of eliminating jar to the tractor and injury to the road surface when traveling on high-ways. Figure 358 shows a special road band for transporting a tractor on a highway. It protects the road surface and, since the effective wheel diameter is greater, the tractor will travel faster.

Fig. 360.—Front-wheel skid ring or guide band.

Extension rims (Fig. 359) are frequently used to improve wheel traction in soft, wet, or loose soil.

Front steering wheels ordinarily are equipped with skid rings or guide bands (Fig. 360), for the purpose of overcoming the tendency of the

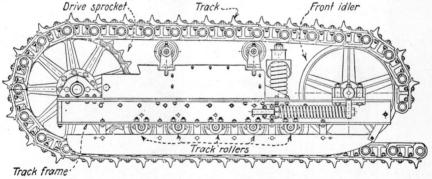

Fig. 361.—Track assembly showing principal parts.

tractor to continue to move straight ahead when pulling a heavy load, even though the front wheels are turned. In other words, these bands produce a more responsive and positive turning action, particularly in loose ground with a heavy load. They also prevent skidding and keep the tractor moving straight ahead on hillsides.

335. Track Mechanisms.—The traction mechanism of a track-type tractor (Fig. 361) consists of (1) frame, (2) drive sprocket, (3) front idler, (4) track rollers, and (5) track. The track frames are built of structural

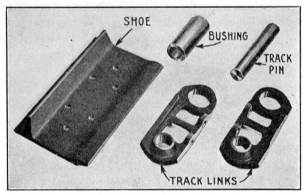

FIG. 362.—Track construction and parts.

steel. The sprocket and front idler drive and support the track. The track rollers on the under side of the track frame act as supports between the machine and that part of the track in contact with the ground. In

FIG. 363.—Method of making track-tension adjustment.

fact, these rollers might be considered as small wheels rolling and conveying the tractor over a stationary track.

The track itself is built up of forged-steel shoes bolted to forged-steel heat-treated links, with alloy-steel heat-treated bolts (see Fig. 362).

Since any track mechanism must have so many wearing points, which are continually exposed to dirt and grit, three fundamentally important considerations must be observed by the manufacturer, namely: (1) the parts themselves must be made of high-grade materials that have been carefully heat-treated and hardened; (2) any important bearings, such as track rollers, front idlers, and so on, must be well enclosed and properly lubricated; and (3) there must be some provision for taking up the slack and maintaining the proper track tension.

Figure 363 shows how the track tension is ordinarily adjusted. The heavy springs located on each side of the idler also provide an automatic safety release, thus permitting the idler to move back temporarily in case any obstruction gets between the track and the sprocket.

336. Slippage and Traction Efficiency.—During the past few years considerable investigation has been made in connection with the factors involved in the effective and efficient operation of tractor wheels, wheel equipment, and other traction devices.

The power applied to any traction device, either round wheel or track, is consumed largely in about four ways, namely, by (1) rolling resistance, (2) wheel slippage, (3) the action of the lugs on the soil, and (4) the tractor-drawbar resistance. Obviously the most efficient device is one in which the first three factors named are low, so that the net power available at the drawbar is as high as possible. In other words, the problem of securing efficient traction is dependent upon the reduction of these apparent power losses. Rolling resistance varies with the soil type and conditions and the weight upon the tractor. A certain amount of weight is essential for traction, but too much weight produces a high rolling resistance and therefore may reduce the net power output of the wheel. The second factor, wheel slippage, is likewise apt to prove excessive under certain conditions and result in low tractive efficiency.

The most important factors affecting the efficiency of a tractor wheel and its equipment are (1) wheel diameter, (2) wheel width, (3) weight on wheel, (4) type of wheel lug or tread surface, (5) speed of travel, (6) soil type and condition, (7) grades, and (8) height of hitch. These factors likewise apply to track-type tractors.

The great variation in soil types and conditions is perhaps the most outstanding handicap encountered in the solution of the problem of tractor traction. It seems quite impractical to attempt to provide a distinct type of equipment for every condition, but it would be rather desirable to have one type of equipment that is adapted to as many conditions as possible. Some of these are:

1. Unplowed grain or corn stubble having a firm surface.
2. Turf or sod which, in addition to being firm, is matted with roots, which assist traction.

3. Plowed land or soil having a similar condition of tilth.
4. Very loose soil such as sand.
5. Firm but sticky soil.
6. Slippery surface, but firm underneath.
7. Hard, firm surfaces, such as highways, pavements, etc.
8. Tall weeds or high cover crops.
9. Terraces, checks, levees, ridges, and ruts.

Studies made of tractor traction devices in general lead to certain definite conclusions as follows:

1. The efficiency is generally lower in loose soils.
2. The higher the rolling resistance, weight, and slip, the lower the efficiency.
3. An extension rim makes a higher tractive pull possible with the same weight on the wheel, but does not increase the efficiency because of the increase in rolling resistance.
4. Maximum efficiency is attained with only enough weight on the wheel or track to prevent excessive slip with a high drawbar pull, a low percentage of slip, and a low rolling resistance.

337. Wheels vs. Tracks.—Tests that have been made of tractor wheels and tracks indicate that a wheel and lugs will give from 35 to 75 per cent efficiency; a track will give from 40 to 85 per cent.

The outstanding advantage of the track type is, no doubt, its ability to secure traction under conditions that will not permit the wheel type to operate with any degree of success. Not only can the track tractor usually secure good footing under most adverse conditions, but there will likely be less loss from slippage under average operating conditions. This is clearly brought out by the Nebraska tractor tests. A checkup of the tests made on current models of tractors, shows that 24 wheel-type machines, when developing their rated drawbar horsepower, gave an average slippage of 4.75 per cent. When developing their maximum drawbar horsepower, the average slippage was 5.73 per cent. Twenty-eight track-type tractors, under the same conditions, showed an average slippage of 1.82 per cent at rated load and 2.16 per cent at maximum load.

RUBBER TIRES FOR TRACTORS

The first tractors to be equipped with solid or pneumatic rubber tires were those used for industrial purposes around factories, airports, and the like and for highway maintenance, the reason being that satisfactory traction was secured and jarring, vibration, and damage were reduced even though the machine traveled relatively fast over the packed gravel, pavement, or similar hard surface. About 1931 investigations were begun concerning the possibility of using low-pressure pneumatic tires for agricultural tractors. Extremely favorable results were immediately observed and a number of advantages in favor of rubber tires over steel wheels and lugs were disclosed. Within a period of 3 or 4 years a pronounced trend developed in favor of rubber tires, particularly for all-purpose farm tractors. Today rubber-tired tractors are used extensively in practically all types of farming, and it is reported that in some sections of the United States there is but a limited demand, if any, for steel-wheel equipment.

338. Tire Construction—Size—Mounting—Inflation Pressures.—Any tractor requires two sizes and types of tires, a large four- or six-ply casing with a special traction-producing tread for the rear or driving wheels, and a smaller casing—usually four-ply—for the front wheels. Rear-wheel tires vary in size, depending upon the size of the tractor and original wheel equipment, from 7.50 to 13.50 in. in cross section, and from 24 to 42 in. in rim size. A casing of a given cross-sectional size may be made in as many as four rim sizes. Front tires vary from 4.00–15 to 9.00–10. Inflation pressures for rear tires vary from 8 to 20 lb.; for front tires they vary from 16 to 40 lb. The pressure recommended depends upon the tire size and operating conditions and is discussed in more detail later on.

A number of methods of equipping a tractor with rubber tires are used as follows:

1. Install completely new wheels with hub, spokes and rim in one piece.
2. Install completely new wheels with demountable rims and lugs.
3. Using the old wheels, cut off the steel spokes and weld the rim (drop-center type) directly to them.
4. Using the old wheels, cut off the steel spokes and weld on felly bands to which demountable rims are clamped.

The integral construction with full drop-center rim is simpler and usually less expensive. However, the use of demountable rims and clamps makes the work of mounting the casing somewhat easier and also permits shifting the tire from one tractor or machine to another.

339. Weights for Rubber Tires.—Since the wheels and tires alone are relatively light and hence do not provide sufficient traction under most conditions, it is necessary to add weight in some manner. This may be done by attaching special iron or concrete weights to the wheels or by partially filling the inner tube with water or some other suitable liquid. The amount of additional weight needed is determined largely by the size and power of the tractor and the traction conditions. For small tractors, 100 to 200 lb. per wheel is usually sufficient. For larger tractors, as much as 400 to 600 lb. per wheel may be needed to obtain the maximum drawbar pull. In general, only sufficient weight should be added to obtain good traction without undue slippage. As little weight as possible should be used in harrowing, planting, drilling, and cultivating land, in order to reduce soil packing.

Figure 364 shows the effect upon drawbar pull of adding wheel weights. Investigations show that the increased drawbar pull will be approximately 50 per cent of the weight added to the wheels.

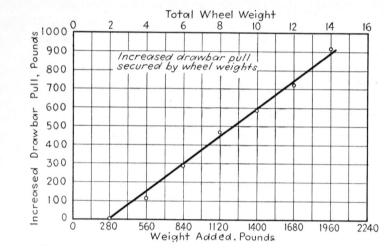

FIG. 364.—Effect upon drawbar pull of adding weight to rear wheels.

Weighting rubber-tired tractor wheels by partly filling the inner tube with water or some other suitable liquid is now common practice and seems to be preferable to using external weights for the following reasons:

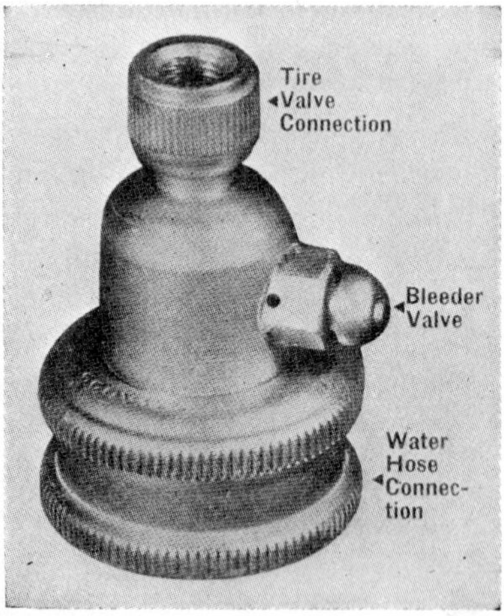

1. Liquids are usually easy to obtain and inexpensive.

2. Better cushioning and shock-absorbing effects are obtained, thus improving riding qualities.

3. The weight is always located in the lower portion of the tire and directly over the tread, thus providing the most effective traction possible.

FIG. 365.—Valve adapter for placing water in inner tube.

The quantity of liquid to use depends upon the tire size and the extra weight desired. As a rule, for smaller tires, a quantity of liquid is used equal to one wheel weight (150 lb.); for medium-size tires, an amount equal to two wheel weights; and for large tires, an amount equal to three wheel weights.

In colder climates where freezing may give trouble, a calcium chloride solution may be substituted for water, although no harm is done by water freezing inside a tire if the tractor is kept out of service and the wheels are jacked up. Specific information concerning the preparation of a calcium chloride solution and putting it into the tire should be obtained from the tractor or tire dealer or the manufacturer.

Water is placed in a tractor tire by one of four methods, namely, (1) by line pressure, (2) by gravity, (3) by a hand force pump, or (4) by a pressure tank. In any case, a valve adapter (Fig. 365) is used. The usual procedure is as follows:

1. Place tube valve at top of wheel.

2. Remove valve core and allow all air to escape.

3. Attach hose adapter to valve to hold valve in place.

4. Jack up tractor wheel so that tire is not deflected and maximum amount of liquid can be admitted.

5. Attach water hose to adapter.

6. Turn on water and allow it to run until tire is filled to desired level which can be determined by rotating the wheel and listening for change in sound made by entering stream of water or by turning off water and pressing special by-pass valve in special hose connection. If tire is to be filled to capacity, keep valve at top; if half filled, place valve in horizontal position.

7. Turn off water, remove adapter, and replace valve core.

8. Remove jack and inflate tire with air to recommended pressure.

9. Check pressure after 1 hr. of operation and check frequently and regularly.

PNEUMATIC TIRES VS. STEEL WHEELS

Numerous investigations relative to the comparative performance of pneumatic tires and steel wheels for farm tractors have been made by various experiment stations. Some of the important problems studied were:

1. Traction effects and efficiency as influenced by:
   a. Traction surface and soil condition.
   b. Inflation pressure.
   c. Weight on wheels.
   d. Travel speed.
   e. Tire size.
   f. Chains and special traction devices.
2. Drawbar horsepower.
3. Fuel consumption.
4. Adaptability to various field operations.
5. Ease of handling and riding qualities.
6. Tractor wear.

340. Rolling Resistance.—According to Davidson[1] *et al.*

Rolling resistance is the drawbar pull or its equivalent required to move the tractor over a given surface. The tractor in field work passes over soft traction surfaces (Fig. 366). The wheels or tracks, in sustaining the weight of the tractor, sink into the surface. Therefore the tractor is virtually climbing an incline as

---

[1] *Iowa Agr. Exp. Sta., Research Bull.* 189.

it moves forward. In addition, rolling resistance includes resistance due to friction in traction members and losses incurred in obtaining adhesion.

Tractors with steel wheels are equipped with lugs to increase the adhesion between the wheels and the traction surface. On soft ground surfaces, for maximum adhesion, long and sharp lugs are desirable to penetrate well into the soil. These long lugs cause considerable energy loss due to soil disturbance. Tractors with smooth-faced wheels have little loss from adhesion, which, however, depends wholly upon friction between the tractor wheel and the ground. Adhesion for field conditions is inadequate. With rimless traction wheels, practically all the

Fig. 366.—Effect of lug and lug shape upon soil disturbance.

adhesion is obtained by the penetration of the lugs into the surface and compaction of the soil back of the lugs. Desired traction or adhesion without some compression of the soil below and back of the lugs is impossible in soft soils. Where adhesion is not good, slippage or failure of the tractor to travel a distance equal to the circumference of the drivers for each revolution may occasion serious loss.

TABLE XVI.—ROLLING RESISTANCE (IN POUNDS) OF STEEL TRACTOR WHEELS
(Total weight on rear wheels, 3,890 lb.)

Wheel diameter, in.	Pulverated soil		Oat stubble	
	Bare wheels	4-in. lugs	Bare wheels	4-in. lugs
38	655	789	362	777
42	610	734	359	746
46	615	690	337	720
50	585	658	313	688
54	528	642	323	660
58	561	628	308	654

Table XVI[1] shows the effect of wheel diameter and lugs upon the rolling resistance of a wheel tractor.

Table XVII[2] shows the comparative rolling resistance of steel wheels and pneumatic tires for both sod and plowed ground. Relative to this, McCuen and Silver[2] state:

A tractor, when equipped with rubber tires, has a much less rolling resistance than when equipped with steel wheels and spade lugs. This is true both on sod and on plowed ground. Furthermore, the difference in rolling resistance between the two types of wheels is greater on sod than it is on plowed ground, and the

[1] *Iowa Agr. Exp. Sta., Research Bull.* 189.
[2] *Ohio Agr. Exp. Sta., Bull.* 556.

rolling resistance of a rubber-tired tractor on plowed ground is less than that of the steel-wheel tractor on sod.

TABLE XVII.—COMPARATIVE ROLLING RESISTANCE OF STEEL WHEELS AND PNEUMATIC TIRES ON SOD AND ON PLOWED GROUND

Wheel	Tractive surface	Miles per hour	Force required to pull tractor, pounds	Hp.
Steel.....................	Sod	2.16	827	5.24
Steel.....................	Sod	3.05	911	7.64
Steel.....................	Sod	4.21	878	10.81
Rubber-tired..............	Sod	3.14	273	2.53
Rubber-tired..............	Sod	4.44	265	3.45
Steel.....................	Plowed ground	2.14	1042	6.46
Steel.....................	Plowed ground	2.67	1150	9.08
Steel.....................	Plowed ground	3.56	1102	11.54
Rubber-tired..............	Plowed ground	2.24	557	3.66
Rubber-tired..............	Plowed ground	2.95	592	5.13
Rubber-tired..............	Plowed ground	3.72	739	6.93

341. Pneumatic Tires vs. Steel Wheels for Various Field Operations.—Figure 367 shows the comparative fuel consumption and acre-hour capacity of pneumatic-tired and steel-wheel-and-lug tractors as reported by Smith and Hurlbut.[1] With further reference to these tests they state:

1. A rubber-tired tractor is harder to hold on listed corn ridges when cultivating the first time over than a tractor equipped with steel wheels and spade lugs. The difference is less noticeable if the front tires are inflated to 25 lb. or higher and if the installation of the rear rubber-tired wheels has not changed the tread of the tractor.

2. There was little difference in the ease of handling or of riding qualities between tractors equipped with rubber tires and those equipped with steel wheels when cultivating corn, except in handling when on ridges.

3. The riding qualities of a tractor equipped with rubber tires are very much better than those of one equipped with steel wheels and lugs when going to and from fields and when traveling on the road.

4. The rubber-tired tractor is considerably better all around for haying operations than the tractor equipped with steel wheels and spade lugs. It does not punch hay into the ground. It does not tear up the ground when approaching the stacker with a sweep. In general, the riding qualities of a rubber-tired tractor are much better on a meadow. However, on meadows that are very rough, an oscillating, bouncing motion is given to the rubber-tired tractor that not only affects the driver, but, when mowing, is reflected in a rough job due to the up-and-down movement of the sickle bar.

[1] *Neb. Agr. Exp. Sta., Bull.* 291.

5. A rubber-tired tractor has less traction on firm soil than a tractor with steel wheels and spade lugs; but in loose, sandy soil it has more.

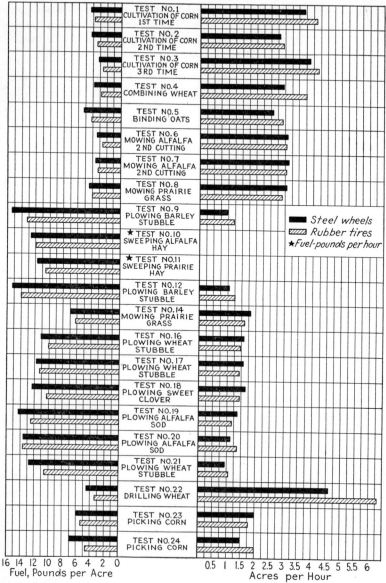

FIG. 367.—Comparative fuel consumption and acre-hour capacity of steel-wheel and pneumatic-tired tractors when performing various field operations.

6. A saving in time and fuel can be made on most farm operations by using a pneumatic-tire-equipped tractor. This saving becomes less significant, and may become negative, as drawbar pulls increase, necessitating the use of low gear.

This saving increases for those operations which make a relatively light load and permit the use of higher gears. Necessarily, the tractors best adapted to rubber tires are those having sufficient speeds to utilize this advantage.

7. Field conditions were encountered following rains that caused a tractor equipped with pneumatic tires to slip when pulling a combine and also when pulling a plow. Although steel wheels and lugs gave some trouble under the same circumstances, they were more satisfactory than the rubber tires.

342. Inflation Pressures.—As previously stated, inflation pressures for the rear tires of a tractor usually range from 8 to 20 lb., depending upon various factors involved. Smith and Hurlbut[1] give the following conclusions as a result of tests comparing inflation pressures of 8 and 16 lb.:

1. Changing the inflation pressure carried in tractor tires causes no appreciable change in the fuel consumption of the tractors on which tires are used.

2. Increasing tire inflation pressure causes an increase in speed of the tractor on which the pneumatic tires are used.

3. Travel reduction (essentially slippage) is less for a tractor equipped with pneumatic tires when the tires are inflated to 16 lb. pressure than when inflated to lower pressures, and this difference increases as the maximum drawbar pull is approached.

4. For the conditions of these tests, greater drawbar pulls were obtained with an inflation pressure of 16 lb. than with lower inflation pressures.

343. Traction Efficiency of Pneumatic Tires.—The traction efficiency of a tractor can be expressed as the ratio of the drawbar horsepower to the engine horsepower, and, regardless of the type of wheel equipment or tread, it is affected by certain factors as previously discussed. In order to determine the relative traction efficiency of steel wheels and pneumatic tires, Smith and Hurlbut[1] made some tests of a standard four-wheel tractor weighing 4,545 lb. with steel wheels and 5,000 lb. with rubber tires and weights. Figure 368 shows the results of these tests with respect to drawbar pull as affected by speed. Their conclusions were:

1. Under favorable operating conditions, pneumatic-tired wheels transform a greater proportion of the engine horsepower into drawbar horsepower than do steel wheels and spade lugs.

2. Under favorable conditions, steel wheels and spade lugs attain a greater drawbar pull in low gear than can be attained with rubber tires.

3. There are two ways of increasing the drawbar horsepower of a tractor: namely, by increasing the drawbar pull and by increasing the speed of travel. The drawbar pull of pneumatic tires is limited by traction in the first three gears. The drawbar pull of a steel-wheeled tractor is usually limited by engine horsepower in any gear.

[1] *Neb. Agr. Exp. Sta., Bull.* 291.

4. The maximum drawbar horsepower of a rubber-tired tractor in various gears covers a much wider range of values than that of a steel-wheel-and-lug

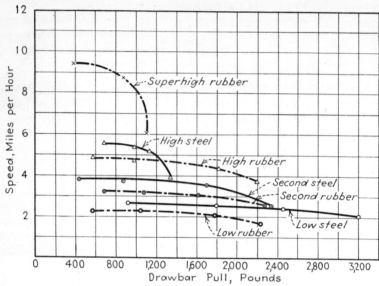

Fig. 368.—The relation of speed to drawbar pull.

tractor. Horsepower values for rubber tires at high speeds exceed any derived for steel wheels and lugs. At low speeds, the horsepower for rubber tires is less than for steel wheels and lugs.

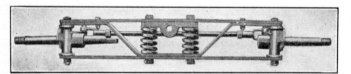

Fig. 369.—Built-up-type front axle.

5. Maximum economy was secured for the rubber-tired tractor when the drawbar pull was slightly more than 50 per cent of the static weight of the rear of the tractor.

6. Maximum drawbar horsepower values were secured for the rubber-tired tractor when the travel reduction was about 16 per cent. No such point is found

Fig. 370.—One-piece forged front axle.

for steel wheels and lugs. The percentage of travel reduction is less for steel wheels and lugs than for rubber tires and decreases as speed increases.

344. Life of Pneumatic Tires.—No definite information is available relative to the life of pneumatic tractor tires. The life will depend upon

such factors as (1) abrasive wear, (2) chipping, (3) punctures and blow-outs, (4) chemical decomposition, and (5) general care given. Abrasive
wear is dependent upon the soil surface
condition and the speed of travel. It
seems logical to assume that the travel
speed will have much less effect on
tractor tires than on auto and truck
tires. Truck tires are usually operated
on hard surfaces at high speeds. The
abrasive action due to these high speeds
has a wearing tendency that is almost
entirely absent in slow-speed operation,
particularly on soft surfaces where a
tractor operates. A high-grade truck
tire operating in heavy-duty work and
on pavements frequently goes 20,000
miles before being discarded. A tractor
tire used 800 hr. per year at a speed of
three miles per hour would cover 2,400
miles per year. Assuming it would
give the same total mileage as the truck
tire, its life would be about eight years.

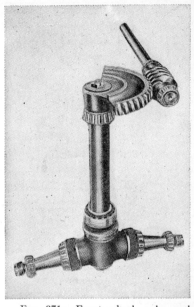

Fig. 371.—Front wheel, axle, and steering mechanism for an all-purpose tractor.

345. Chains for Tractor Tires.—
Special tractor tire chains are available
but are seldom necessary except for adverse conditions or for securing a
slightly greater maximum pull in low gear.

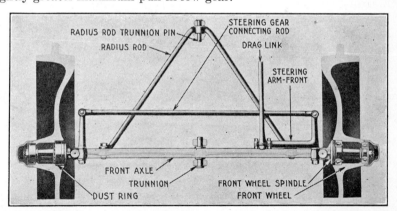

Fig. 372.—Construction of a tractor front axle and steering mechanism. (*Courtesy of Ford Motor Company.*)

346. Front Axles.—Front axles are of either the built-up type (Fig. 369), or one piece (Fig. 370). The latter type is more common. In all cases the connection to the engine frame is made by means of a large pin at the center of the axle which permits a hinging action to provide proper flexibility in uneven or rough ground.   Radius rods, extending

Fig. 373.—Fifth-wheel method of front-axle mounting and steering

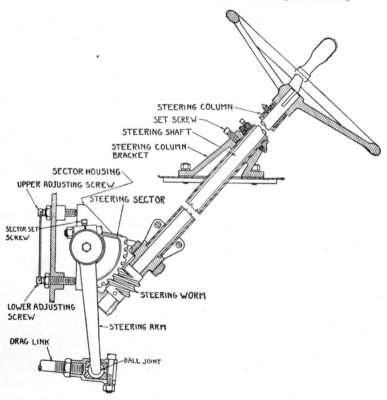

Fig. 374.—Worm-and-sector type of steering mechanism.

from the axle back under the frame, maintain the axle in an exactly crosswise relation to the frame.

Figure 371 shows the front-wheel construction and steering mechanism of an all-purpose tractor. A heavy-duty antifriction bearing

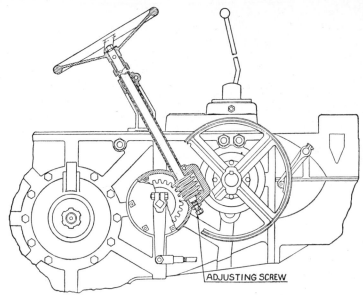

Fig. 375.—Worm-and-wheel steering mechanism.

supports the front end of the tractor and permits easy turning.

Front wheels are equipped with either plain or antifriction bearings. The latter are preferable because wear can usually be taken up. Figure 372 shows the usual front-wheel mounting.

347. Steering Mechanisms.—The fifth-wheel arrangement (Fig. 373) was used on the first tractors. Since they were very heavy and slow moving, it worked reasonably well. As shown, the axle is pivoted at the center and is rotated horizontally about a vertical axis by means of chains attached at each side. The chains are wound about an iron roller, which in turn is actuated by a steering wheel and worm.

With the introduction of smaller and faster tractors, a more positive means of steering was necessary, and the automobile type (Fig. 372) was adopted. The most important part of a steering mechanism is the gear

Fig. 376.—One type of worm-and-nut steering gear.

that transmits the motion from the steering wheel to the drag link and steering rods. There are a number of different types as shown by Figs. 374, 375, and 376. These gears are always enclosed in dirt-proof

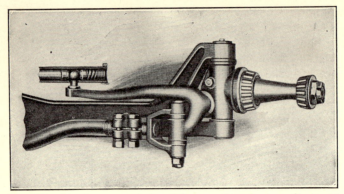

Fig. 377.—Steering-arm and drag-link connection showing adjustment.

housings and operate in oil or grease. Most of them have a simple and convenient means of taking up wear. Figure 377 shows the usual

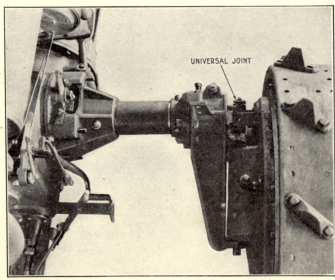

UNIVERSAL JOINT

Fig. 378.—Axle and mounting construction when tractor is guided through the drive wheels.

method of connecting the steering gear to the spindle arms, and a take-up device on the drag link.

Some tractors, particularly four-wheel-drive machines, are driven as well as steered through the front wheels. This involves a similar but somewhat more complicated connection as shown by Fig. 378. A universal joint must be inserted in the countershaft that transmits the power to the wheel, and the spindle mounting must be considerably heavier.

348. Turning Radius.—The turning radius of a tractor (Fig. 379) is the radius of the smallest circle in which it will turn, measuring from the center to the outermost point on the outside-wheel track. It varies from about 6 ft. for the small-size crawler tractors to 20 ft. or more for large wheel machines. All-purpose tractors have a turning radius of 8 to 10 ft.

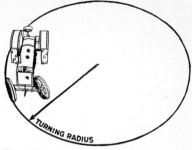

Fig. 379.—Turning radius of a tractor.

References

A Comparative Study of Pneumatic Tires and Steel Wheels on Farm Tractors, *Univ. Neb., Agr. Exp. Sta., Bull.* 291.

A Method of Studying Soil Stresses, *Trans. Amer. Soc. Agr. Eng.*, Vol. 19.

An Analysis of Soil Dynamics Factors Affecting the Operation of Tillage and Tractor Machinery, *Trans. Amer. Soc. Agr. Eng.*, Vol. 17.

Apparatus for Testing the Efficiency of Tractor Wheel Lugs, *Trans. Amer. Soc. Agr. Eng.*, Vol. 19.

A Review of Users' Experience with Rubber-tired Farm Tractors, *Agr. Eng.*, Vol. 16, No. 2.

Compaction of Soil Due to Tractors, *Trans. Amer. Soc. Agr. Eng.*, Vol. 10, No. 1.

Effect of Drawbar Pull upon the Effective Weight on Front and Rear Wheels of Farm Tractors, *Trans. Amer. Soc. Agr. Eng.*, Vol. 22.

Effect of Tractor Tire Size on Draw-bar Pull and Travel Reduction, *Agr. Eng.*, Vol. 17, No. 4, and Vol. 18, No. 2.

Efficiency Tests of Tractor Wheels and Tracks, *Agr. Eng.*, Vol. 14, No. 2.

Methods of Investigating Slippage of Traction Wheels in Tractors, *Trans. Amer. Soc. Agr. Eng.*, Vol. 16.

Methods of Testing Drive Wheels and Tracks, *Agr. Eng.*, Vol. 16, No. 2.

Pneumatic Tires for Tractors (three reports), *Agr. Eng.*, Vol. 14, No. 2.

Pneumatic Tractor Tires on Listed Crop Ridges, *Agr. Eng.*, Vol. 16, No. 2.

Relation of Lug Equipment to Traction, *Trans. Amer. Soc. Agr. Eng.*, Vol. 22.

Reports of Studies and Tests of Pneumatic Tractor Tires (fifteen reports), *Agr. Eng.*, Vol. 15, No. 2.

Rubber-tired Equipment for Farm Machinery, *Ohio Agr. Exp. Sta., Bull.* 556.

Tests of Tractor Wheel Equipment, *Agr. Eng.*, Vol. 9, No. 10.

Tests on Use of Rubber Tires and Steel Wheels on a Farm Tractor, *Mont. Agr. Exp. Sta., Bull.* 339.

Torque Dynamometer for Tractor Drive Wheels, *Trans. Amer. Soc. Agr. Eng.*, Vol. 22.

Tractive Efficiency of the Farm Tractor, *Iowa State Coll., Agr. Exp. Sta., Research Bull.* 189.

Tractor Lug Studies on Sandy Soil, *Trans. Amer. Soc. Agr. Eng.*, Vol. 20.

Tractor Lug Studies on Sandy Soil, *Trans. Amer. Soc. Agr. Eng.*, Vol. 21.

CHAPTER XXV

POWER AND ITS MEASUREMENT—FUEL CONSUMPTION— ENGINE EFFICIENCY

In the study of farm power it is important that one has a clear and definite understanding of the exact technical meaning of such terms as *horsepower*, *energy*, *efficiency*, and so on, as applied to mechanical devices. To obtain such an understanding, one must consider, first of all, certain fundamental physical terms, definitions, and units.

349. Force.—A *force* is an action, exerted upon a body, that changes or tends to change its natural state of rest or uniform motion in a straight line. It is thus observed that a force may or may not be effective in producing motion in the body acted upon. Likewise, if a body is in motion, a force may be applied that may or may not change its direction of movement. The unit of measurement of a force is the pound weight.

The force required to move a body may not necessarily be the same as the weight of the body. Only when the latter is moved vertically, with respect to the earth's surface, will this be true. Under certain conditions the force required to move an object might be greater than its weight.

350. Work.—If a force is applied to a body and its state of motion is changed, that is, it is made to move from a condition of rest or, if already in motion, its rate or direction of travel is changed, then *work* is done. In other words, if a force acts on a stationary body but does not produce motion, no work is done. Work is measured by determining the force in pounds and the distance through which it acts in feet. The product of the two gives the work done in foot-pounds. That is, the unit of work is the foot-pound.

351. Energy.—The energy possessed by a body is defined as its capacity for doing work. The energy possessed by an object or body by virtue of its position is known as *potential energy*. The energy possessed by the body by virtue of its motion is called *kinetic energy*. For example, the water stored in an elevated tank possesses potential energy. If this water is now discharged through a pipe, that water which is in motion in the pipe possesses kinetic energy.

Energy, like work, is measured in foot-pounds. The potential energy of a body, with respect to a given point or surface, is equal to the product of the weight in pounds and the vertical height in feet through which it has been lifted above this point or surface.

The kinetic energy of a body is dependent upon its weight, and its velocity or rate of travel. For a body moving at a uniform velocity,

$$\text{Kinetic energy} = \frac{WV^2}{2g}$$

where W = weight of body in pounds.
      V = velocity in feet per second.
      g = acceleration of a freely falling body.
        = 32.2 ft. per second per second.

352. Power.—Power is defined as the rate at which work is done. In other words, power involves the time element. For example, if a force of 100 lb. acts through a distance of 50 ft., 5,000 ft.-lb. of work are done. If, in one case, this 100-lb. force requires 1 min. to move the object 50 ft. and, in another case, the same force consumes 2 min. to move the object this distance, then twice as much power is required in the first case as in the second because the same work is done in one-half the time.

353. Horsepower.—The term horsepower is defined as a unit of measurement of power and is equal to doing work at the rate of 33,000 ft.-lb. per minute or 550 ft.-lb. per second. There is no real reason why this unit should have this particular value. However, it was fixed some time during the eighteenth century as a result of observations made of the work done by a horse in England in hoisting freight. It was estimated from these observations that the average horse was able to lift vertically a load of 150 lb when traveling at the rate of $2\frac{1}{2}$ m.p.h. Calculating as follows:

$$\frac{150 \times 2.5 \times 5{,}280}{60} = 33{,}000 \text{ ft.-lb. per minute}$$

Therefore, 33,000 ft.-lb. per minute was chosen as the rate at which the average horse could work and, consequently, was termed 1 hp.

In calculating the horsepower developed or required by a machine, it is only necessary to determine the total foot-pounds of work done or required per minute and divide this total by 33,000.

RELATION BETWEEN MECHANICAL AND ELECTRICAL POWER UNITS

354. Electrical Work.—Electrical work is measured in joules, a joule being defined as the amount of work done by a current of 1 amp. flowing for 1 sec. under a pressure of 1 volt, that is

$$\text{Electrical work} = \text{volts} \times \text{amperes} \times \text{seconds} = \text{joules}$$

355. Electrical Power—the Watt.—Since power is the rate of doing work, the power of an electric current would be the electrical work it is capable of doing per time unit (second).   In other words,

$$\text{Electrical power} = \frac{\text{electrical work}}{\text{time}}$$

$$\text{Electrical power} = \frac{\text{joules}}{\text{seconds}}$$

$$= \frac{\text{volts} \times \text{amperes} \times \text{seconds}}{\text{seconds}}$$

The unit of electrical power, known as the watt, is the power required to do 1 joule of electrical work per second, that is,

$$1 \text{ watt} = 1 \text{ joule per second}$$

$$= \frac{1 \text{ amp.} \times 1 \text{ volt} \times 1 \text{ sec.}}{1 \text{ sec.}}$$

$$= 1 \text{ amp.} \times 1 \text{ volt}$$

or

$$\text{Watts} = \frac{\text{joules}}{\text{seconds}}$$

$$= \frac{\text{amperes} \times \text{volts} \times \text{seconds}}{\text{seconds}}$$

$$= \text{amperes} \times \text{volts}$$

It has been found by experiment that if mechanical work is done at the rate of 1 ft.-lb. per second, then 1.356 watts of electrical power will be required to do the same work; that is,

$$1 \text{ ft.-lb. per second} = 1.356 \text{ watts}$$

but

$$1 \text{ hp.} = 550 \text{ ft.-lb. per second}$$

therefore,

$$1 \text{ hp.} = 550 \times 1.356$$

$$= 746 \text{ watts}$$

and

$$\text{Horsepower} = \frac{\text{watts}}{746}$$

$$= \frac{\text{volts} \times \text{amperes}}{746}$$

Since

$$1 \text{ kw.} = 1{,}000 \text{ watts}$$

then

$$1 \text{ kw.} = \frac{1{,}000}{746}$$

$$= 1.34 \text{ hp.}$$

HORSEPOWER OF ENGINES

356. Indicated Horsepower.—The indicated horsepower (i.hp.) of an engine is the power generated in the cylinder and received by the piston.

357. Belt or Brake Horsepower.—The belt or brake horsepower (b.hp.) of an engine is the power generated at the belt pulley and available for useful work. Several methods are used for measuring brake horsepower as described later.

358. Friction Horsepower.—The friction horsepower of an engine is the power that it consumes in operating itself at a given speed without any load. That is, it is the power required to overcome friction in the moving parts of the engine. The i.hp. minus the b.hp. equals the friction hp.

359. Drawbar Horsepower.—The drawbar horsepower of a pulling machine such as a tractor is the power developed at the hitch or drawbar and available for pulling, dragging, or similar tractive effort. In a tractor, for example, the drawbar horsepower would be equal to the b.hp. less the power consumed in moving the tractor itself. Methods of measurement are described later.

360. Rated Horsepower.—The rated horsepower is the amount of power that the manufacturer states that the engine will generate. It is stamped on the engine along with the rated crankshaft speed (r.p.m.), model designation, and serial number. Tractors are usually given both a b.hp. rating and a drawbar horsepower rating. The manufacturer's rating may or may not be the maximum horsepower that the engine will develop. As a rule, manufacturers rate their engines rather conservatively so that they will develop their rated power at the rated speed without being overloaded.

361. Measurement of Horsepower.—In the measurement of the power of an engine it must be kept in mind that the fundamental problem is to determine specific values for the physical factors involved, namely, the force acting, the distance through which the force acts, and the time it is acting. Knowing these, the rate of

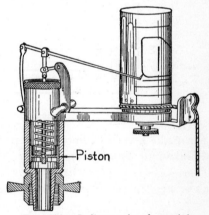

Fig. 380.—Indicator for determining the mean effective pressure of a gas engine.

power generation in foot-pounds per minute can be calculated and, by dividing by 33,000, the generated horsepower is obtained.

362. Measurement of Indicated Horsepower.—The power generated in the cylinder of an engine owing to the explosion pressure acting on the piston is termed the *indicated horsepower*, because a device known as an *indicator* (Fig. 380) is necessary in determining it. The indicator is attached to the cylinder at the closed end. The combustion chamber is connected to the small indicator cylinder by means of a small passage so that the pressure existing in the engine cylinder at any time reacts upon the indicator piston and spring. As this piston moves, it actuates a recording arm bearing a pencil point, and the latter records the pressure graphically on the paper-covered drum. This drum is connected to the crankshaft of the engine so that it revolves back and forth with the crank and piston movement. The pencil, therefore, will have a tendency

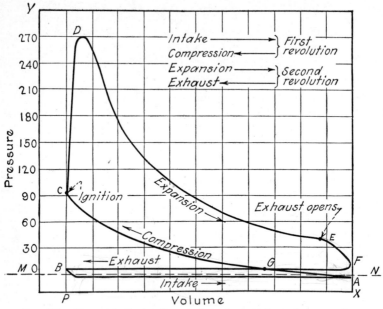

FIG. 381.—Typical indicator diagram for a four-stroke-cycle gas engine.

to make a vertical line on the paper due to the pressure reaction, and a horizontal line due to the drum movement. The result of the two movements will be what is known as an *indicator diagram* (Fig. 381). A specially prepared indicator form or card is used.

Referring to the diagram, the vertical line *OY* represents the pressure, and the horizontal line *PX* represents the piston movement and corresponding volume. The horizontal line *MN* represents atmospheric pressure. Examining the curve, the irregular line *BA*, representing the action in the cylinder on the suction or intake stroke, shows that the existing cylinder pressure is slightly below atmospheric. This, of course, is natural because the cylinder volume is increasing. At the end of the intake stroke, the piston starts back on compression and the pressure increases rather uniformly to about 90 lb., as shown by the curve *AC*. At the point *C*, slightly before the completion of the stroke, the spark ignites the charge and the pressure jumps quickly to about 270 lb., as shown by the curve from *C* to *D*. The curve *ACD*, therefore, represents the action during the entire compression stroke, with *C* indicating the point of ignition. The curve *DE* represents the action on the expansion or power stroke. As the piston moves outward, the cylinder volume increases and the pressure drops rapidly. The exhaust valve opens at E just before the expansion stroke ends and the pressure drops nearly to atmospheric pressure. The exhaust stroke, shown by the horizontal line FB, indicates that there is little change of pressure during this event. At B, the piston is back at its starting point and ready to begin the intake stroke again.

It is thus observed that two irregular closed areas are described by the indicator pencil during one complete cycle. The one is the area $FCDE$ and the other the area ABG. The first or larger area represents the energy applied to the piston in producing power and might be called the positive area. The second or smaller area represents energy tending to retard the movement of the piston, or negative energy. Since this latter area is extremely small, it can be neglected in making calculations.

The next step is to determine, from this indicator diagram, the average working pressure existing in the cylinder during the cycle. This is known as the mean effective pressure (m.e.p.). The procedure consists first in determining the exact area of $FCDE$ in square inches. For this purpose a special device, known as a planimeter, is used. Then, dividing the area by its length—horizontal distance PX—in inches will give the average height in inches. The height multiplied by the scale of the indicator spring (pounds required to deflect the spring 1 in.) will give the average or m.e.p. per square inch of cylinder cross-sectional area.

In order to calculate the i.hp. of an engine, its operating r.p.m., bore of cylinder in inches, and length of piston stroke in feet must be known, in addition to the m.e.p. just explained. Having determined these values, the power may be calculated by means of the following formula:

$$\text{I.h.p.} = \frac{PLANn}{33{,}000 \times 2} \text{ (for four-stroke-cycle engine)}$$

or

$$\text{I.h.p.} = \frac{PLANn}{33{,}000} \text{ (for two-stroke-cycle engine)}$$

where P = m.e.p.

L = length of piston stroke in feet.

A = area of cylinder in square inches

= (bore)^2 × 0.7854.

N = r.p.m.

n = number of cylinders.

For multiple-cylinder engines, the i.hp. of one cylinder is multiplied by the number of cylinders because it is assumed that the m.e.p. of all cylinders is alike.

Example.—What is the i.hp. of a four-cylinder four-stroke-cycle engine if the m.e.p. is 90 lb., cylinder dimensions 4 × 5 in., r.p.m. 1,200?

$$\text{I.hp.} = \frac{90 \times \frac{5}{12} \times 4^2 \times 0.7854 \times 1{,}200 \times 4}{33{,}000 \times 2}$$

$$= 34.27$$

363. Measurement of Belt Horsepower.—Devices used for the determination of the belt or brake horsepower of engines are known as *dynamometers;* dyna or dynamo meaning force or power, and a meter

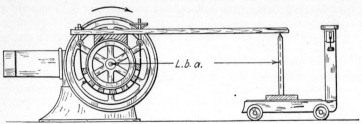

Fig. 382.—Use of Prony brake in determining the horsepower of an engine.

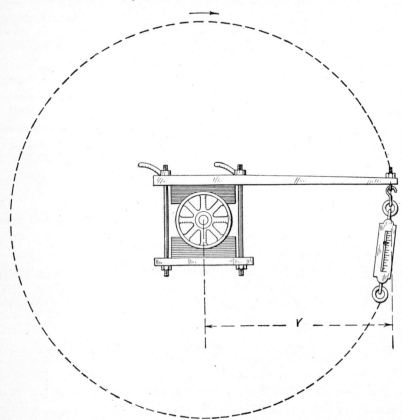

Fig. 383.—Sketch showing principle of operation of a Prony brake.

being a measuring device. Therefore, a dynamometer is a power-measuring device. There are a number of different types of dynamometers, such as the Prony brake (Fig. 382), the electric (Fig. 386), the hydraulic, and the fan type. Of these, the first two named are used most extensively.

364. Prony Brake Dynamometer.—The Prony brake consists of an adjustable friction band that fits on the belt pulley or a special pulley keyed to the engine crankshaft, and an arm from 2 to 5 ft. in length, which rests on or is supported by scales (Fig. 382). To explain its operation more clearly, let us assume that the engine crankshaft is locked so that it cannot turn and that a pair of spring scales are fastened to the outer end of the brake arm (Fig. 383). Now suppose that the brake band is tightened on the pulley but that the brake is rotated on the latter by grasping the scales and applying the rotative force at the outer end of the brake arm. The scales will register a certain number of pounds, depending upon the tension of the brake in addition to the weight of the arm itself. Likewise, as the arm rotates, the point of application of the rotating force will describe a circle whose circumference is equal to $2\pi r$ (Fig. 383). The work done in one turn in foot-pounds is equal to the product of the force in pounds registered by the scales (less the weight of the brake arm) and the circumference in feet of the circle described. By counting the total revolutions made in 1 min., the total foot-pounds of work per minute can be calculated. Dividing by 33,000 gives the horse-power developed. In other words:

$$\text{Hp.} = \frac{\text{pounds on scales (due to friction only)} \times 2\pi r \times \text{r.p.m.}}{33,000}.$$

In actual practice, the brake arm is held in a fixed position as shown by Fig. 382, and the engine pulley turns in the band. This will not alter conditions for the reason that the friction contact will still be the same and will, therefore, create a like pressure on the scales. By measuring the distance r in feet, better known as the length of brake arm (l.b.a.), and obtaining the engine r.p.m., the horsepower output can be calculated as follows:

$$\text{Hp.} = \frac{\text{net load (lb.)} \times \text{l.b.a. (ft.)} \times 2\pi \times \text{r.p.m.}}{33,000}.$$

The net load on the scales is determined by first weighing the brake arm and support, with the brake loose and engine not running, and then subtracting this so-called tare load from the total load on the scales when engine is running under test; that is, the gross load minus the weight of arm and stand (tare load) equals net or friction load. The crankshaft r.p.m. is determined by means of a speed counter or indicator such as that shown in Fig. 385.

Very often it is inconvenient or unsatisfactory to place the Prony brake directly on the pulley as, for example, in the case of a tractor. Under such conditions the brake can be mounted on a special frame with

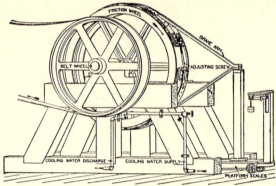

FIG. 384.—Belt-driven Prony brake for testing tractors and other large engines.

a second or belt pulley mounted on the brake-pulley shaft. The engine to be tested drives the brake by means of a belt, as shown in Fig. 384.

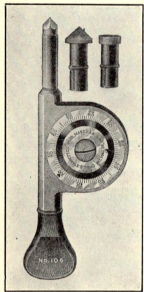

FIG. 385.—Revolution counter or speed indicator.

By obtaining the net load, l.b.a., and brake r.p.m., the horsepower can be calculated as before. The principal objection to a belt-driven Prony brake is that, at heavy loads, some belt slippage is apt to occur that will result in giving a horsepower output slightly less than the engine is actually developing.

The Prony brake is an inexpensive apparatus to construct and is simple and easy to operate as compared to other types of brake dynamometers. However, the friction between the brake and the pulley surface generates considerable heat, depending upon the size and speed of the pulley. This heat has a tendency to cause the frictional contact and the resulting brake load to increase or vary. To eliminate or remedy this undesirable load variation it is often necessary to apply some cooling medium such as water or oil to the surfaces. It is better to apply the cooling liquid continuously in small quantities than intermittently in large amounts.

365. Electric Dynamometer.—For precise power measurements of large, or multiple-cylinder, variable-speed engines, the electric dynamometer is preferred. The complete outfit consists of the generating unit (Fig. 386), the resistor unit (Fig. 387), and the control and instrument board (Fig. 388).

The outer field frame of the direct-current generator is mounted with ball bearings on two heavy iron pedestals as shown, so that it is possible for the entire generator unit to rotate. This is prevented, however, by the torque arm and scales. The generator armature rotates independently in the field frame as in any other generator. The engine to be tested is mounted beside the generator and its crankshaft connected

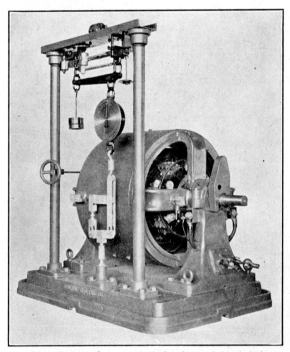

Fɪɢ. 386.—Electric-type dynamometer for determining belt horsepower.

directly to the armature; or, where such is impractical, a pulley and belt drive may be used as in testing a tractor. The load or pressure on the scales is created by the electromagnetic action between the field frame and the armature, which exists in any such machine. That is, as the armature rotates, the electromagnetic field set up tends to cause the field frame to rotate with the armature. This pulling or torque action, as it is called, does make the frame rotate a certain amount and creates a pressure on the scales attached to the torque arm. The electromagnetic field can be closely adjusted so as to increase or decrease the load on the engine by any desired amount. In other words, the stronger the current supplied to the field windings, the greater the load on the engine and the resulting pull or torque. In other respects this device is very much like the Prony brake. That is, by observing the r.p.m. of the armature, the length of the torque arm, and the net load on the scales, the power output can be calculated by the same formula: namely,

$$Hp. = \frac{\text{net load (lb.)} \times \text{l.b.a.} \times 2\pi \times \text{r.p.m.}}{33,000}$$

The electric dynamometer is much more expensive to install than the Prony brake and requires more experience and care in operation. However, where a large amount of accurate and precise engine testing is done, it is the most desirable apparatus to use.

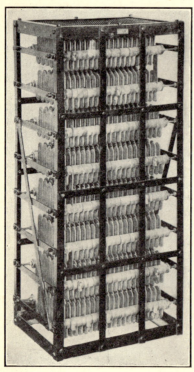

FIG. 387.—Current absorber or resistor for electric dynamometer.

FIG. 388.—Control unit for an electric dynamometer.

366. Determination of Drawbar Horsepower.—It is often desirable to know the pulling power of a tractor under various conditions. The usual procedure is to hitch the machine to some heavy object or load to be pulled or dragged, which will require not more than the maximum pulling power of the tractor. Between the tractor and the load is placed the dynamometer, which must at least record or indicate the pulling effort in pounds required to drag the load at a certain rate of travel. Such a device is shown in Fig. 389. The rate of travel can be determined by observing the time required to cover some definite measured distance such as 500 ft. Then, the horsepower developed can be calculated as follows:

$$\text{Hp.} = \frac{\text{average lb. pull} \times \text{rate of travel (ft. per min.).}}{33,000}$$

Figures 390 and 391 illustrate special types of self-recording drawbar dynamometers. The one (Fig. 390) records directly a certain value from

Fig. 389.—Plain dial-type, indicating drawbar dynamometer.

Fig. 390.—An integrating and recording drawbar dynamometer.

which the average pull over a given distance, usually 50 ft., can be determined. In the other (Fig. 391), the average pounds pull can be determined very accurately by checking the chart record according to a certain recommended procedure. In addition to the pull, the rate of travel is also accurately recorded on the chart, and it is unnecessary to measure off a given distance and observe the time required to cover this distance.

367. Operating Characteristics of Multiple-cylinder Engines.— Figure 392 shows typical horsepower and torque curves for a multiple-cylinder engine. It is noted that the horsepower generated varies almost directly as the crankshaft speed, up to a certain point when it drops off. The reason for this is that, as very high piston speeds are reached, the fuel charge per stroke decreases slightly owing to the effect of the inertia

FIG. 391.—Hydraulic-type recording drawbar dynamometer.

and friction of the gas as it rushes through the manifold and valves into the cylinders.

The meaning of torque can best be understood by reference to the Prony brake (Fig. 382). Here we note that the outer end of the brake

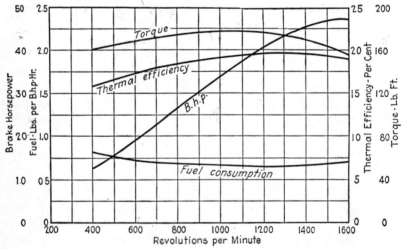

FIG. 392.—Curves showing typical operating characteristics of a multiple-cylinder gasoline engine.

arm exerts a certain pressure on the scales, depending upon the tension of the brake, the length of the arm, and the engine speed. The product of the brake-arm length in feet and the pounds pressure exerted is known as torque and is expressed in pounds-feet. This should not be confused with the unit of work or foot-pound.

Referring again to the torque curve in Fig. 392, it is observed that the torque varies somewhat with the engine speed but is greatest at a comparatively low speed. The dropping off is again due to the smaller fuel charge taken in with increasing piston speeds. However, the torque does not decrease sufficiently with increased engine speed to affect the horsepower developed. In other words, even though the torque begins to drop off at a comparatively low engine speed, it drops very slowly as compared to the rate of increase in r.p.m. Consequently, the horsepower continues to increase until a much higher engine speed is reached.

368. Mechanical Efficiency.—The mechanical efficiency of an engine is the ratio of its b.hp. to its i.hp., that is,

$$\text{Mechanical efficiency (per cent)} = \frac{\text{b.hp.}}{\text{i.hp.}} \times 100.$$

The mechanical efficiency of internal-combustion engines varies from 75 to 90 per cent, depending upon the load, speed, and other factors.

369. Thermal Efficiency.—The thermal efficiency of an engine is the ratio of the output in the form of useful mechanical power to the power value of the fuel consumed; that is,

$$\text{Thermal efficiency (per cent)} = \frac{\text{b.hp.}}{\text{power value of fuel}} \times 100.$$

In order to determine the thermal efficiency of an engine, the quantity of fuel consumed and the power generated in a given time must be measured. Then this power and fuel must be converted into a common form; that is, the power must be converted into heat-energy units, or the fuel into mechanical power units. The heat unit used is the British thermal unit (B.t.u.). It has been determined that 1 B.t.u. of heat is equivalent to 778 ft.-lb. of work. This is known as the mechanical equivalent of heat. Since 1 hp. equals 33,000 ft.-lb. per minute,

$$1 \text{ hp.} = \frac{33,000}{778}$$
$$= 42.42 \text{ B.t.u. per minute}$$

and 1 hp. generated for an hour is equal to 60 × 42.42, or 2,545 B.t.u.

Again the heat value in B.t.u. has been determined for the various engine fuels. Gasoline, kerosene, and other petroleum fuels contain approximately 20,000 B.t.u. per pound (see Table VII). A gallon of gasoline (6.2 lb.) contains 6.2 × 20,000 or 124,000 B.t.u. Therefore, if an engine were 100 per cent efficient in its operation, that is, if the entire heat value of the fuel burned were converted into useful power, then, a gallon of gasoline would produce 124,000/2,545 or 48.7 hp. for 1 hr.

Such, however, is obviously impossible because every engine wastes a large quantity of heat through the exhaust, the cylinder walls and the piston, by friction, and so on.

The thermal efficiency of internal-combustion engines varies from 12 to 25 per cent, depending upon the type of engine, speed, load, design, and other factors. For example, the ordinary stationary engine uses about 0.8 lb. of fuel per horsepower generated per hour. The input is

$$0.8 \times 20,000 = 16,000 \text{ B.t.u. per hour.}$$

The output is

$$1 \text{ hp. per hour} = 2,545 \text{ B.t.u.}$$

Then,

overall

~~Thermal~~ efficiency $= \dfrac{\text{output}}{\text{input}}$

$$= \dfrac{2,545}{16,000} \times 100$$

$$= 15.9 \text{ per cent.}$$

Certain Diesel-type engines often burn as low as 0.5 lb. of fuel per horsepower per hour. Calculating as above:

$$0.5 \times 20,000 = 10,000 \text{ B.t.u. per hour}$$

$$\text{Thermal efficiency} = \dfrac{2,545}{10,000} \times 100$$

$$= 25.5 \text{ per cent.}$$

370. Brake Mean Effective Pressure.—The brake mean effective pressure of an engine is that portion of the total or indicated mean effective pressure that is actually consumed in generating the useful or brake horsepower. Therefore it is equal to the indicated mean effective pressure multiplied by the mechanical efficiency and is calculated by making the necessary substitutions for b.hp. and i.hp. and solving the formula

Brake mean effective pressure

$$= \text{indicated mean effective pressure} \times \dfrac{\text{b.hp.}}{\text{i.hp.}}$$

371. Factors Affecting Fuel Consumption and Efficiency.—One of the fundamental objectives in the design, construction, and operation of any internal-combustion engine is to secure the greatest possible efficiency without interfering with other considerations involved in practical adaptability to a particular purpose. As already explained, efficiency in an engine means obtaining the greatest possible power out of it with the lowest possible fuel cost—not necessarily fuel consumption. For example, an engine might burn either of two fuels satisfactorily but use slightly less of one than of the other, indicating that the one is the better to use.   However, if the second fuel costs considerably less per gallon, it might be the more economical fuel to use in the engine.

The actual fuel consumption of engines is usually expressed as "lb. per hp.-hr." The total fuel consumption of engines varies, of course, according to the size, the power generated, and the length of time in operation. However, when reduced to the "lb. per hp.-hr." basis, the fuel consumption of two engines of entirely different size and type may be very nearly the same.

In general, the fuel consumption of all gasoline- and kerosene-burning engines, such as stationary farm engines, automobile engines, tractor engines, airplane engines, and so on, is about the same when reduced to the "lb. per hp.-hr." basis, provided they are all operated under a one-half to full load.   Such engines seldom burn less than 0.6 lb. of fuel per hp.-hr.   The average is around 0.8 lb. and may run as high as 0.9 lb.   Diesel-type and similar high-compression heavy-duty oil-burning engines often show a fuel consumption of 0.4 to 0.5 lb. per hp.-hr., and seldom use more than 0.6 lb.

Given a certain type of engine burning a given fuel, the most important factors affecting its economical and efficient operation are:

1. Normal operating compression pressure.
2. Operating load: light, medium, or heavy.
3. Mechanical condition:
   a. Ignition correctly timed.
   b. Valves correctly timed.
   c. Fuel mixture properly adjusted.
   d. Piston rings and cylinder not badly worn.
   e. Bearings properly adjusted.
   f. Properly lubricated.

372. Compression and Efficiency.—Referring to Fig. 393, it is observed that, theoretically, the greater the compression pressure, the higher the thermal efficiency.   It is difficult and impractical to take advantage of this, however, in the carbureting type of engine, such as the farm, automobile, or tractor engine, for the reason that too high a compression causes preignition or detonation of the fuel mixture.   In other words, the high compression produces a higher cylinder temperature and the mixture ignites and explodes too early.   In the Diesel and similar types of heavy-duty oil engines, high compression is practical because the fuel charge does not enter the cylinder until the piston is ready to receive the explosion.   Engines of this type, therefore, are somewhat more efficient than carbureting engines.

373. Effect of Load on Efficiency.—The curves (Fig. 394) show that any engine, when operating at a very light load, will use more fuel per horsepower-hour. As the load increases, the fuel consumption decreases until, at about nine-tenths of the maximum power developed by the

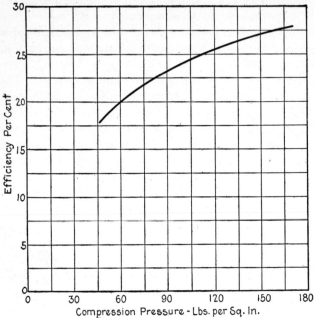

Fig. 393.—Curve showing effect of compression pressure on efficiency.

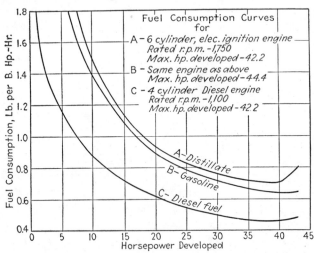

Fig. 394.—Fuel consumption curves for tractor engines.

engine, it gives the most economical results. It is obvious that a certain amount of fuel is required to operate the engine itself, that is, to supply the power necessary to overcome friction. Furthermore, this friction power and the amount of fuel required to overcome it remain practically constant, regardless of the load on the engine. Therefore, as the load increases, this quantity of fuel required to overcome friction becomes less and less in proportion to the total amount burned.

Engines operating at, or very near, their maximum output show a slightly greater fuel consumption than at nine-tenths load for the reason that the cylinder temperature becomes too high to permit the engine to derive the greatest possible power from the fuel. This explains the straightening of the curves (Fig. 394) at the full load.

This characteristic of engines to give the best fuel economy at medium to heavy loads means that it is important always to use an engine that fits the job. That is, do not use a 5-hp. engine to operate a machine requiring only 1 hp. On the other hand, it is not good practice to use an engine that is too small and will be overloaded.

FORMULAS FOR CALCULATING HORSEPOWER

A number of formulas have been suggested for calculating the horsepower output of a given engine using certain cylinder dimensions and other values. Although none of these should be considered as reliable and accurate when applied to any engine because of the great variation in certain factors involved, such as piston speed and compression pressure, they can often be used to advantage to check up on the engine's rating or to determine its approximate power output.

374. S.A.E. Formula.—This formula, originally adopted by the American Licensed Automobile Manufacturers (A.L.A.M.), and later by its succeeding organization, the Society of Automotive Engineers (S.A.E.), for determining the approximate brake horsepower of automotive engines and engines of similar type is

$$\text{Hp.} = \frac{D^2 \times N}{2.5}$$

where D = bore of cylinder in inches.

N = number of cylinders.

The S.A.E. formula is a simplification of the i.hp. formulas previously explained. It assumes that the mean effective pressure is 90 lb. per square inch, that the piston speed is 1,000 ft. per minute, and that the mechanical efficiency is 75 per cent.

References

A Drawbar Dynamometer and Its Use in Soil Tillage Experiments, *Univ. Mo., Agr. Exp. Sta., Research Bull.* 226.

Cornell Univ., N. Y. State Coll. Agr., Ext. Bull. 85.

ELLIOTT: "Automobile Power Plants."

FAVARY: "Motor Vehicle Engineering—Engines."

HELDT: "The Gasoline Automobile," Vol. I.

Iowa State Coll., Agr. Exp. Sta., Bull. 240.

JUDGE: "The Testing of High Speed Internal Combustion Engines."

KERSHAW: "Elementary Internal Combustion Engines."

Mechanical Tests on Tractor Farming Equipment, *Mont. State Coll., Agr. Exp. Sta., Bull.* 243.

Some Power Studies through the Use of the Ohio Recording Belt Dynamometer, *Trans. Amer. Soc. Agr. Eng.,* Vol. 17.

CHAPTER XXVI

TRACTOR POWER TRANSMISSION—RATINGS—
NEBRASKA TESTS

The present-day farm tractor is designed to deliver power in three distinct ways, namely, (1) by a pulley and belt, (2) by pulling effort at the drawbar, (3) by means of a power take-off.

375. Power Rating.—As a rule, tractor sizes are designated according to the power capable of being generated at the pulley and at the drawbar. Obviously, the maximum power that can be generated by the power take-off will be practically the same as the belt power. However, when a tractor is operating a machine through the power take-off, the machine is usually being pulled. Therefore, only a certain fraction of the engine power is available at each point.

The power of tractors is expressed as so many horsepower, either belt or drawbar, or both. Formerly a manufacturer designated the various models according to their approximate power output; for example, a 15-30 tractor would deliver about 15 hp. at the drawbar and 30 hp. at the belt. This system is going out of use, however, and certain models are now designated by a letter as "Model W." If a manufacturer makes but one size, the trade name only is sufficient. This does not mean that modern tractors are not given a power rating. Most manufacturers give their machines such a rating by stamping it on the serial-number plate attached to the machine. If the power is not stamped on the machine it can usually be found in the catalogue specifications.

Owing to the crude construction and excessive weight, early tractors seldom delivered more than one-half of the engine power at the drawbar. Therefore the rating was always expressed as 10-20, 15-30, and so on. With the adoption of higher grade, lighter weight materials, enclosed transmissions running in oil, antifriction bearings, and improved methods of workmanship, this power loss has been greatly reduced until at the present time it is not uncommon for a tractor, to deliver 75 per cent of its belt power at the drawbar. For example, the Nebraska tractor tests[1] show that 26 current-model wheel-type machines delivered 73.4 per cent of the maximum developed belt horsepower to the drawbar. Twenty-five crawler machines delivered 84.4 per cent. In addition to those mentioned above, such factors as wheel or track construction, lug

[1] *Neb. Agr. Exp. Sta., Bull.* 292, 296, and 304.

equipment, slippage, and soil conditions also materially affect the drawbar power output.

The American Society of Agricultural Engineers has adopted the following standard specifications for farm-tractor rating and testing:

TRACTOR RATING SPECIFICATIONS

Belt Rating.—The belt-horsepower rating of the tractor shall not exceed 85 per cent of the maximum corrected horsepower which the engine will maintain by belt at the brake or dynamometer for 2 hr. at rated engine r.p.m., the test to be carried out as specified herein.

Drawbar Rating.—The drawbar rating of the tractor shall not exceed 75 per cent of the maximum corrected drawbar horsepower developed at a rate of travel recommended for the ordinary operation of the tractor, under conditions of testing as specified herein.

TESTING PROCEDURE

Nature of Tests.—The following rating tests are to be conducted in the order given on three or more tractors picked at random from factory stock, run by the engineer or engineers conducting the test; the averages of all tests are to be used in determining the results.

Test A (Limbering-up Run).—Before a test is undertaken it is important that the tractor shall have been in operation for a sufficient length of time to attain proper operating conditions throughout so that the results of the test shall express the true working performance. The tractor or tractors to be tested shall therefore be submitted to "limbering-up" runs on the drawbar of 12 or more hours' duration. Drawbar loads of approximately one-third, two-thirds, and full load shall be pulled by the tractor during the runs, each load being drawn for approximately an equal length of time, the lighter loads being used first.

Test B (Maximum Brake-horsepower Test).—The tractor engine is to be tested in the belt with the governor set to give full opening of governor valve, and the carburetor set to give maximum power at rated engine crankshaft speed. (The rated speed is that which the manufacturer recommends for the engine under normal load.) The test shall begin after the temperature of the cooling fluids and other operating conditions have become practically constant. The duration of this test shall be 2 hr. of continuous running with no change in engine adjustments. If the speed should change sufficiently during the test to indicate that the operating conditions had not become constant when the test was started, the test will be repeated with the necessary change in load. (The term "load," as used in this connection, means pounds on dynamometer or brake scale.) Minor changes in load to be made to maintain rated speed, and the average load and average speed for the period shall be used in computing the horsepower. All belt-horsepower tests must be made with an electric dynamometer, or with an accurately tested Prony brake or other accurate power-measuring device. Correction shall be made for temperature and altitude effect on horsepower output on maximum test only. Standard conditions of barometric pressure of 29.92 in. Hg. and a temperature of 60°F., or 520°Abs. shall be used.

The following correction formula shall be used (on maximum test only):

$$BHPc = BHPo \times (Ps \div Po) \times \sqrt{To \div Ts}$$

where $BHPc$ = corrected brake horsepower.
   $BHPo$ = observed brake horsepower.

Po = observed barometric pressure in inches of mercury.
Ps = standard barometric pressure in inches of mercury.
To = observed absolute temperature in degrees Fahrenheit.
Ts = standard absolute temperature in degrees Fahrenheit.

Test C (*Maximum Drawbar Horsepower*).—After the tractor has attained proper working conditions it shall be subjected to a series of drawbar tests, preferably on level ground, and of such a nature that it will provide a firm footing and offer tractive resistance simulating that of grain stubble ground in good plowing condition. The loads shall be successively increased or decreased to a point where the tractor can sustain a constant pull for a distance of not less than 1,000 ft. with the average engine r.p.m. maintained within 5 per cent of its rated full-load speed and a wheel slippage of not more than 10 per cent, the wheel slippage to be secured as follows: The tractor shall be driven without any drawbar load for a distance sufficient to give 10 revolutions of the drive wheels. This distance is then accurately measured and is used as the basis for computing wheel slippage. The number of revolutions of all drivers shall be counted for the entire distance through which the load is pulled. The drawbar pull shall be measured by means of an accurately calibrated draft dynamometer or draft-measuring device placed between the tractor and the load. The actual distance traveled shall be used in calculating the horsepower, no allowance being made for wheel slippage. During this test the tractor shall be run in the gear recommended for plowing under favorable conditions.

FUELS

These tests will be made on the lowest commercially available grade of fuel which the manufacturer recommends.

LUBRICANTS

The lubricants used in these tests shall be such as are regularly recommended by the manufacturer for the tractor.

BELTS

The belt or belts used in these tests shall be such as the manufacturer recommends for use with the tractor in ordinary operation. No allowance will be made for belt losses.

376. Tractor Horsepower Testing.—The belt horsepower of a tractor can be conveniently tested with a Prony brake (Fig. 384) or an electric dynamometer as described in Chap. XXV. The drawbar power is determined by placing a drawbar dynamometer in the hitch between the tractor and the load to be pulled, as explained in Chap. XXV.

NEBRASKA TRACTOR TESTS

A law known as the Nebraska Tractor Law, which was enacted in that state and put into effect in 1919, has been of far-reaching effect, particularly in the design, construction, and operation of the farm tractor. The provisions of the law, the tests involved, and the results are outlined in detail in the discussion that follows.

I. Nebraska Tractor Law

Full text of the law as amended by Senate File 280, 45th Session (in force March, 27, 1929), is as follows:

Sec. 1. That on and after July 5, 1919, no tractor or traction company shall be permitted to sell or dispose of any model or type of gas, gasoline, kerosene, distillate, or other liquid fuel tractor engine in the State of Nebraska, without first having said model tested and passed upon by a board of three competent engineers who are or shall be under the control of the State University management. Each and every tractor presented to the State University management for testing, shall be a stock model and shall not be equipped with any special appliance or apparatus not regularly supplied to the trade. Any tractor not complying with the provisions of this section shall not be tested under this act, nor the results certified. Provided, no official tractor tests shall be conducted during the months of December, January, and Feb-

Fig. 395.—Tractor testing at the University of Nebraska—track-type tractor being tested for drawbar horsepower.

ruary. Applications for the test of a tractor shall be made to the testing board of engineers and shall be accompanied by specifications of the tractor required by said board of engineers, and by the fee specified in Sec. 5535, Compiled Statutes of Nebraska for 1922. If the application with specifications and fee is submitted during December, January, February, or at any other time when the test cannot be started at once, the State Railway Commission, with the approval of the said board of engineers, may issue a temporary permit for the sale of tractors of the model specified in the application for test, the date on which said temporary permit shall terminate to be fixed by the said board of engineers. All temporary permits shall be conditioned upon such tractor as is covered thereby being tested at the earliest available date, and the tractor company to which a temporary permit has been issued shall submit a tractor for test which conforms to the specifications filed with the application, said tractor to be delivered for test at any time specified by the board of engineers. Upon failure so to do, all such fees deposited by said companies shall be forfeited to the State of Nebraska, and in addition such companies shall be liable to the penalties prescribed by Sec. 5540, Compiled Statutes of Nebraska for 1922, and shall never thereafter be issued any temporary permit whatever. Provided, however, that all sales of tractors upon which a temporary permit has been issued, shall be made subject to the final official test and approval of the model.

Sec. 2.   Such tests to consist of endurance, official rating of horsepower for continuous load, and consumption of fuel per hour or per acre of farm operations; the results of such tests to be open at all times to public inspection.

Sec. 3.   The State University management after having duly tested any model of liquid fuel traction engine shall certify the results to the State Railway Commission. Prior to the issuing of a permit by said commission to any liquid fuel traction engine company to do business in the State of Nebraska, the official tests of the University shall be compared with the representations of the tractor company as to horsepower rating for not less than ten consecutive hours of continuous load, fuel used for developing such horsepower, and any other representation such company shall make, and in case any such representations shall be found false, the State Railway Commission shall deny the company manufacturing or assembling such tractor, the right to do business in the State of·Nebraska, except as hereinafter provided.

Sec. 4.   Likewise, the State Railway Commission shall deny to any liquid fuel tractor company the right to do business in the State of Nebraska which shall be found, on complaint of two or more *bona fide* customers residing within the State, to fail to maintain an adequate service station with full supply of replacement parts within the confines of the State and within reasonable shipping distance of said customers.

Sec. 5.   Application to the State University management for test of a tractor shall be accompanied by a fee of two hundred and fifty dollars ($250.00) paid to the State University management as a partial reimbursement for the cost of such test. Payment shall also be made upon presentation of the proper statement for costs of all fuel and oil consumed in making said test.

Sec. 6.   Provided further, that the failure of any model of tractor to come up to the representations of the company manufacturing or assembling same, shall not prevent said company from placing on the market other models of tractors that do comply with specifications and ratings.   Any model of tractor that fails in the official test to come up to the company's own specifications may be retested after alteration and remodeling.

Each and every permit issued under this act shall specify the model or models included in such permit to sell.

Sec. 7.   The report of the official tests herein provided shall be posted in the Agricultural Engineering Department of the State University and in such other places as may be designated by the State University management.   The same shall be incorporated in the annual report of the State Railway Commission.

Sec. 8.   Provided, that no tractor company shall use the results of said tests in such manner as would cause it to appear that the State University management intended to recommend the use of any given type or model of tractor in preference to any other type or model.   It shall be unlawful for any tractor company operating in the State of Nebraska, to publish extracts from such official tests for advertising purposes without publishing the entire report.

Any infraction of the foregoing provisions shall result in the suspension or denial to any company so offending the permission to do business in the State at the discretion of the State Railway Commission.

Sec. 9.   Tractors shall be tested by the State University management in the order in which they are presented for such tests, and no discrimination shall be made for or against any tractor company in any manner whatsoever.   Complaints against the violation of this provision shall be heard and adjusted by the State Railway Commission.

Sec. 10.　After July 15, 1919, any gas, gasoline, kerosene, distillate, or other liquid fuel tractor or traction company selling or offering for sale in the State of Nebraska or any automobile, implement or other company or individual operating in behalf of such tractor company or on their own behalf who shall after the date specified, sell or offer for sale in the State any model of liquid fuel tractor engine, without having in his possession a permit from the State Railway Commission to sell such model of tractor as he is offering for sale, the same shall be deemed guilty of a misdemeanor.

On conviction such misdemeanor shall be punished by a fine of not less than $100.00 nor more than $500.00 for each offense, in the discretion of the court.

Sec. 11.　The State Railway Commission shall have full authority to enforce the provisions of this act, both by denial of permit to do business in the State and by due process of law to compel compliance therewith.

II. The Test

PROCEDURE—EQUIPMENT

Application for Test.—Tractor companies desiring to have tractors tested are required to make application on a form supplied by the Department of Agricultural Engineering, College of Agriculture, Lincoln, Nebraska.　This form calls for detailed specifications of the tractor.

Dates for test are given in the order in which the specifications and application blanks, mentioned above, together with the testing fee of $250.00, are received at the Department of Agricultural Engineering, College of Agriculture, Lincoln, Nebraska. Actual test depends on date of arrival of tractor and suitable weather conditions.

It is assumed that the application for test is made by the tractor company with the knowledge that it is not an experimental, but an official service test to be run under specifications submitted with the application and under the testing rules in force.

There shall be no revision of specifications after the test starts, nor any change of parts or equipment except to make replacement for that which may be defective or broken.

Experimentation desired by the manufacturer must be deferred until after the official test is completed and is then contingent on available time.　Any data and results thus attained are for the information of the manufacturer and will not be included in the official report.

1. All tractors will be delivered for test to the Agricultural College, Lincoln, Nebraska.

2. *Fuels.*—All tests are made on the lowest grades of fuels sold throughout Nebraska on which the tractor manufacturer claims that his tractor will operate; that is, if the manufacturer claims that his tractor will operate only on gasoline, it is tested on the lower grades of gasoline sold within Nebraska.　If the manufacturer claims that the tractor will operate on two or more distinct fuels (for instance, gasoline and kerosene) it is tested on the least volatile of these fuels, namely, kerosene.

Fuels are supplied by the University and charged to the manufacturer at current prices.

All fuels are measured in accordance with Art. 7.

3. *Lubricants.*—Manufacturers are requested to specify the kinds and grades of lubricant to be used on the different parts of each tractor.　These lubricants are supplied by the University and charged to the manufacturer at current prices.

All engine oils are measured in accordance with Art. 7.

4. *Manufacturers' Representatives and Tractor Operators.*—Representatives of manufacturers are welcome and urged to be present during the test but none of them

are permitted to handle the tractor except on the limber-up run. They are not permitted to manipulate any testing equipment during the test except by authority from the engineer-in-charge or some person designated by him.

The manufacturer should have one representative present during the test whom he shall designate as his representative with authority to act in an official manner should the necessity arise.

5. *Belt Tests.*—These tests are made with belt-driven electric dynamometer.

Tractors are given credit only for the power delivered to the dynamometer through a suitable belt. The following belts are supplied by the University:

> One 5-in. single first-quality leather belt, 50 ft. long
> One 6-in., 5-ply friction surface rubber belt, 60 ft. long
> One 7-in., 4-ply friction surface rubber belt, 70 ft. long
> One 8-in., 4-ply friction surface rubber belt, 80 ft. long
> One 10-in., 5-ply rubber belt, 100 ft. long.

The tractor manufacturer may select any of the belts mentioned above or may supply his own belt.

6. *Drawbar Tests.*—These tests are made on a half-mile track. This track is not level, but has several short grades, none of which is greater than $3\frac{1}{2}$ per cent. The condition of the track is kept as nearly uniform as possible by dragging and rolling when necessary.

To provide a drawbar load that can be kept constant during the test and which can be adjusted to suit different sizes of tractors, a dynamometer car or loading machine is used. The drawbar pull is registered by a hydraulic traction dynamometer, the draft unit of which is attached between the tractor and the load pulled.

7. *Measuring Fuel, Oil, and Water.*—The quantity of fuel used in each part of the test (except Art. 8a) is determined by weight and the quantity reduced to gallons at 60°F. For brake tests a tank is placed on a scale and set at the same height as the tank on the tractor. Fuel is drawn from this tank on the scale during the tests.

For the 10-hr. drawbar test either of the following methods may be used:

First Method: Fill the tank to a measured level at the beginning of the test. Fill to the same level at the end of the test, weighing the fuel put in.

Second Method: Drain the tank at the beginning of the test. Fill the tank, weighing the fuel put in. Drain the tank again at the end of the test, weighing the fuel drawn out.

The quantity of oil used is determined with standard gallon, quart, and pint measures, or by weight when more convenient. The quantity used in the complete test is determined as accurately as possible. In most cases it is not practical to determine the amount of oil used in each separate part of the test.

Quantity of water used in each part of the test (except Art. 8a) is determined by measuring the height of water in the radiator or tank at the beginning of the test and filling to the same level at the end of the test, weighing the water added.

RULES AND DETAILED DESCRIPTION OF TEST

8a. *Limbering-up Test.*—The principal object of this test is to take out the stiffness, likely to be found in a new machine, and give an opportunity to check the condition of the tractor and ascertain if all parts are working normally. Beginning with this run, a record is kept of repairs, adjustments, oil, loads carried, and the actual running time. The tractor crankcase is drained and supplied with fresh oil and tank supplied with fuel and water at the beginning of this test.

Tractors in this test are used to pull drags and rollers in maintaining the testing track and for any other drawbar work that may be convenient. The loads pulled are: Approximately one-third load for 4 hr., two-thirds load for 4 hr., full load for 4 hr., a total of 12 hr. actual running time. Reasonable additional time will be given for this test if desired by the manufacturer.

During this test only, the tractor is operated by an employee of the manufacturer and he shall equip it with such accessories as lugs, extension rims or other equipment to be used with, but not assembled on, the tractor when shipped. On all subsequent tests the tractor is operated by a University employee.

9*b. Maximum Brake-horsepower Test.*—The object of this test is to check or establish ratings and to determine adjustments for operating conditions. The tractor is run under full load until it is thoroughly warmed up and has reached a condition of constant operating temperatures.

All adjustments are made to secure maximum output from the engine at rated r.p.m. The governor is set to hold the throttle valve to the extreme open position, the spark is set to give best results and the carburetor is carefully adjusted to that point at which no further additional amount of fuel gives an increase in power output at the rated speed of the engine. After uniform operating conditions are reached, this test is continued for 2 hr.

The record of this test will not appear in the published report.

The purpose of all tests after *b* is to determine performance both in horsepower and economy, with adjustments not made for spectacular performance, but in agreement with the best engineering opinion and field practice.

9*c. Operating Maximum Brake-horsepower Test.*—The object of this test is to determine the maximum power developed and the fuel consumption at a carburetor setting that is practicable for field operations.

The idling adjustment, if present, is first checked. The carburetor is so adjusted that the rated r.p.m. of the engine is maintained with wide-open throttle and at a setting which is to be used on this and all subsequent tests. The manufacturer may specify the carburetor setting to be used. In case he does not so specify, it will be made for approximately 97 per cent of the load determined in test *b*. The duration of this test shall be 1 hr.

10*d. Brake-horsepower Test at Rated Load.*—The object of this test is to determine if the tractor will carry its rated load in the belt and to secure a record of fuel consumption and other operating data.

The tractor is given, as nearly as possible, its rated load. The governor is set to run the engine at its rated speed. The same setting of the carburetor is used as in test *c*, Art. 9. The duration of this test shall be 1 hr.

11*e. Brake-horsepower Test at Varying Load.*—The object of this test is to show the governor control of the speed, also fuel consumption when the load varies.

All adjustments are as in test *d*, Art. 10.

Time and loads (pounds on scale) are as follows:

Twenty minutes rated load.
Twenty minutes no load.
Twenty minutes one-half load.
Twenty minutes maximum load.
Twenty minutes one-fourth load.
Twenty minutes three-fourths load.

The duration of this test is 2 hr. continuous running with no change of adjustments. Any special attention required or unsatisfactory performance of the tractor noted during any part of this test will be mentioned in the report.

12f. Drawbar Test at Maximum Load.—The object of this test is to determine the maximum horsepower the tractor will develop in each gear (except reverse) with all adjustments as in test *c*, Art. 9. During this test the drawbar pulls are secured over distances of 1,000 ft. on the testing track.

Where no brake tests are possible a preliminary test is made on the drawbar, similar to the preliminary test described in test *b*. The procedure then followed is as outlined in test *c*, Art. 9.

13g. Drawbar Test at Rated Load.—The object of this test is to determine if the tractor can pull its rated load continuously and to secure a record of fuel consumption on drawbar work.

The governor is adjusted to give rated speed of the engine with the tractor pulling rated load, and in the gear recommended by the manufacturer for plowing or ordinary farm work. The carburetor has the same adjustments as in test *f*.

The duration of the test is 10 hr. actual running time, as nearly continuous as possible, with constant load. A record is made of the time and reason for each stop; also of any adjustments or repairs made on the tractor.

OPERATION OF ONE-CARBURETOR-SETTING PLAN[1]

At the beginning of the tractor-testing season of 1928 a new procedure was incorporated in the method of testing tractors. This plan is as follows:

After the limber-up run of 12 hr. is completed, the tractor is given a series of brake-horsepower tests, the first being a 2-hr. test with the engine developing maximum horsepower at rated engine speed. This maximum-horsepower test is carried out according to the provisions of the tractor-rating code developed by the American Society of Agricultural Engineers and the Society of Automotive Engineers. The purpose of this test is to check the manufacturer's rating of the tractor and in case this rating is higher than the code rating, attention is called to this fact in the official report. This 2-hr. maximum test is not included in the report of the official test because duplication of the results attained would scarcely be possible outside of the laboratory.

Upon completion of the 2-hr. maximum test, a series of experimental runs of 20 min. duration is made. Starting with the 100 per cent maximum, all adjustments are made to bring about maximum horsepower at the rated engine speed. The load on the dynamometer is then changed to give rated horsepower at rated engine speed and a run of 20 min. is made. After completing this rated-load run at maximum carburetor settings, the adjustments are all changed back to the 100 per cent maximum test and former readings verified. The load on the dynamometer scales is next reduced 1 per cent and the carburetor adjusted to a leaner mixture until this load is carried at rated engine speed. A run is then made at rated load with the above mentioned carburetor setting. The same procedure is carried out for 98, 97, 96 per cent, and so on, of the ultimate maximum.

Figures 396 and 397 present in curves the data from two such preliminary tests on two different tractors that appeared for test during 1928.

After the completion of the preliminary run and the plotting of the data in curves, the manufacturer is asked to select the particular carburetor setting that he desires to be used in the official tests. All official tests are then made on the setting chosen by him.

[1] *Neb. Agr. Exp. Sta. Bull.* 233.

Referring to the curves (Figs. 396 and 397), it is observed that 97 per cent of the ultimate maximum is the most economical setting in both cases; however, experience has shown that this point is usually too lean to give satisfactory performance on the

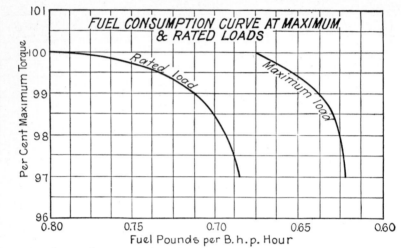

FIG. 396.—Curves showing relation of carburetor setting to fuel economy and power generation for a certain tractor.

drawbar tests and particularly so if the weather is cool. While pulling a maximum drawbar load an engine operating at or near the most economical fuel mixture is rather sensitive to temperature changes of either the cooling fluid, fuel, or air. The

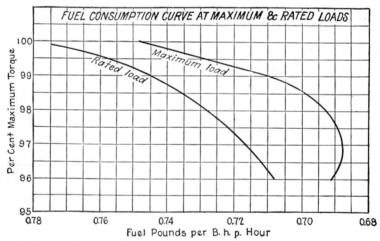

FIG. 397.—Curves showing relation of carburetor setting to fuel economy and power generation for a certain tractor.

drawbar load also fluctuates to a much greater extent than the brake loads. On the maximum drawbar tests it is necessary to stop at the end of each run to take certain readings and also to calculate the data to determine the necessary changes in load for the next trial. The following procedure in obtaining the maximum drawbar pull is used. All runs are made with the butterfly or throttle valve wide open over a 400-ft. course. Therefore, the controlling factor of the speed is the load carried. To arrive

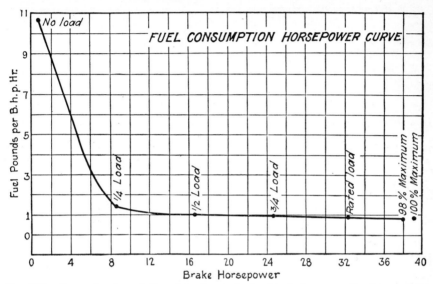

Fig. 398.—Curve showing relation of tractor-engine load to fuel consumption for a certain tractor.

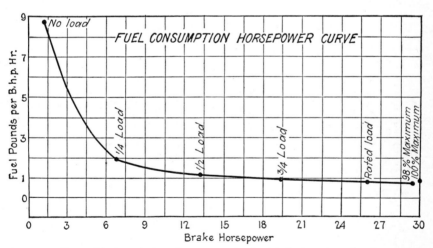

Fig. 399.—Curve showing relation of tractor-engine load to fuel consumption for a certain tractor.

at the maximum horsepower at rated engine speed is of necessity a cut-and-try process. That is, if at the end of a trial, the engine speed is found to be high, another trial is made with a greater load. Hence, due to such stops, the temperature of the cooling fluid and of the motor as a whole drops. If the fuel-air mixture is at or very near the critical mixture, the above mentioned cooling of the engine changes the gasifying point of the fuel to the extent that the actual mixture is too lean to support maximum or full load of the engine.   Experience shows that if stable operation is to be attained, it is safer to select some point on the richer side of the most economical point of the curve as shown in Figs. 396 and 397.

The curves (Figs. 398 and 399) give the fuel consumption at varying loads.   All tests were made on a one-carburetor-setting basis.   Particular attention is called to the curve in Fig. 398.   It is to be observed that the load carries well beyond the half load and almost to the one-fourth load before an excessive increase in fuel consumption takes place.   This curve shows quite clearly the possibilities and advantages of a carburetor designed to operate economically over a wide range of loads without changing its adjustment.   A tractor equipped with this carburetor would show good economy under field conditions where the load varies through quite a wide range.

Results of tests indicate that the manifolding must be very carefully worked out and should be verified by actual tests before final adoption.   The beneficial results of a good carburetor may be almost lost if it is used in conjunction with a poor manifold.   Some manufacturers are fully aware of this problem and are giving it due consideration.

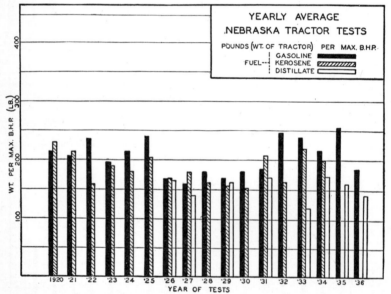

Fig. 400.—Relation of tractor weight to maximum b.hp. (yearly average).

ANALYSIS OF NEBRASKA TESTS

In an analysis of the Nebraska Tractor Tests, Smith[1] reports as follows:

During the 17 years in which Nebraska tractor tests have been made, enough data have accumulated to show some relationship between tractor weight and

[1] Cooperative Tractor Catalog, 22d ed.

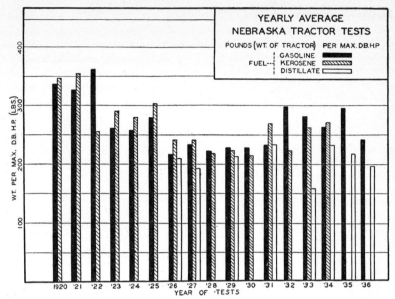

Fig. 401.—Relation of tractor weight to maximum db.hp. (yearly average).

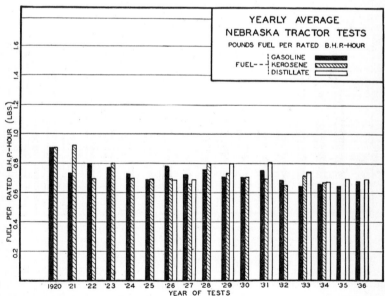

Fig. 402.—Yearly average fuel consumption (lb. per b.hp.) of tractors as shown by Nebraska tests.

horsepower.　For example, Fig. 400 indicates that from 1920 to 1925 inclusive, the tractors weighed approximately 200 lb. per brake horsepower; from 1926 to 1930 inclusive, the averages were all below 175 lb.; and from 1931 to 1936 inclusive, the spread in the averages for the different fuel-burning tractors became more noticeable, the averages for the gasoline-burning tractors being very decidedly above the average for the distillate-burning tractors.　Figure 401 shows much the same trend as Fig. 400.　However, it is apparent that, in recent years, considerably more drawbar horsepower was delivered for an equal amount of weight.

Figure 402 shows the yearly average of the Nebraska tractor tests in pounds of fuel per rated brake horsepower-hour.　It is observed that in 1920, the average fuel consumption on the belt was about 0.90 lb. per horsepower-hour.　From 1922 to 1931, practically all averages were below 0.80 lb., and from 1932 to 1936,

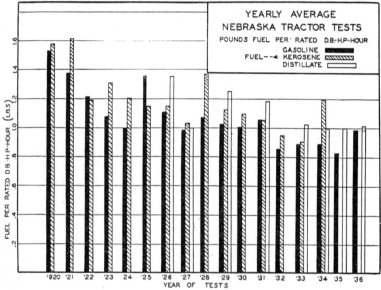

Fig. 403.—Yearly average fuel consumption (lb. per db.hp.) of tractors as shown by Nebraska tests.

the averages are largely below 0.70 lb. of fuel per brake-horsepower-hour.　This shows a very definite gain in fuel economy during these years.　If 0.90 lb. is taken as an approximate average for 1920 and 0.70 lb. for the last few years, there is a gain of 0.20 lb. or 22 per cent.　Making a similar analysis of Fig. 403 shows an average fuel consumption in 1920 of over 1.50 lb. per drawbar horsepower-hour. The trend has been down until that for 1936 is approximately 1.00 lb. per drawbar horsepower-hour.　This represents an increase in efficiency of 33.33 per cent.

The following is a sample of a test report as issued upon completion of a test:

UNIVERSITY OF NEBRASKA—AGRICULTURAL ENGINEERING DEPARTMENT
AGRICULTURAL COLLEGE, LINCOLN
Copy of Report of Official Tractor Test No. 290

Dates of test: October 4 to 12, 1937.
Name and model of tractor: M-M TWIN CITY ZT
Manufacturer: Minneapolis-Moline Power Implement Company, Minneapolis, Minnesota.
Manufacturer's rating: NOT RATED

BRAKE HORSEPOWER TESTS

Hp.	Crank-shaft speed, r.p.m.	Fuel consumption			Water consumption per hour, gal.			Temp., °F.		Barometer in. mercury
		Gal. per hr.	Hp.-hr. per gal.	Lb. per hp.-hr.	Cool-ing	In fuel	Total	Cool-ing med.	Air	

Test B.—Maximum Load. Two Hours

Hp.	Crank-shaft speed, r.p.m.	Gal. per hr.	Hp.-hr. per gal.	Lb. per hp.-hr.	Cool-ing	In fuel	Total	Cool-ing med.	Air	Barometer in. mercury
26.84	1,498	2.905	9.24	0.750	0.000	0.000	0.000	189	61	29.150

Test C.—Operating Maximum Load, One Hour

Hp.	Crank-shaft speed, r.p.m.	Gal. per hr.	Hp.-hr. per gal.	Lb. per hp.-hr.	Cool-ing	In fuel	Total	Cool-ing med.	Air	Barometer in. mercury
25.20	1,500	2.495	10.10	0.686	0.000	0.000	0.000	189	60	29.175

Test D.[1]—One Hour

Hp.	Crank-shaft speed, r.p.m.	Gal. per hr.	Hp.-hr. per gal.	Lb. per hp.-hr.	Cool-ing	In fuel	Total	Cool-ing med.	Air	Barometer in. mercury
23.62	1,502	2.349	10.06	0.689	0.000	0.000	0.000	190	60	29.180

Test E.—Varying Load. Two Hours (20-min. runs; last line average)

Hp.	Crank-shaft speed, r.p.m.	Gal. per hr.	Hp.-hr. per gal.	Lb. per hp.-hr.	Cool-ing	In fuel	Total	Cool-ing med.	Air	Barometer in. mercury
23.72	1,502	2.355	10.07	0.688	.....	.....	.....	190	60	
0.80	1,621	1.147	0.70	9.938	.....	.....	.....	179	59	
12.27	1,582	1.745	7.03	0.985	.....	.....	.....	186	60	
25.22	1,500	2.489	10.13	0.684	.....	.....	.....	190	61	
6.22	1,602	1.416	4.39	1.577	.....	.....	.....	186	60	
17.53	1,562	1.991	8.80	0.787	.....	.....	.....	188	61	
14.29	1,562	1.857	7.70	0.901	0.000	0.000	0.000	186	60	29.190

DRAWBAR HORSEPOWER TESTS

Hp.	Draw-bar pull, lb.	Speed, m.p.h.	Crank-shaft speed, r.p.m.	Slip on drive wheels, %	Fuel consumption			Water used, gal. per hr.	Temp. °F.		Barom-eter, in. mercury
					Gal. per hr.	Hp.-hr. per gal.	Lb. per hp.-hr.		Cool-ing med.	Air	

Test F.—100% Maximum Load. Third Gear

Hp.	Draw-bar pull, lb.	Speed, m.p.h.	Crank-shaft speed, r.p.m.	Slip on drive wheels, %	Fuel consumption			Water used, gal. per hr.	Cool-ing med.	Air	Barom-eter, in. mercury
20.53	2,270	3.39	1,500	4.51	Not Recorded				194	69	29.200

Test G.—Operating Maximum Load

Hp.	Draw-bar pull, lb.	Speed, m.p.h.	Crank-shaft speed, r.p.m.	Slip on drive wheels, %	Fuel consumption			Water used, gal. per hr.	Cool-ing med.	Air	Barom-eter, in. mercury
19.82	3,262	2.28	1,495	8.31	Not Recorded				188	61	29.180
20.37	2,691	2.84	1,500	5.14	Not Recorded				188	68	29.180
19.53	2,149	3.41	1,503	4.20	Not Recorded				189	70	29.225
17.55	1,302	5.05	1,502	3.75	Not Recorded				188	65	29.180

Test H.[1]—Ten Hours. Third Gear

Hp.	Draw-bar pull, lb.	Speed, m.p.h.	Crank-shaft speed, r.p.m.	Slip on drive wheels, %	Gal. per hr.	Hp.-hr. per gal.	Lb. per hp.-hr.	Water used, gal. per hr.	Cool-ing med.	Air	Barom-eter, in. mercury
15.98	1,768	3.39	1,497	4.34	2.145	7.45	0.931	0.016	187	50	29.070

[1] Formerly called RATED LOAD; see *Remarks* 4, page 386.

Fuel, Oil, and Time:

Fuel _____ distillate. _____ Weight per gallon ⎰Belt _____ 6.93 _____ pounds.
⎱Drawbar _____ 6.94 _____ pounds.

Oil: S.A.E. No. _____ 30. To motor _____ 3.405 _____ gal. Drained from motor _____ 3.780 _____ gal.

Total time motor was operated _____ 46 _____ hr.

Brief Specifications:

Advertised speeds, miles per Hour: First _____ 2.18. Second _____ 2.62. Third _____ 3.13

Fourth _____ 4.57. Fifth[1] _____ 14.3. Reverse _____ 1.0.

Belt pulley: Diameter _____ 14 in. Face _____ 7 in. R.p.m. _____ 786.

Clutch: Make _____ Twin Disk. Type _____ Single-plate, dry. Operated by _____ Hand.

Seat: _____ Pressed steel.

Total weight as tested (with operater): _____ 4,280 _____ lb.

Motor:

Make: _____ Own. Serial No. _____ 40973. Type _____ 4 cylinder, vertical.

Mounting _____ Crankshaft lengthwise. Lubrication _____ Pressure.

Bore and stroke: _____ $3\frac{5}{8}$ by $4\frac{1}{2}$ in. Rated R.p.m. _____ 1,500.

Valves: Position _____ In block, horizontal over piston.

Port diameter: Inlet _____ 1.25 in. Exhaust _____ 1.25 in.

Magneto: Make _____ Fairbanks-Morse. Model _____ RV-4.

Carburetor: Make _____ Schebler. Model _____ TRX 12. Size _____ 1 in.

Governor: Make _____ Own. Type _____ Variable-speed, centrifugal.

Air cleaner: Make _____ Donaldson. Type _____ Oil-washed, wire filter and pre-cleaner.

Chassis:

Type: _____ Tricycle. Serial No. _____ 560228. Drive _____ Enclosed gear.

Tread width: Rear _____ 54–76 in. Front Top _____ 13 in. Bottom _____ 7 in.

Drive wheels: Type _____ Standard. No. _____ 2. Diameter _____ 50 in. Face _____ 8 in.

Lugs: Type _____ Spade. No. per wheel _____ 24. Size: _____ 4 in. high by 3 in. wide.

Extension rims: Face _____ 5 in. Lugs per rim _____ 12. Size _____ 4 in. high by 3 in. wide.

Front wheels: Type _____ Standard. No. _____ 2. Diameter _____ 25 in. Face _____ $4\frac{1}{2}$ in.

Repairs and Adjustments:

No repairs or adjustments.

Remarks:

1. All results shown on page 385 of this report were determined from observed data and without allowances, additions, or deductions. Tests *B* and *F* were made with carburetor set for 100 % maximum horsepower, and data from these tests were used in determining the horsepower to be developed in tests *D* and *H*, respectively. Tests *C*, *D*, *E*, *G*, and *H* were made with an operating setting of the carburetor (selected by the manufacturer) of 93.9 % of maximum horsepower.

[1] Not recommended for steel wheels.

2. Observed maximum horsepower (tests *F* & *B*): Drawbar: 20.53.   Belt: 26.84.
3. Sea level (calculated) maximum horsepower: Drawbar: 21.21.   Belt: 27.56.
   (based on 60°F. and 29.92″ Hg.)
4. Seventy-five per cent of calculated maximum
   drawbar horsepower and eighty-five per cent
   of calculated maximum belt horsepower (for-
   merly A.S.A.E. and S.A.E. ratings):   Drawbar: 15.91.   Belt: 23.43.

   We, the undersigned, certify that the above is a true and correct report of official tractor test No. 290.

Carlton L. Zink   E. E. Brackett
Engineer-in-charge
                                               Ivan D. Wood

                                               L. W. Hurlbut
                                               Board of Tractor Test Engineers

References

Considerations Affecting Belt Speeds, *Trans. Amer. Soc. Agr. Eng.*, Vol. 16.

Nebraska Tractor Tests, *Univ. Neb., Agr. Exp. Sta. Bull.* 177, 200, 212, 220, 224, 233, 242, 255, 265, 277, 285, 292, 296, 304 and 313.

Nebraska Tractor Tests in 1917, *Trans. Amer. Soc. Agr. Eng.*, Vol. 11.

Recent Changes in Tractors as Noted from the Nebraska Tractor Tests, *Trans. Amer. Soc. Agr. Eng.*, Vol. 22.

Standardization of Gas Tractor Ratings, *Trans. Amer. Soc. Agr. Eng.*, Vol. 10, No. 2.

Testing Draft Horses, *Iowa State Coll., Agr. Exp. Sta., Bull.* 240.

Testing the Power of Horses, *Trans. Amer. Soc. Agr. Eng.*, Vol. 17.

The Iowa Recording and Integrating Traction Dynamometer, *Trans. Amer. Soc. Agr. Eng.*, Vol. 4.

The Nebraska Tractor Law, *Univ. Neb., Agr. Exp. Sta., Circ.* 13.

Tractor Testing, *Trans. Amer. Soc. Agr. Eng.*, Vol. 13.

Tractor Testing in Nebraska, *Trans. Amer. Soc. Agr. Eng.*, Vol. 14.

CHAPTER XXVII

TRACTOR HITCHES—FIELD OPERATION

Frequently, when a tractor or a field implement is condemned owing to the unsatisfactory results obtained, the machine is really not at fault, but the trouble can be traced to the way it is "hooked up"—in other words, to the hitch. It is not always possible to devise a perfect hitch, but a little study and a simple analysis of the factors involved will often lead to at least partial elimination of the trouble.

377. True Line of Pull and Draft.—Referring to Fig. 404a, the true line of pull of an ordinary tractor passes through a point in the hitch midway between the drivers and is parallel to the direction of travel. The true line of draft of any drawbar-operated implement passes through that

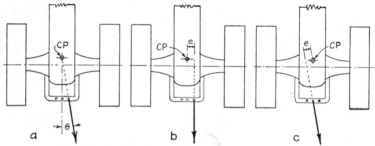

Fig. 404.—Hitch and pull conditions and their effect on side draft.

point on the implement to which one can hitch and make it move squarely forward in the direction of travel. In other words, the true line of draft of an implement is a line parallel to the direction of travel and passing through its center of resistance.

378. Side Draft.—In all drawbar work the object should be to make the true line of pull of the tractor and the true line of draft of the implement coincide as far as possible. If they do not, more or less side draft results. In other words side draft exists in a tractor whenever the direction of pull on the implement is not parallel to the direction in which the implement is moving. It is unnecessary when the load is such that it can be trailed straight back from the center of the drawbar.

According to Clyde[1]: In a tractor with two drive wheels connected by the usual differential, the pull exerted by the wheels is approximately equal. Friction

[1] *Pa. Agr. Exp. Sta., Bull.* 343.

in the differential gears is all that permits one wheel to pull slightly more than the other. For all practical purposes a point *CP* (Fig. 404) may be called the center of pull, being on a vertical line slightly ahead of and midway between the wheel centers. It corresponds in some ways to the center hole of a double tree. The point *CP* is farther ahead on a 4-wheel drive and track-laying tractors.

Side draft exists if the pull is not straight back from *CP*. There are, however, three distinct types of side draft with different effects, as shown in Fig. 404 *a*, *b*, and *c*. The important point is the angle of pull ⊖, and its offset from *CP*, if any, rather than where the clevis is attached to the drawbar. For example, when a disk is attached to the center of a rigid drawbar, an offset-angled pull exists in turning (Fig. 404*c*). A swinging drawbar pivoted near *CP* would make the tractor much easier to turn with a load.

Side draft is encountered largely with plows and, to some extent, with harvesting machines such as mowers and binders. In these cases the path of the tractor with respect to the implement is somewhat restricted,

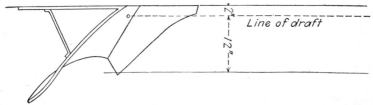

Fig. 405.—Sketch showing location of center of resistance and line of draft for a moldboard plow bottom.

for example, by the edge of the furrow wall, by standing grain or hay, or by rows of trees. Plows give more trouble in this respect because of their heavy draft and narrow swath.

Side draft in tractor plowing is affected by a number of factors such as (1) whether tractor runs in the furrow or entirely on the unplowed land, (2) width and tread of drive wheels, (3) type of plow, and (4) width of cut and number of plow bottoms.

In plowing, practically all the small and medium sizes of wheel tractors are operated with the right-hand wheels in the furrow. This simplifies steering and eliminates the use of a special guiding device. Some large-size wheel tractors and all track-type machines operate on the unplowed land.

The effect of these different factors on side draft is clearly shown by the accompanying illustrations and explanation. For example, for a 14-in. moldboard plow bottom, the center of resistance passes through a point about 2 in. to the right of the land side as shown by Fig. 405. Therefore, for two 14-in. bottoms, the line of draft would fall midway between the lines of draft of the individual bottoms or at a distance of 19 in. inward from the furrow wall (Fig. 406*A*). For a three-bottom

14-in. plow, the line of draft would coincide with that of the middle bottom (Fig. 406B). For a four-bottom, it would be 33 in. inward from the furrow wall (Fig. 407A), and so on.

Assuming the average two-plow tractor to be 60 in. wide and to have 10-in. drive wheels (Fig. 406A), the line of pull with the wheel in the furrow would be 20 in. from the furrow wall and would fall within one

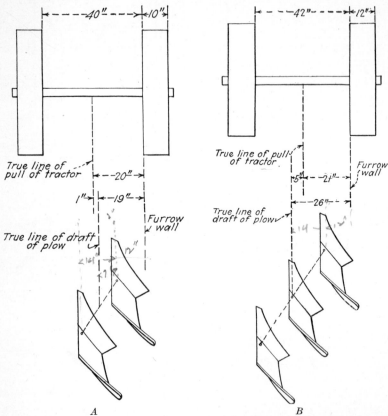

A   B

Fig. 406.—Relation of line of hitch of tractor and line of draft of plow for (A) two 14-in. plows and (B) three 14-in. plows.

inch of the line of draft of the plow. With narrower plow bottoms or a wider-tread tractor a greater difference would exist resulting in some side draft.

Again assuming the average three- and four-plow tractor to be 66 in. wide and to have 12-in. drive wheels, with the machine running in the furrow and pulling three 14-in. bottoms (406B), there will be a variation of 5 in. in the two lines. With the same tractor pulling four bottoms (Fig. 407A and B), the variation will be 12 in. when running in the furrow and 8 in. when on the unplowed land.   In this case the latter method will work more satisfactorily.   It is thus evident that as the total effective width of the plow approaches the tractor width the lines of pull and draft fall closer together with the tractor running on the unplowed land.

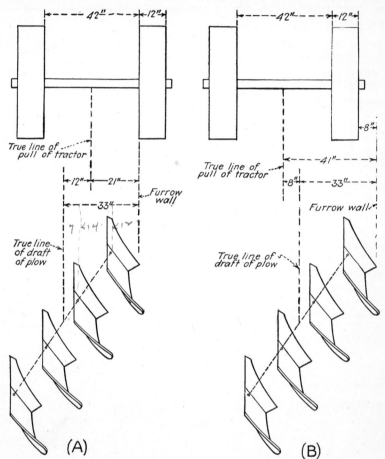

Fig. 407.—Effect on line of draft of operating a four-plow tractor with (*A*) one wheel in furrow and (*B*) both wheels on unplowed land.

379. Effects and Control of Side Draft.—With further reference to side draft, its effects, and correction, Hoffman[1] states that:

It causes increased friction especially on the tractor-wheel bearings.   This means rapid wear, more repairs, and shorter life for the tractor.   It makes the tractor less able to pull its full load because a larger part of the power developed in the engine is needed just to move the tractor itself.   Side draft makes the trac-

[1] *Calif. Agr. Exp. Sta., Bull.* 349.

tor harder to steer.   In some cases it is found absolutely impossible to turn the
tractor, except in the direction of the side draft, without first removing the load.
The result is crooked furrows and, in general, poor, uneven work.   Side draft
effects are practically the same for wheel-type and track-type tractors.

To detect side draft Hoffman[1] suggests as follows:

A rope, chain, hook, loose clevis, or any other object capable of transmitting
a pull will, if free to swivel at both ends, set itself in the direction of the force
it is transmitting.   Hence, if there is somewhere between the tractor or drawbar
and the plow a link or clevis free to turn at both ends, its position when the force
is applied will show the direction of the force.   Whenever such a free link is
not parallel to the direction of motion but makes an angle with it (in the horizontal
plane), there is side draft.

He says further that a study of the underlying principles shows
clearly how we may reduce or even entirely prevent side draft.   Assuming
a rear-drive four-wheel tractor and the hitch to drawbar center unless
otherwise specified:

1. Reduce the angle of offset to nothing, and there will be no side draft.
2. Make the angle of offset very small, and the side draft will be correspondingly
small.
3. Reduce the total draft required, and the side draft will be reduced in the same
proportion.
The angle of offset may be reduced in three ways:
  a. By running one driver in the furrow.
  b. By running very close to the furrow.
  c. By lengthening the hitch.

In some cases, dividing the side draft between the tractor and the plow
gives better results.   In regard to this Hoffman[1] says,

By offsetting the point of hitch on the tractor drawbar we may take off of the
plow much, if not all, of the side draft due to the pull of the tractor.   This, however, causes the load to be heavier on the driver nearer the point of hitch.   The
wheel carrying the larger load will slip more easily than the other.   Thus we see,
in the extreme case, the more lightly loaded driver tending to run in a circle
around the more heavily loaded one, shoving over the front of the tractor toward
the furrow.   This means a side force on the front wheels.   Whenever the steering
wheels must be held turned toward the land to make the tractor go straight
forward there will be strong forces pressing the axle collars against the front-wheel
hubs.   More power is therefore needed just to move the tractor and less is left
to do the plowing.   If circumstances compel us to have the center lines of tractor
and plow offset, we may divide the bad effects between tractor and plow by
hitching slightly (2 to 6 in.) off center on the tractor drawbar and adjusting the

[1] *Calif. Agr. Exp. Sta., Bull.* 349.

hitch bars of the plow so that the pull of the tractor will make a small angle with the direction of motion.

In attempting to relieve the side draft with any tractor plowing outfit, care must always be taken to maintain the correct width of cut of the front plow bottom by adjusting the hitch bars (Fig. 408).

380. Vertical Hitch Adjustment.— Reed and Silver[1] state as follows concerning this problem:

Hitching too high on the plow is quite common practice. When a plow in good adjustment is being used under good plowing conditions, attach the hitch bars low enough on the plow to let them point straight from the drawbar of the tractor to the center of resistance of plow, as shown in Fig. 409. In other words, the line of hitch *ABC* should be straight—not broken upward or downward as would be the case if the hitch were too high or too low. If you are using a tractor in which the height of the drawbar is adjustable, both

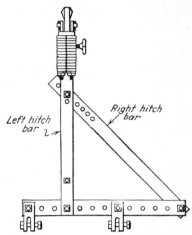

FIG. 408.—Common plow hitch showing horizontal adjustment.

the tractor drawbar itself and the hitch bars should lie in a straight line from the center of resistance of the plow to that point under the tractor to which the drawbar is attached. Keep

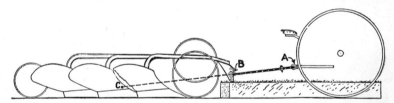

FIG. 409.—Sketch showing proper vertical hitch adjustment for a tractor plow.

the proper amount of down-suck in the shares, keep the shares sharp, maintain the proper heel clearance so that the line of hitch can be as straight as possible.

SPECIAL HITCHES

It is frequently desirable and necessary to devise special hitch arrangements, particularly under the following circumstances: (1) where the tractor to be used is capable of pulling several units of a certain implement and (2) when two or more distinct field operations can be performed at the same time.

381. Multiple-unit Hitches.—Multiple hitches are used when a single unit of the implement is insufficient to properly load the tractor.

[1] *Ohio Agr. Ext. Service, Bull.* 80.

Murdock[1] gives the requirements of such a hitch as follows: (1) sufficient strength and rigidity to transmit the pull from the tractor to the implement; (2) a design to prevent side lashing of tongues or twisting of the implements; (3) noninterference of the implement on turns; (4) a design that will make the implements follow each other in their respective positions on straight pulls, turns, and hillsides. He states further that "hitches that may give complete satisfaction on large level fields may prove inadequate on a hilly farm. In going downhill such implements as combines may have to be held back by the tractor, hence a hitch for this implement would have to be designed to take a forward thrust."

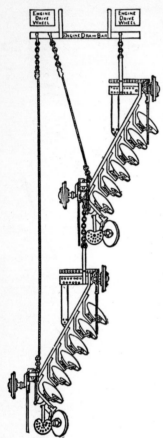

Fig. 410.—Special hitch for pulling two disk-plow units.

Figures 410 and 411 show methods of hooking up two tractor disk-plow units. Figures 412 and 413 show general-purpose hitches for use with two- and three-unit outfits, respectively, such as disk harrows, grain drills, rollers, and so on. Figures 414 and 415 are illustrations of special grain-drill hitches. Figure 416 shows how several peg-tooth harrow sections may be connected together. Figure 417 shows a special hitch for operating two or more mowers.

382. Combination Hitches.—Combination hitches are sometimes desirable if a single implement is insufficient to load the tractor and two or more operations can be done at the same time. As a rule, however, the problem is a rather simple one for the reason that the different implements must trail each other to a large extent. For example, Fig. 418 shows a tractor pulling three different tools. However, the hitch is taken care of almost entirely by chains because all machines follow each other directly behind the tractor.

LAYING OUT FIELDS FOR TRACTOR PLOWING

The proper method to follow in laying out and plowing a field with a tractor depends upon so many factors that no attempt will be made here to give complete and definite information applicable to every possible

[1] *Mont. Agr. Exp. Sta., Bull.* 229.

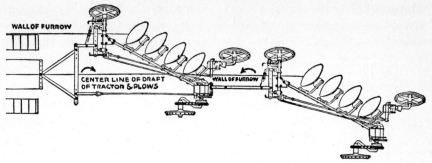

FIG. 411.—Method of hitching two disk-plow units to a tractor.

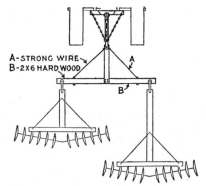

FIG. 412.—Simple hitch for pulling two disk-harrow units.

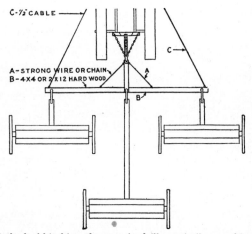

FIG. 413.—Method of hitching three grain drills or similar machines to a tractor.

condition.　In the last analysis, it is largely a matter of judgment, custom, and preference.　In plowing a field, certain fundamental factors must be considered such as (1) saving time, fuel, and labor; (2) doing the

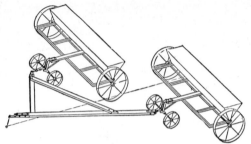

Fɪɢ. 414.—Method of hitching two grain drills to a tractor.

Fɪɢ. 415.—A simple and practical hitch for handling three grain drills.

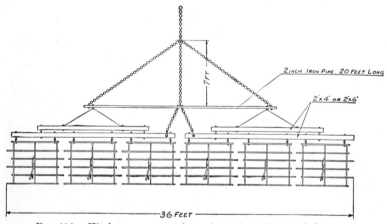

Fɪɢ. 416.—Hitch arrangement for a six-section spike-tooth harrow.

best possible job; (3) elimination of dead furrows, which might start erosion; and (4) proper drainage.

383. Methods Used.—There are two general classes of methods of laying out and plowing fields; namely, those in which the plows are lifted out of the ground at the ends and those where they are not lifted.

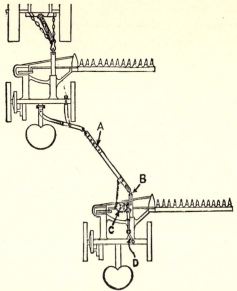

Fig. 417.—Special hitch for pulling two mowers with a tractor. *A*. Adjustable tongue connection. *B*. Flexible joint. *C*. Steering quadrant. *D*. Steering crank.

Fig. 418.—A large tractor performing three operations simultaneously by trailing the different implements directly behind it in proper order.

Methods of the first class involve back furrows, dead furrows, and head-lands for turning, and are better adapted to large, rectangular fields, which are reasonably level and well drained.

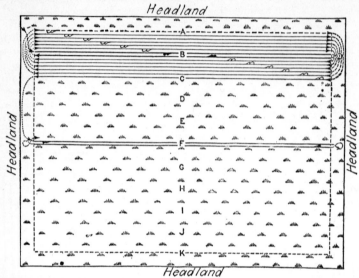

FIG. 419.—**Method 1.**   The body of the field is plowed in successive back-furrow and dead-furrow lands with a headland all around the field.

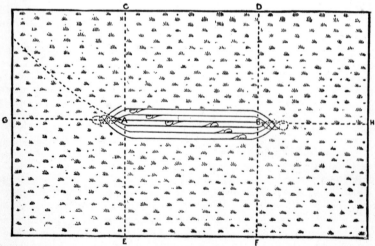

FIG. 420.—**Method 2.**   First stage.   Begin in the center of the field by laying out a back furrow *AB*.   The distance *AC* should be equal to distance *AE*.

384. Method 1.—Referring to Fig. 419, a 20- to 100-ft. headland is first measured off on all four sides of the field.   The field is then staked out into lands of equal size as *A* to *C*, *C* to *E*, and so on.   A back furrow is run across the center of the first land through *B*.   Plowing is continued around the back furrow as shown, until *A* and *C* are reached.   Then lay out another back furrow at *F* whose distance from *C* is equal to $1\frac{1}{2}$ times the width of the land A to C. Plow around this back furrow as far as E and G. Then, the land from C to E, still unplowed, will be of the same width as each of the two lands already plowed. This is then plowed by turning to the left from the furrow through E, to the furrow through C until the strip is plowed out to a dead furrow at D. In a similar manner, there would be a back furrow at J and a dead furrow at H. Finally, the headlands are plowed out.

If this method is used continuously, year after year, the back furrows will, in time, form ridges and the dead furrows will form ditches. This can be prevented, to some extent, by varying the width of the lands and headlands each time the field is plowed. A still better way is to lay out the field just as described, but have a back furrow in place of a dead furrow, and a dead furrow in place of a back furrow.

385. Method 2.—This method involves beginning the work at the center of the field and working toward the outside, leaving the corners round as shown by Fig. 420.

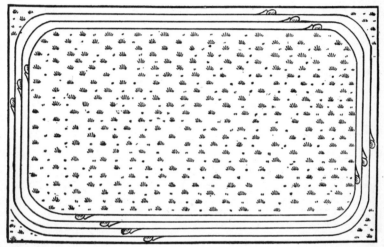

Fig. 421.—**Method 3.** Plowing around a rectangular field by turning the furrows toward the fence, rounding the corners enough to permit turning without lifting the plows.

The plows are lifted only for the first few short turns until the land becomes wide enough to leave them in the ground. The distance A to G should be made enough shorter than distance A to C so that, when the land is rounded off at the ends and plowing entirely around is begun, the furrow across the end will be the same distance from the edge of the field as are the furrows down the sides.

386. Method 3.—In this method the operator starts plowing at the outside of the field, as shown in Fig. 421, throwing the furrow toward the fence and turning to the left at the corners without lifting the plow. A rectangular field like that shown is plowed in a single land and with one dead furrow. The corners will have to be rounded to a certain extent on the first trip around the field and kept this way throughout the plowing so as to permit the tractor to make the turns without encroaching too far on the plowed ground or getting the furrows irregular and crooked near the corners. The plows will be pulled away from the last open furrow to a certain extent in making the turns, and the diagonal strips running from the ends of the dead furrow to the corners of the field will usually have to be replowed. A field with slightly irregular or crooked boundaries can be plowed very conveniently this way. This method is very popular with many tractor operators and has been adopted almost exclusively where disk plows are used.

387. Method 4.—This is like the preceding method except that the plows are lifted each time at the corners in plowing the body of the field, and the diagonals are left entirely unplowed until the finish of the field. Care must be taken to get the width of all the diagonals the same, that is, from C to D, E to F, and so on (Fig. 422). The width should be ample for turning the outfit and getting it in line with the furrow, before the point where the plows are to be put into the ground again is reached. It will be better to make an extra round in plowing out the diagonals than to be cramped for space at every turn in plowing the body of the field.

388. Irregular Fields.—Irregular fields have such a variety of shapes and present such a variety of conditions that it is impossible to give any definite directions applicable to all. If the field is comparatively level,

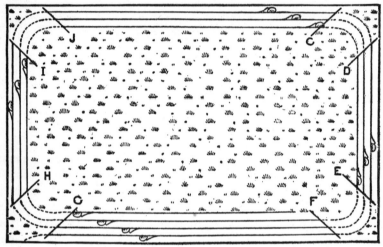

Fig. 422.—**Method 4.** Plowing around a rectangular field as in Method 3 but lifting the plows at the turns.

and the irregularities are confined to the boundaries on one or two sides, some one of the methods described for rectangular fields can usually be adopted.

References

Hitches for Field Machinery, *S. D. State Coll., Agr. Exp. Sta., Bull.* 297.
Laying out Fields for Tractor Plowing, *U. S. Dept. Agr., Farmers' Bull.* 1045.
"Plow Manual," J. I. Case Company.
Plows and Good Plowing, *Ohio State Univ., Agr. Ext. Service Bull.* 80.
Plows and Plowing, *Univ. Saskatchewan, Agr. Ext. Bull.* 32.
RADEBAUGH: "Repairing Farm Machinery and Equipment."
Sidedraft and Tractor Hitches, *Univ. Calif., Agr. Exp. Sta., Bull.* 349.
SMITH: "Farm Machinery and Equipment."
"The Care and Operation of Plows and Cultivators," John Deere Plow Company.
The Tendency of Tractors to Rise in Front, *Univ. Calif., Agr. Exp. Sta., Circ.* 267.
Tractor Hitches, *Univ. Mont., Agr. Exp. Sta., Bull.* 229.
Using the Tractor Efficiently, *Pa. State Coll., Agr. Exp. Sta., Bull.* 343.

CHAPTER XXVIII

TRACTOR TYPES—ROW-CROP AND GARDEN TRACTORS

389. Classification of Tractor Types.—Tractors as now manufactured can be classified as follows:

A. According to method of securing traction and self-propulsion:
1. Wheel tractors.
   - *a.* Three wheels.
   - *b.* Four wheels.
2. Track-type tractors.

B. According to utility:
1. General purpose.
2. All-purpose cultivating type.
3. Orchard.
4. Garden.
5. Industrial.

390. Wheel Tractors.—The wheel-type tractor is the predominating type of machine, particularly for agricultural purposes. Wheel tractors are made either with three wheels (Fig. 423), or with four wheels (Fig. 424). Figure 425 shows a four-wheel tractor in which all four wheels drive the machine. With the established success of the row-crop type of tractor and its greater adaptability, the three-wheel tractor now probably predominates in agriculture. The usual arrangement consists of two rear-drive wheels and a front steering member as in Fig. 426. The drive wheels are, as a rule, spaced farther apart than in the ordinary four-wheel machine to secure stability on sloping land. The steering member, although usually consisting of two wheels, acts as a single wheel, because the two are placed very close together. By using two wheels in this manner more rigid construction and easier steering are secured. At the same time, the steering wheels can easily work in the narrowest rows, thus eliminating straddling the row with the front axle.

391. Track-type Tractors.—The traction mechanism in the track-type tractor (Fig. 427) consists essentially of two heavy, endless, metal-linked devices known as *tracks*. Each runs on two iron wheels, one of which bears sprockets and acts as a driver. The other serves as an idler (see Fig. 361). Steering is accomplished through the tracks themselves by reducing the movement of one track below the speed of the other.

The use of the track-type tractor was quite limited up to the World War period. The great tanks that were developed at that time and proved so successful in traveling over the battle-swept areas did more perhaps than anything else to demonstrate the adaptability and utility of such a traction arrangement.   Consequently, during the past few

Fig. 423.—All-purpose tractor with two-row cultivating attachment.

Fig. 424.—Four-wheel general-purpose tractor.

years, the development of successful track-type tractors in both small and large sizes for every known use has been rather rapid.   In fact, they have practically supplanted the wheel tractor for road construction and maintenance and for many industrial jobs requiring tractor power.

392. Tractor Types According to Utility.—A *general-purpose tractor* is one of more or less standard or conventional design, such as an ordinary four-wheel machine (Fig. 424), or a track-type (Fig. 427).   It is made to perform only the usual tractor jobs including both field and belt work,

FIG. 425.—Four-wheel-drive all-purpose tractor.

FIG. 426.—Wide-tread rear-drive all-purpose tractor.

such as plowing, harrowing, road grading, threshing, silo filling, and the like.

An *all-purpose cultivating* type is a light-weight tractor (Fig. 426), designed to handle practically all the field and belt jobs on the average farm, including the planting and intertillage of row crops. The most important requirements of such a tractor are: (1) greater clearance both

Fig. 427.—Track-type tractor pulling a terracing machine.

Fig. 428.—General-purpose tractor equipped for orchard work.

vertical and horizontal, (2) adaptation to the usual row widths, (3) quick, short-turning ability, (4) convenient and easy handling, and (5) quick and easy attachment and removal of field implements and attachments.

Orchard tractors are small or medium-size, general-purpose machines of either the wheel or crawler type (see Fig. 428), so constructed and equipped as to be operated to better advantage around the trees.   Such

Fig. 429.—Industrial tractor.

tractors are often built lower with as few projecting parts as possible and with special fenders.

Industrial tractors are machines of any size or type specially constructed for various industrial operations and for heavy hauling about

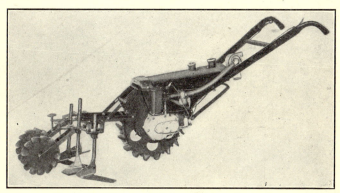

Fig. 430.—The small, light-weight garden tractor.

factories, freight depots, airports, and so on.   They are equipped with rubber-tired wheels and special high-speed transmission, as shown by Fig. 429.   Others have hoisting, excavating, power-loading, and similar attachments built on them.

Garden tractors are the smallest machines manufactured. Their use is limited to garden and truck farms or similar farms of limited acreage. They are made in three types: (1) a small light size (Fig. 430), (2) a

Fig. 431.—A medium-size garden tractor.

Fig. 432.—A large-size garden tractor.

medium size (Fig. 431), and (3) a large size (Fig. 432). The small machines are used primarily for the planting and cultivation of gardens and small truck farms. The medium and large sizes may be used for plowing, harrowing, and belt work, as well as for planting and cultivating. In most cases, the operator of a garden tractor walks and guides the machine by means of handles as he would a walking plow or planter.

393. Make-up of a Tractor.—Any tractor is made up of three distinct parts or assemblies as follows:

1. Power unit—engine and all accessories necessary for its operation.
2. Power-transmitting system—clutch, change-speed gears, differential, final-drive mechanism, belt pulley, power take-off.
3. Chassis—frame, wheels, and steering mechanism.

394. Power Unit.—The power unit of a tractor consists of the engine and all essential parts and accessories, such as ignition system, fuel-supply and carburetion system, cooling system, lubrication system, and governing mechanism. The various types of engines and their construction and operation, as well as the detailed operation and construction of the related systems mentioned above, are discussed in detail in the succeeding chapters.

395. Power-transmitting System.—Certain parts and units are necessary for the transmission of the engine power to the rear wheels, to the belt pulley, and to the power-take-off shaft. The arrangement, construction, and operation of the different units and parts making up the complete transmission system in the various makes and types of tractors are discussed fully in preceding chapters.

396. Tractor Chassis.—An automobile is said to consist of two main parts, the chassis and the body. The chassis consists of the entire mechanical make-up of the car, including the engine, transmission, frame, wheels, and steering mechanism. Tractors do not have a body. Therefore, the chassis is considered as including only the frame, wheels, and steering mechanism. A more complete discussion of these parts will be found in Chap. XXIV.

ALL-PURPOSE TRACTORS

The definition and general requirements of an all-purpose tractor have been stated in Par. 392. Other terms applied are *general-purpose, row-crop, all-crop*, etc. With this type of tractor being introduced and adopted in all sections of the country, the principal problem of the manufacturers has been to build one machine and the necessary field equipment that will operate satisfactorily with any crop and soil condition. This has proven somewhat difficult, if not impossible, and, as a result, some manufacturers supply all-purpose tractors in two or even three sizes. In most cases it is necessary to adapt the tractor and equipment to present crop-growing practice, rather than to adapt the crop to the tractor.

397. General Design.—Although there is considerable variation in the design and construction of the individual machines and equipment, the trend appears as follows:

1. Type: The majority of the machines on the market are wheel machines of the tricycle type. However, Fig. 433 shows a track-type all-purpose tractor with a four-row cultivator attachment.
2. Power: Three sizes having power ranges of:
   a. Small: 8 to 11 drawbar hp., 12 to 18 belt hp.
   b. Medium: 12 to 20 drawbar hp., 20 to 30 belt hp.
   c. Large: 20 to 25 drawbar hp., 28 to 40 belt hp.
3. Weight: 3,000 to 6,000 lb. depending on size and power rating.

Fig. 433.—Special wide-tread, track-type tractor with four-row cultivating attachment.

4. Speeds: Four speeds forward with a low of about 2.5 m.p.h.; second, 3.25 m.p.h.; third, 4.25 m.p.h.; and high, 5 to 10 m.p.h.
5. Field-tool and Belt Capacity:
   a. Small size: One 16-in. or two 10-in. moldboard plow bottoms or a two-disk plow; one-bottom lister plow; 6-ft. tandem disk harrow; 8- to 10-ft. grain drill; two-row planter or cultivator; 7-ft. mower; 8- to 10-ft. grain binder; small feed mill or ensilage cutter.
   b. Medium size: Two 14-in. moldboard plow bottoms or a 3-disk plow; two-bottom lister plow; 7- to 8-ft. tandem disk harrow; 12- to 14-ft. grain drill; four-row planter or cultivator; two 7-ft. mowers; one 10-ft. power binder; single-row corn picker; 6- to 10-ft. combine; 22-in. cylinder thresher; medium-size feed mill or ensilage cutter.
   c. Large size: Three to four 12- or 14-in. moldboard plow bottoms or four- to five-disk plow; three-bottom lister plow; 10-ft. tandem disk harrow; two 12-ft. grain drills; four-row planter or cultivator; two 7-ft. mowers; two grain binders; two-row corn picker; 26-in. cylinder thresher; large feed grinder or ensilage cutter.

398. Load Factor.—Common sense as well as experience tells us that efficient results cannot be secured from a tractor that is not sufficiently loaded. On the other hand, topographic, soil, and crop conditions may serve as limiting factors with respect to the size of field tools that it is possible to use. For example, a tractor capable of pulling two 14-in. plows ought to handle a four-row cultivator in most soil conditions. However, the topography of the land, the size of fields, or the crop itself may make the use of a four-row cultivator impractical, although a two-row will operate satisfactorily. This situation has led to the manufacture of both two-row and four-row equipment for the same tractor.

FIG. 434.—All-purpose tractor with four-row planting attachment.

399. Row Widths.—In a paper entitled "The Requirements of the General-purpose Farm Tractor,"[1] Heitshu states that:

A major factor in the cultivating-tractor problem is the varying widths of row spacings used in different parts of the country. The distance between rows varies from 12 to 72 in. Crop, soil, climatic, and economic conditions allow but little argument in favor of changing the established widths.

An analysis of the data secured in relation to the width of rows (Fig. 435) indicates that the normal width of rows falls between 36 and 48 in. To adapt the tractor fully to interrow tillage operations, however, it is necessary that the tractor work in rows from 30 to 48 in. in width. For rows smaller than 30 in. or larger than 48 in., some combination of widths can be arranged to adapt the crop to tractor cultivation.

400. Clearance.—With reference to clearance Heitshu[1] says:

[1] *Agr. Eng.*, Vol. 10, No. 5.

The height and width of plants at the time of the last cultivation varies even more than does the width of row. The maximum height of the cultivated plants

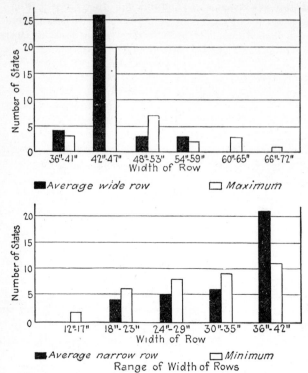

Fig. 435.—The range of row widths of cultivated crops in various states.

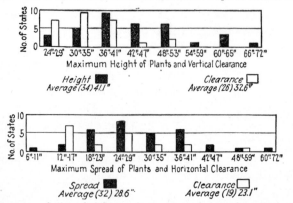

Fig. 436.—The range of plant heights and spreads and desired vertical and horizontal clearances in various states.

in the different states ranges from 24 to 72 in., the 36- to 41-in. classification covering the major group (Fig. 436). The average height of the time of the last cultivation is 41.1 in., with corn representing the major crop in this class. The maximum width is even less fixed, extending from 6 to 60 in. The average maximum width of plants is 28.6 in. In twelve states, potatoes have the maximum spread; in eight states, corn; and in eight other states, cotton.

That the majority of crops can be cultivated with less clearance than the actual size of a plant demands is shown in the average desirable clearance, which is 32.6 in. vertically, and 23.1 in. horizontally. The clearance necessary in row-crop cultivation can be chosen only by arbitration. All the data available show that 30-in. vertical clearance, and 24-in., or even less, horizontal clearance should be sufficient for practically all crops.

401. Steering and Control.—Ease of handling and control of an all-purpose tractor and its attached field equipment are essential. The steering mechanism should (1) permit short, quick turning, (2) require the minimum of effort in its operation regardless of whether the machine is moving or stationary and (3) permit precise and accurate control of the attached units, particularly planters and cultivators.

The implements and attachments should be as simple as possible, in both construction and operation. They should be capable of attachment or removal in a few minutes' time without any appreciable dismantling. The adjusting levers should be within convenient reach of the tractor operator.

402. Power Lift.—A positive, easily controlled power lift (Fig. 350) is an important accessory for lifting cultivators, planters, and similar tools at the ends of the rows or whenever necessary. This arrangement permits raising and lowering the tool with the tractor either stationary or moving.

GARDEN TRACTORS

Garden tractors have never come into extensive use except in certain sections where the farms are comparatively small and devoted largely to truck crops. It seems likely that, as this type of tractor is improved in design and construction, it will find considerably wider favor.

A. A. Stone, in a booklet entitled "Garden Tractors on Long Island," states that although there are several different sizes and types of garden tractors on the market, there is no sharp line of distinction between them and farm tractors. He considered a garden tractor as a machine capable of handling one 12-in. plow or less.

Regarding the development of the garden tractor, Stone gives the following information:

At the outset designers of garden tractors attempted a general-purpose machine—one that could be used for light plowing and cultivating as well as numerous other jobs, and also afford a source of belt power. Many problems were met with, similar to those that occurred in the development of the farm tractor. The chief difficulty lay in producing a machine with proper power for good plowing, and yet small and light enough for close cultivation of narrow-row crops of truck farms, market gardens, florists, nurseries, and so on.

The first machines averaged about 5 belt hp. and weighed approximately 700 lb. The majority of these early machines were rated to pull a 10-in. plow. They proved to be too large for general use in many of the market garden crops where cultivation was the principal operation. More demand existed for a small power cultivator than for a small plowing outfit.

The production of smaller units began shortly after 1920. These machines had motors of from 1½ to 2 hp. and weighed from 200 to 400 lb. In general, they were fairly well adapted to light cultivation and many of the better models met with fair success. They could not, however, be successfully used for light plowing.

The larger size (5 to 10 belt hp.) shows a decrease in number of models manufactured after 1920, until 1924 the average of all garden-type tractors produced shows a belt horsepower of 2.65.

Beginning with 1925 the average belt horsepower and the weight of all models of garden tractors increased, until 1928 the average belt horsepower of all models is 3.23 and the weight 483 lb. This increase in power and weight has made such tractors capable of better and deeper cultivation, and also offers the possibility of operating an 8-in. plow under average conditions without any material sacrifice in the ease of handling or control. It appears to be a step toward a garden tractor which is more nearly a general-purpose type. Stronger and more efficient tools and implements, especially adapted for tractor work, were also produced during this period.

403. Classification of Garden Tractors.—Concerning garden-tractor types, Stone states as follows: "The term 'garden tractor,' although not a well-chosen expression, has come to be widely used. This class of tractors includes three sizes or types, although the limits of each are difficult to define."

Stone suggests the following classification:

A. Large Garden Tractors (Fig. 432).
 The capacity of machines of this class is indicated by the following approximate specifications and ratings, which are the result of observation of their use under Long Island conditions:
 1. Five- to twelve-horsepower motors (usually two cylinders).
 2. Riding models.
 3. Pull one 12-in. plow to a depth of 7 in., and at the rate of about two acres per day under favorable conditions.
 4. Operate three 30-in. lawn mowers.
 5. Operate 5-ft. field mower.
 6. Operate 5-ft. single-disk harrow.
 7. Gasoline consumption 4 to 6 gal. per day (approximate).
B. General-purpose Garden Tractors (Fig. 431).
 1. Three- to four-horsepower motor, usually single cylinder.
 2. Walking models usually.
 3. Pull 8-in. plow at rate of one acre per day (approximate).

4. Operate 3½-ft. field mower and cut at rate of three to four acres per day.
5. Operate 20- to 22-in. wood saw.
6. Operate 30- to 36-in. lawn mower.
7. Gasoline consumption 2 to 3 gal. per day (approximate).
C. Small Power Cultivators (Fig. 430).
1. One- to two-horsepower motors; usually single cylinder.
2. Walking models.
3. Operate 22- to 30-in. lawn mower.
4. Pull 4- to 6-in. plow (used mostly for hilling up or opening rows for seed, although some models do light garden plowing in a very creditable manner).
5. Pull one section (3 ft.) spike-tooth harrow.
6. Pull one 3-ft. Acme harrow.
7. Gasoline consumption 1½ to 2 gal. per day (approximate).

Stone[1] is of the opinion that the most successful garden tractor should be of the general-purpose type and conform as far as possible to the following requirements:

1. Sufficient power to operate a 10- to 12-in. plow.
2. Interchangeable lug and wheel equipment to secure traction.
3. Variable tread, adaptable for multiple-row work.
4. Adaptability to cultivating narrow rows, as well as the wider rows of truck and market garden crops.
5. Lateral stability.
6. Ample vertical clearance.
7. Vine turners or plant-shielding devices.
8. Adequate depth adjusting, spacing, and tool-lifting mechanisms.
9. Accurate, quick-acting cultivator control.
10. Easy interchangeability of attachments.
11. Easy handling, short turning, and reverse gear.
12. Power take-off and belt-pulley equipment.

Sauve,[2] as the result of a study made of the use of garden tractors in Michigan, gives the following advantages and disadvantages:

Advantages:
1. The overhead cost of a garden tractor when not in use is low compared to other forms of power.
2. The operating cost of a garden tractor is low.
3. The garden tractor when properly handled does not destroy the plants.
4. The garden tractor is a labor and time saver.
5. The garden tractor may be adapted to belt work.
6. The small sizes are especially adapted to narrow-row crops.
Disadvantages:
1. Lacks traction in sandy soil.
2. Difficult turning at end of rows in some cases.
3. May have poor service on repairs.
4. High initial cost.

[1] *Agr. Eng.*, Vol. 11, No. 12.
[2] *Mich. Agr. Exp. Sta., Quart. Bull.*, Vol. 11, No. 3.

Sauve states further:

Most of the garden-tractor owners who operate small acreages, and these are in the majority, desire that the tractor should perform the operations of plowing, fitting, and cultivating. With the more experienced users, it is found that the small tractor is too small for preparing the seed bed, and the larger garden tractors are too large for efficient cultivation of narrow-row crops. Since cultivating is the major function of the garden tractor, it would seem that this should be given first consideration. Usually, where a small garden tractor is used, plowing and fitting of the land are done with other forms of power.

He suggests that the following points be considered in selecting a garden tractor.

1. Availability of quick service is the first and most important consideration in the selection of a garden tractor.
2. The initial investment should be second in importance to service.
3. The tractor should be well balanced and easily controlled by the operator.
4. Ability to make short turns with little manual effort is desirable.
5. Buy only from reputable concerns.

References

Farm Requirements of the Small All-purpose Tractor (three reports), *Agr. Eng.*, Vol. 18, No. 5.
Field Requirements of Garden Tractors, *Agr. Eng.*, Vol. 15, No. 3.
Garden Tractor Development and Application, *Agr. Eng.*, Vol. 11, No. 12.
General-purpose Tractor Design, *Agr. Eng.*, Vol. 12, No. 3.
Possibilities of the All-purpose Tractor, *Trans. Amer. Soc. Agr. Eng.*, Vol. 16.
Quick-on and Quick-off Power Farming Equipment, *Agr. Eng.*, Vol. 18, No. 9.
Requirements and Design of Cultivating Equipment for the General-purpose Tractor, *Agr. Eng.*, Vol. 11, No. 2.
Requirements of a General-purpose Agricultural Tractor, *Agr. Eng.*, Vol. 14, No. 10.
Results of a Garden Tractor Survey, *Agr. Eng.*, Vol. 12, No. 5.
Spacing of Row Crops in the United States, *Agr. Eng.*, Vol. 14, No. 9.
Stone: Bulletin entitled "Garden Tractors on Long Island."
The Garden Tractor in Michigan, *Agr. Eng.*, Vol. 15, No. 3.
The General-purpose Farm Tractor, *Trans. Amer. Soc. Agr. Eng.*, Vol. 17.
The Requirements of the General-purpose Farm Tractor, *Agr. Eng.*, Vol. 10, No. 5.

CHAPTER XXIX

GAS-ENGINE AND TRACTOR TROUBLES—
OVERHAULING TRACTORS

Any internal-combustion engine, no matter how large or how simple it may be, will start and operate satisfactorily only when given reasonable handling and care according to the manufacturer's instructions. An engine, in order to give dependable service, must be kept in good condition and proper adjustment at all times. Negligence and abuse are the greatest gas-engine trouble producers. Occasionally, of course, even the best cared-for engines give trouble. But, as a rule, the time given to keeping an engine properly lubricated, tightened up, correctly adjusted, and clean will mean less delay and loss of time in the middle of an important job, as well as a longer life and period of service for the engine.

It is impossible for any one to prepare a trouble chart, lay down a complete set of ironclad rules, or devise any other scheme by which any and all gas-engine troubles may be diagnosed and corrected in a minimum period of time by an inexperienced operator. Internal-combustion engines are subject to hundreds of troubles. Many of these are simple to the well-informed and experienced operator who can quickly detect and correct them. In other words, a thorough knowledge of the fundamental principles of construction and operation, combined with sufficient practical training and experience, is more reliable than any trouble chart or set of rules that can be prepared.

The following discussion and outline of gas-engine troubles are more or less general and apply primarily to the simpler types of engines such as those used for stationary farm power, and in tractors and automobiles. No attempt is made to cover all the possible troubles, their diagnosis, and remedy. The primary object is to present, in a general way, a systematic method of procedure in diagnosing troubles, together with a list of the more common troubles and their causes and remedies.

404. Classification of Troubles.—As a rule, certain conditions are essential in order to start an engine and have it operate normally These are:

1. Each cylinder should have reasonable compression.
2. A correct fuel mixture must be supplied to each cylinder.
3. The ignition system must function properly and ignite the fuel mixture at the right time.

4. The valves must be correctly timed.

5. The moving parts must be properly lubricated.

6. The cooling system must maintain the usual operating temperature.

In short, there are six general classes of troubles, namely, (1) compression, (2) fuel and carburetion, (3) ignition, (4) timing, (5) lubrication, (6) cooling.

Again there are (1) starting troubles and (2) running troubles. An engine may start hard or fail to start at all, or it may start readily but run only a short time and stop.

STARTING TROUBLES

405. Engine Fails to Start.—When an engine fails to start or show any signs of firing, check it up as follows:

1. See that there is fuel in the tank and that it is turned on and flowing to the carburetor. Starting troubles are often caused by low-grade fuels, water in the fuel or carburetor, or a fuel that has stood in the tank long enough to permit the more volatile portions to evaporate. Fresh fuel should be used for starting in cold weather.

2. See that a fuel mixture is being drawn into the cylinder but not in such quantities as to flood the engine. Too much fuel in the cylinder is just as troublesome as too little.

3. See that the switch is closed and the ignition system is producing a good spark at the correct time.

4. See that there is reasonable compression.

5. See that the valves are properly timed.

406. Engine Hard to Start.—Difficult starting and slow pickup in an engine may be due to one or more of a number of conditions, some of which are:

1. Poor compression.

2. Poor grade of fuel or water in the fuel, carburetor, or cylinder.

3. Carburetor improperly adjusted so that fuel mixture is too lean or too rich.

4. Weak spark or ignition system poorly timed.

5. Valves poorly timed.

6. Engine stiff, due to tight bearings and pistons, cold weather, poor lubrication, or overheating.

407. Engine Starts Easily but Stops Immediately.—If an engine starts readily but runs only a few minutes at the most, check up the following:

1. The fuel flow from the tank to the carburetor may be slow or irregular due to a clogged fuel line or strainer, stuck float and valve, or a low fuel level in the tank.

2. There may be water in the fuel.

3. There may be a loose connection or a broken wire in the ignition system.

4. The engine may be poorly cooled or lubricated and, therefore, heats up rapidly.

408. Causes of Poor Compression.—Any engine must have reasonable compression in order to start readily and operate efficiently. Compression depends largely upon the care and attention given an engine and the period of service since its initial operation or last overhaul. Some of the common causes of poor compression are:

1. Badly worn piston, piston rings, or cylinder.
2. Poorly fitted piston or rings.
3. Scored or damaged cylinder walls.
4. Leaks around cylinder head, spark plug, or igniter.
5. Worn, damaged, stuck, or poorly adjusted valves, or weak or broken valve spring.
6. Carbon or other foreign matter under valves.
7. Open priming cock.
8. Poor piston and cylinder lubrication.

409. Fuel and Carburetor Troubles.—The fuel-supply and carburetion system of an engine probably ranks next to the ignition system as a source of trouble, both in starting and in operating. Most of the more common troubles are listed below but no attempt is made to include every possible trouble or to suggest a diagnosis and remedy for each one. Some of the usual troubles are:

1. Fuel flow cut off or restricted, due to closed valve, clogged fuel line or strainer, low fuel supply, stuck carburetor float and valve.
2. Excessive fuel flow, due to stuck float and valve, dirt under valve, punctured or waterlogged float.
3. Fuel mixture too lean or too rich, due to improperly adjusted carburetor.
4. Water in the fuel or carburetor.

410. Ignition Troubles.—The various systems of electric ignition with which the more common types of engines are equipped are perhaps the greatest source of trouble. The generation and production of a good electric spark in the combustion chamber at the correct time in the cycle depend upon the almost perfect functioning of a number of more-or-less delicate parts and devices, even in the simplest system. A particle of metal, a bit of moisture, or a single loose connection may disrupt the ignition system entirely, which would mean considerable delay and loss of time.

Ignition troubles are usually indicated by the engine failing to fire when cranked, or by its suddenly going "dead" while apparently running smoothly. However, certain ignition troubles, such as loose connections or parts which intermittently open and close the circuit due to vibration and jars, cause the engine to fire irregularly or unevenly.

The troubles as listed below are divided into two groups depending on whether the source of ignition is a battery or a magneto.

A. Battery Ignition Troubles.
1. Open switch.
2. Battery weak or discharged.

   3. One or more weak cells in a battery.

   4. Loose or corroded connections.

   5. Broken wire.

   6. Short circuit due to bare wire, moisture on coils, spark plugs, or connections.

   7. Coil winding short-circuited or burned out.

   8. Coil points rough, worn, or improperly adjusted.

   9. Timer dirty or making poor contact.

   10. Breaker points rough, wet, oily, poorly adjusted or sticking.

   11. Spark plug or igniter fouled and short-circuited by carbon.

   12. Spark plug damp or wet on outside.

   13. Spark-plug insulator cracked or broken.

   14. Spark-plug points too close or too far apart.

   15. Igniter points rough or improperly adjusted.

   16. Movable igniter point stuck and does not work freely.

   17. Spark out of time.

   18. System not wired up correctly.

  B. Magneto Ignition Troubles.

   1. Brushes dirty or broken.

   2. Armature dragging due to worn bearings.

   3. Breaker points rough, dirty, oily, or improperly adjusted.

   4. Movable breaker point stuck and does not work freely.

   5. Winding short-circuited.

   6. Cracked or broken insulation.

   7. Magneto grounded.

   8. Magneto not in time with engine.

   9. Magnet poles incorrectly arranged—like poles should be on same side.

   10. Magnets weak.

   11. Distributor gear or breaker point opening not in time with armature.

411. Timing Troubles.—Timing troubles involve the correct opening and closing of the valves and occurrence of the spark with relation to the piston travel. Incorrect timing seldom prevents the engine from starting and running reasonably well, but often reduces the power output and causes overheating and high fuel consumption. Some of the common timing troubles are:

   1. Cam gear improperly timed with crankshaft gear.

   2. Valve clearance improperly adjusted.

   3. Range of operation of valve incorrect.

   4. Ignition-timing mechanism such as timer, timer contact, igniter trip arm, breaker points, or distributor poorly adjusted or improperly connected.

412. Overheating.—Overheating of an engine may be caused by the following:

   1. Late spark.

   2. Late exhaust opening.

   3. Too rich or too lean fuel mixture.

   4. Engine poorly lubricated.

   5. Cooling system not functioning properly, water low, fan belt broken or loose, water pump not working, excessive scale deposits in water jacket, water passages clogged.

6. Excessive carbon deposits.
7. Piston, piston rings, and bearings tight.

413. Lack of Power.—Some of the usual causes of lack of power are:

1. Improperly adjusted carburetor.
2. Valves out of time.
3. Spark coming too early or too late.
4. Engine stiff or poorly lubricated.
5. Poor compression.
6. Engine running too hot.

414. Knocking and Pounding.—In general, knocks are caused by preignition of the fuel mixture or by loose parts. The former are readily distinguished from the latter by the characteristic "ping." Preignition knocks are caused by:

1. Spark advanced too far.
2. Engine overheated.
3. Excessive carbon deposits.
4. Incorrect fuel mixture.
5. Failure to use water with kerosene mixture.

The usual engine knocks may be due to:

1. Loose or worn bearings.
2. Loose or worn piston pins.
3. Worn timing gear.
4. Excessive valve clearance
5. Badly worn pistons.

415. Backfire—Afterfire—Kickback.—*Backfire* is the term applied to the popping and sputtering that usually take place in the carburetor and air intake when the fuel mixture is too lean. It is caused by the lean mixture in the cylinder burning so slowly that when the intake valve again opens, the fresh incoming mixture is ignited back in the manifold and carburetor.

Afterfire is the term applied to an explosion occurring in the muffler of an engine. It is usually an indication of ignition trouble such as a loose connection or poor contact. The explanation is that the spark is suddenly cut off while the engine is running but comes on again before the engine stops. During this brief interval of no spark several fuel charges pass on through the engine unburned and lodge in the muffler. When the engine fires again it ignites this fuel in the muffler and an explosion results.

An engine is said to *kickback* when, during cranking, the charge is ignited and explodes before the piston reaches compression dead center, causing a sudden reverse movement of the piston and backward rotation of the crankshaft. A kickback usually results from failure to retard the spark when starting an engine.

TRACTOR CARE AND TROUBLES

The life and service given by the farm tractor or any similar machine are dependent largely upon the care and treatment accorded it by the operator. Tractors are now made of high-grade materials; greater precision and better workmanship are used in their manufacture; and their design in every respect is improved and refined. Furthermore, the widespread use of both the automobile and the farm tractor has given operators more experience and a better knowledge of the working of the machine.

No attempt will be made here to outline any definite procedure or set of rules that, if followed, will insure the satisfactory performance of any machine under all conditions. Practically all tractor manufacturers now supply with their machines a well-prepared set of operating and overhauling instructions. If they do not, they should. The first thing a tractor operator should do is become thoroughly familiar with the machine by studying the instruction book carefully. Second, he should follow these instructions in adjusting, repairing, and operating the machine.

416. Hints on Care and Operation.—The following are some general suggestions to be observed in the handling of any make or type of tractor under all conditions.

1. A new tractor should be broken in carefully by operating it at low speeds and light loads for several days and observing its action closely.

2. Follow the lubrication instructions faithfully—a machine working under high temperatures and heavy strains, as a tractor does, must be kept well lubricated. Check the lubricant frequently for correct level, in both the engine and the transmission.

3. Keep the air cleaner cleaned out and in proper working order at all times. It is much easier and cheaper to check up an air cleaner frequently than it is to put in new pistons and rings and tighten bearings.

4. Keep everything tightened up. This includes engine bearings, cylinder-head bolts, valve covers, crankcase bolts, transmission-housing bolts, wheels and wheel bearings, steering gear and mechanism, fender bolts, and so on.

5. Check the valve clearance frequently and adjust if necessary.

6. Become thoroughly familiar with the carburetor construction and adjustment and keep it set as closely as possible for best results.

7. Do not let the water in the cooling system get low. Use clean water if possible, preferably rain water.

8. Keep the magneto clean and free from oil, moisture, and dirt accumulation.

9. Remove the spark plugs occasionally, clean and adjust the points, and wipe off the outside of the porcelain insulator.

10. Protect the gears by shifting only with the clutch completely disengaged.

11. Save the clutch by keeping it in complete engagement as much as possible. Do not try to move a heavy load or get through a hard pull by engaging the clutch quickly and making the tractor jump.

12. Do not change the governor setting or speed the engine up. It only shortens the life of the machine.

13. Avoid overloading either at the belt or on the drawbar except for very short periods of time if necessary.

417. Common Troubles.—The average tractor will seldom give trouble even in the hands of a "green" operator, if the care and operation suggestions previously enumerated are properly observed. Practically all tractor troubles are confined to the engine. Since a tractor engine does not differ fundamentally from other types, the troubles are essentially the same. All tractors are subject to a few occasional troubles, primarily starting troubles, ignition troubles, and carburetion troubles.

418. Starting.—Starting troubles are caused largely by incorrect fuel mixtures. In general the following rules should be observed:

1. See that there is fuel in the tank, that it is turned on, and getting to the carburetor.

2. Be sure that gasoline is being used for starting instead of kerosene.

3. See that the carburetor needle valve is opened from one to two turns.

4. Always have the hand throttle wide open and see that governor rod and butterfly are not stuck.

5. See that magneto is not short-circuited and that impulse starter is engaged.

6. Prime or choke as sparingly as possible. Getting too much fuel into the cylinders is just as bad as not having enough.

7. After engine starts and gets warmed up, place it under a load and adjust the carburetor closely.

8. Hot engines frequently give more trouble in starting than cold ones because a very slight excess of fuel chokes them easily. The best plan is to try starting them first without priming or choking. If unsuccessful, then choke sparingly.

9. To reduce starting troubles in cold weather:
   a. Drain out the lubricating oil the night before while it is warm, and warm it before putting it back in.
   b. Fill the radiator with warm but not hot water.
   c. Have spark plugs clean and carefully adjusted.
   d. Use high-test gasoline for priming.

419. Carburetion Troubles.—These troubles have been discussed in detail. However, in tractors such things as dirt and trash in the fuel line or under the float valve, water in the fuel, clogged jets, and so on, often develop. It is good practice to remove, partly disassemble, and clean the carburetor at least twice a year. More or less corrosion always takes place and forms scale that interferes with proper carburetor performance.

420. Ignition Troubles.—Since tractors use high-tension magnetos almost exclusively for ignition, such troubles are largely in the magneto itself. As stated elsewhere, tractor ignition troubles can be avoided by:

1. Keeping the magneto clean and free from dust, dirt, and moisture.

2. Avoiding the use of too much lubricating oil in the magneto.

3. Keeping the breaker points smooth and properly adjusted.
4. Keeping the spark plugs clean and properly set.

Information concerning other tractor troubles, such as overheating, knocking and pounding, loss of power, slipping clutches, and so on, may be found on preceding pages.

OVERHAULING THE TRACTOR

A farm tractor ought to be cleaned up and overhauled after every 500 to 1,000 hr. of operation. Since most manufacturers supply rather detailed instructions for doing this work, including making certain adjustments and replacing worn parts, no attempt will be made here to

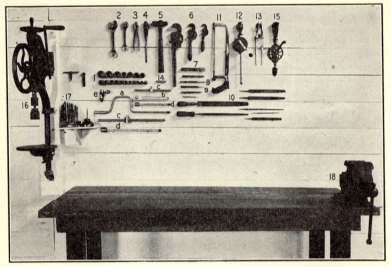

Fig. 437.—Tools and equipment necessary for repairing and overhauling a tractor.

cover the subject in such detail as to enable one to completely overhaul any make or size of machine. Only those suggestions will be given that are applicable in general to all tractors.

421. Workshop and Tools.—Every tractor owner should have a suitable shed or shop in which to house the machine and work on it if necessary. A concrete or tight wooden floor is desirable because bolts, nuts, washers, springs, and similar small parts when dropped in loose dirt are hard to find. Ample space with plenty of light and heating facilities to permit working in cold weather are also important.

The tools regularly supplied with a tractor are seldom adequate for doing a complete overhaul job conveniently and with the least loss of time. There are usually certain special tools included which, of course, should be kept with the machine and not misplaced. However, the money spent for a complete set of high-grade tools is a good investment.

Figure 437 shows a tool layout that includes the following:

1. Socket wrench set with sockets varying from $\frac{7}{16}$ to $1\frac{1}{2}$ in. and the following handles:
   a. Speeder.
   b. Ratchet.
   c. Sliding handle.
   d. Extensions.
   e. Universal.
2. Adjustable end wrenches.
3. Pliers.
4. Screwdriver.
5. Machinist's hammer.
6. Pipe wrenches.
7. Cold chisels.
8. Plain punches.
9. Center punch.
10. Files.
11. Hack saw.
12. Valve grinder.
13. Valve lifter.
14. Thickness gage.
15. Hand drill.
16. Post drill.
17. Drills $\frac{1}{16}$ to $\frac{1}{2}$ in.
18. Work bench and vise.

Practically any overhaul job can be handled readily with such an outfit.

422. General Overhaul Procedure.—The first thing is to plan and carry out the job carefully according to the following steps:

1. Clean up the tractor thoroughly on the outside, removing mud from wheels with water, and dirt and grease from other parts with kerosene. Old paint brushes are good for this work.
2. Disassemble, clean up, and overhaul the engine, replacing worn parts with new ones.
3. Examine and adjust clutch, transmission, steering mechanism, front-wheel bearings, and similar parts.
4. Reassemble and replace engine.
5. Start engine and operate at least an hour without load, until loosened up.
6. Check everything over carefully.
7. Give entire machine at least one coat of paint.

423. Engine Overhaul.—The complete overhaul of an engine involves the following:

1. Cleaning the carbon from the cylinder head, pistons, and valves.
2. Grinding the valves, and replacement of badly worn or pitted ones.
3. Replacement either of (a) piston rings, (b) pistons and rings, or (c) pistons, rings, and cylinder block or liners depending upon the degree of wear and the general condition of the engine.
4. Replacement of piston pins; also the bushings if old pistons or connecting rods are used.
5. Taking up all main crankshaft and connecting-rod bearings.
6. Cleaning up and checking oil pump and lines, fuel and carburetion system, water pump, magneto, governor, and similar accessories.

424. Cleaning Carbon.—The carbon that accumulates in the cylinder head and on the pistons, valves, and so on is usually quite hard. It is best removed with an old flat file that is ground off like a wood chisel on the end. A small scraper or putty knife will also be found useful.

The loose carbon should be kept out of the cylinder-block bolt holes by keeping the bolts in the holes.

425. Grinding Valves.—To do a good job of valve grinding requires experience and patience as well as the proper tools. A common brace grinder is satisfactory but slower than a regular tool (Fig. 437). Any well-known brand of commercial grinding compound will give good results. Use a fine compound in preference to the coarse. The latter is only for grinding out pits and bad places in the valve face or seat. If the valve and seat are carefully cleaned with a rag and gasoline from time to time, as the grinding proceeds, one can readily tell when to stop. A light spring placed under the valve permits a better and a quicker job. Grind-

Fig. 438.—Proper method of removing piston rings to prevent breakage.

Fig. 439.—Fitting a piston ring to its groove.

ing the valves always changes the clearance. Therefore, it must be checked and readjusted when the engine is reassembled.

426. Pistons, Rings, and Cylinder.—In most engines the pistons are removed by disconnecting the connecting rods from the crankshaft and pushing the pistons out through the head end of the block. Care must always be taken not to let the piston slip downward enough to release the lower ring from the cylinder and let it catch on the lower edge inside the crankcase. If this happens, it is very difficult to compress the ring and get it back into the cylinder. Usually it is necessary to break the ring.

Piston rings can be removed and replaced readily without breaking by using three or four stiff strips of sheetmetal (Fig. 438). The grooves will always contain more or less hard carbon, especially the top ones. This must be completely removed.

New piston rings must be so fitted that they are free in the grooves and have the correct joint clearance. To fit the ring to the groove, roll it around as shown by Fig. 439. If it is too tight, it can be ground down slightly by rubbing on a piece of fine sandpaper or emery cloth tacked on a smooth flat board, as shown by Fig. 440.

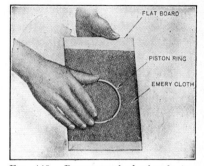

Fig. 440.—Proper method of reducing the width of a piston ring.

Correct joint clearance is very essential for a piston ring. To check this, the ring is inserted in the cylinder and squared up with one of the pistons. The space at the joint is then checked with a thickness gage (Fig. 441). This clearance should be about 0.001 in. per inch of cylinder diameter for the lower ring, and 0.002 in. per inch of diameter for the top ring.

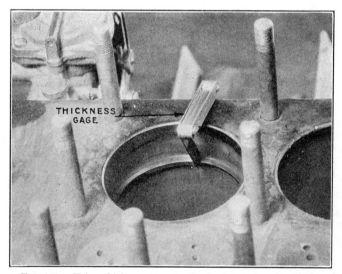

Fig. 441.—Using thickness gage to check piston-ring clearance.

When the rings have all been fitted, they are placed on their respective pistons. When the latter are inserted in the cylinder, the ring joints should be staggered around the piston and not left in a line.

New pistons are necessary whenever the cylinder is sufficiently worn to permit the old pistons to have too much clearance and to fit too loosely. As a rule, oversize pistons will be required. For tractors having removable cylinder walls (Fig. 58), instead of using oversize pistons, it is recommended that new cylinder liners complete with new pistons, rings, and piston pins be used. These parts can be secured in sets all fitted and ready for assembly in the old block. This means practically a new engine at comparatively light expense.

427. Taking up Bearings.—Most tractor-engine bearings are of the plain type with removable bronze-backed babbitt liners. Shims between the two halves permit taking up any wear. These shims are either separate thin strips or laminated (Fig. 442). The latter consist of a heavy brass liner to which is sweated a number of thin shims about 0.002 in. thick. To adjust a bearing, one or more laminations are peeled off with a knife as shown.

In taking up bearings, the main bearings should be adjusted first. Then, adjust the crank bearings. To start with, all bearings should be

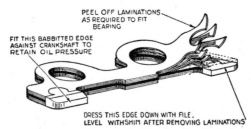

PEEL OFF LAMINATIONS AS REQUIRED TO FIT BEARING

FIT THIS BABBITTED EDGE AGAINST CRANKSHAFT TO RETAIN OIL PRESSURE

DRESS THIS EDGE DOWN WITH FILE. LEVEL WITHSHIM AFTER REMOVING LAMINATIONS

Fig. 442.—Laminated shim for facilitating bearing adjustment.

free and loose. Adjust one bearing until it is just snug but does not bind. Then, loosen it slightly and adjust the next one. Loosen this one and adjust the third, and so on. When all bearings have been properly adjusted, tighten the bolts and try turning the engine over. It should turn rather hard but should not bind enough to prevent hand cranking. When the engine is completely assembled, it should be operated an hour or so at low speed without load, and the bearings examined again.

428. Engine Reassembly.—Before reassembling an engine, be sure that such parts as the oil pump, the fuel and carburetion system, the water pump, the magneto, and the governor have been examined and any worn or defective pieces replaced. As the engine is assembled, wipe off everything, such as cylinder walls, pistons, and bearings, with a clean rag, to remove all grit and dirt. Before putting on the crankcase and cylinder head, squirt some oil on the bearings and cylinder walls. It will make cranking easier.

429. Transmission and Chassis Overhaul.—Tractor transmissions and chassis parts usually do not require anything more than a thorough checkup. Under normal conditions the gear wear will be almost negligible. Some tractors have a means of adjusting the mesh of the main differential gears. If a chain final drive is used, it should be checked and tightened if necessary. Other parts that may need adjusting are the steering mechanism, clutch, and front-wheel bearings.

References

A Field Test of Tractor Wear and Endurance, *Agr. Eng.*, Vol. 10, No. 1.
Cornell Univ., N. Y. State Coll. Agr., Ext. Bull. 108.
Cornell Univ., N. Y. State Coll. Agr., Ext. Bull. 133.
Cornell Univ., N. Y. State Coll. Agr., Ext. Bull. 147.
Cornell Univ., N. Y. State Coll. Agr., Junior Ext. Bull. 40.
"Dyke's Automobile and Gasoline Engine Encyclopedia."
Tractor Repair and Maintenance, *Ill. Univ., Agr. Exp. Sta., Circ.* 425.

CHAPTER XXX

ECONOMICS AND SELECTION OF THE FARM TRACTOR

The farm tractor has come to be recognized as a major source of power in both American and foreign agriculture. The recent introduction of successful all-purpose tractors and tractor-operated equipment that will greatly reduce the man and horse labor requirements for producing such important row crops as corn and cotton is a development that seems likely to usher in a new agricultural era. Mechanical power is now being applied successfully to practically all the operations involved in the production of the major field crops as well as fruits, vegetables, and similar farm commodities.

There is a vast amount of information available concerning the subject of farm power and its utilization under the great variety of existing agricultural conditions. Yet, owing to the rapid changes in tractor design and equipment and in crop-production methods much of this information is more or less obsolete. It is not the purpose of the author to present the detailed results of all the published material on the subject of tractor economics but merely to give such information as will be of general application and answer the more common questions along this line. For the assistance of the reader who desires to make a more complete and intensive study of the subject, a carefully selected list of authoritative references will be found at the end of this chapter.

ECONOMICS OF THE TRACTOR

About 93 per cent of the total horsepower-hours utilized annually on farms in the United States is supplied by animals and heat engines in some form. Animals supply 40 per cent and heat engines 53 per cent. Again about 62 per cent of the annual farm-power consumption in the United States is required for tractive or draft work and 38 per cent for stationary work. It is quite evident, therefore, that the farmer must choose between the tractor and the horse or mule as the proper source of power supply.

430. General Considerations.—Any discussion of the economic phases of tractor use and operation must revolve around certain general questions such as:

1. The factors to be considered in determining whether animal or mechanical power or both can be utilized profitably on a given farm.

428

2. The application, adaptability, and operating costs of tractors in producing the more important crops.

3. The cost of mechanical power as compared with other forms for doing the same work.

431. Factors Determining Kind of Power to Use.—The principal factors to be considered in choosing the most suitable kind of power for a given farm setup are:

1. Size of the farm.
2. Topography of the land.
3. Crops and kind of farm.
4. Soil characteristics.
5. Size of fields.
6. General adaptability of either animal or tractor power to handling the various operations about the farm.

The smallest size of farm on which a tractor can be used profitably cannot be stated definitely because other factors, such as crops raised, size of fields, and so on, also enter into the problem. Before the introduction of the cultivating type of tractor, farms devoted largely to row crops were dependent upon animal power for planting and cultivation. Consequently, the size of the farm on which a tractor could be used profitably was seldom less than 80 acres, and some investigations showed that farms considerably larger than this could be operated about as profitably with horses alone. Now that tractors are available that will successfully perform all operations in row-crop production and thereby eliminate the need of a single horse or mule, it seems reasonable to say that the size of farm on which a tractor can be used with profit might be as low as 40 acres or even less.

The topography of the land is of minor consideration except in extreme cases. A tractor will not operate satisfactorily on what might be considered steep hillsides owing to the tendency to slip and slide down hill. In going directly up the hill most of the power may be consumed in propelling the tractor without a load. Land that is terraced to control soil erosion can usually be handled satisfactorily with tractors.

The particular crops to be grown on a given farm and the acreage of each must be considered along with the other factors mentioned. However, the introduction of the all-purpose tractor has removed certain limitations in the utilization of mechanical power for producing a number of crops and greatly simplified the problem from this standpoint.

With the numerous types and styles of wheels and lug equipment now available for most tractors, and the working parts of the machine well enclosed and protected from dust and dirt, there is no longer any question concerning the adaptability of the farm tractor to the most adverse soil conditions. Wheels are now used that, instead of packing the soil, actually have a pronounced tilling effect.

The size of fields is no longer an important factor in economical tractor operation. It is true that for most operations larger fields mean somewhat less loss of time and more efficient results, but the smaller tractors are now capable of being handled easily in small or irregular fields containing as few as 5 to 10 acres.

In selecting the proper kind of power for a farm, one should consider carefully the numerous jobs, both tractive and stationary, that are likely to develop. If a tractor of a certain type is well suited to handling the majority of these jobs, it should prove a profitable investment. In other words, the greater the amount of time the machine is kept busy, the lower the cost per horsepower-hour of power developed. The possibility of doing a certain amount of custom work for neighbors, provided it is done at a profit, often solves the problem of whether to buy a tractor or not. Frequently certain special tractor jobs develop, such as clearing land, dragging roads, moving buildings, operating snowplows, and the like, which assist in reducing the annual power cost.

According to Fletcher and Kinsman,[1]

Each farm is an individual problem when power requirement and choice of equipment are being considered because of:

1. Variation in the managing and mechanical skill and personal likes and dislikes of the farm operator.
2. Differences in type of farming and in practices followed in caring for and handling the individual crops.
3. Size, topography, and soil conditions of the farm.
4. Plans for future developments.
5. Relative cost of labor and of the various kinds of available power.
6. Kind and availability of hired power and possibility of doing custom work.
7. The ability of any one kind of power to do all the necessary field work.
8. Climatic conditions, such as high temperatures at time of power application.
9. Available capital.
10. Other occupations, if any, of the farm operator.

Since each kind of power has its advantages and disadvantages, each factor should be weighed in proportion to its relative importance as applied to the particular farm in question.

The personal factor is a very important item in the choice of power. Some farm operators are mechanically inclined and much prefer tractor power if the other factors nearly balance, while others have a personal liking for animals and prefer to use that type of power under the same circumstances.

It is generally conceded that for some operations animal power is more practicable than tractors; while for others tractors are the more satisfactory. Animal

[1] *Calif. Agr. Exp. Sta., Bull.* 415.

power is frequently better adapted for working in corners and other cramped or difficult places, and is much more flexible than tractor power because of the ease with which the size and combination of the units may be changed. A horse or mule also has a great overload capacity, that is, the ability to exert, for short periods, a force several times greater than that normally exerted. On the other hand, there are some heavy operations, such as subsoiling, and deep plowing and cultivation, where animal power can seldom be employed satisfactorily.

432. Benefits from Tractor Ownership and Use.—In addition to the general advantages already mentioned, there are certain other benefits to be derived when tractors are substituted for animal power.

At least some, if not all, of the work stock may be displaced, depending upon the type and size of farm and the adaptability of the tractor to handling the various operations. The amount received from horses or mules sold should be credited to the first cost of the tractor.

The purchase of a tractor should always mean a reduction in the man-labor requirements of a farm. The row-crop tractor with its various kinds of attachments will materially reduce the man labor needed for row crops, and the larger tractors with multiple hitches pulling several plows or two or three different tools at the same time permit one man to handle, for example, a large acreage of grain.

The use of a tractor may permit an increase in the size of the farm, owing to the ability of the tractor to do more work in a given time with the same or less man labor. This is true, of course, only where it is possible to add to the farm acreage by renting or buying adjoining or near-by tracts.

Tractor use often materially increases the crop yield, because a better seed bed may be prepared, or the crop may be planted in better season.

With the improved tractors and special implements now available, it is generally agreed that most field operations can be done better as well as faster.

433. Drawbacks to Tractor Ownership and Use.—There are a few possible objections to tractor ownership and use, none of which are of any great consequence, provided the farm is adapted to tractor operation in other respects as already discussed. One possible drawback may be the excessive cost of the tractor and equipment. The purchase of a tractor usually means an expenditure equally as great for implements to go with it. Horse-drawn tools are not adapted to tractor use and, therefore, must be disposed of at a considerable sacrifice. Under such circumstances the size or type of farm may not justify the total outlay involved.

The operator's inability to handle a tractor successfully is sometimes a handicap in securing satisfactory results. Two farmers operating under the same conditions as to size of farm, crops, and make of tractor may obtain decidedly different results simply because of the greater ability of one of them in using the machine. This does not mean that unusual mechanical skill is necessary in order to secure good results with a tractor, but it does mean that one man who likes machinery will frequently have greater success with mechanical power than another who is not so inclined or who is unusually fond of horses. Likewise, dependence upon hired help to care for and operate the tractor is frequently unsatisfactory and may be a determining factor in its purchase. An incompetent or irresponsible operator often means the difference between the success and failure of a machine under otherwise favorable conditions.

SELECTING THE TRACTOR

The selection of the proper type, size, and make of tractor for a given set of conditions is frequently a perplexing problem and may be the one factor that determines the success or failure of mechanical power on a certain farm.

In general in selecting a tractor the three fundamental considerations are:

1. The type.
    a. Should it be a general-purpose or an all-purpose machine?
    b. If general-purpose, should it be a wheel or track type?
2. The size.
3. The make.

434. Choice of Type.—As a rule the acreage and kind of crops raised determine the type of tractor best adapted to a given farm. If such row crops as corn, cotton, or grain sorghums are grown either alone or with wheat, oats, hay, and similar broadcast crops, the all-purpose tractor can likely be utilized to best advantage and with greater efficiency. This type should prove particularly well adapted to dairy and general grain and livestock farms because of the variety of power jobs which arise. For wheat or all-grain farming the ordinary general-purpose tractor of the wheel or track type is most suitable.

435. Wheel vs. Track Type.—According to Fletcher and Kinsman[1] California farmers gave the following reasons for their choice of a wheel or a track-type tractor:

For Wheel Tractors:
    1. First cost lower.
    2. Wheels less complicated than tracks.
    3. Expense of upkeep less on wheel tractor.
    4. Riding easier on rough ground for the operator.
For Track-type Tractors:
    1. Better traction on loose soils.

[1] *Calif. Agr. Exp. Sta., Bull.* 415.

2. No packing of the soil.
3. Quicker and shorter turning.
4. Less damage to road surfaces.

It is evident that both types have their merits, and other factors such as the kind of work to be done, the maximum justifiable investment, dealer's service facilities, and even personal preference may likewise be involved.

436. Size of Tractor.—The choice of the correct size of tractor is important, particularly if the farm setup justifies the purchase of an all-purpose tractor. As previously stated, three distinct sizes of this type of tractor are now available. In general, for small farms or for large farms made up of small fields, the small or two-row size will prove most satisfactory. On the other hand, for large farms having large fields, a larger tractor capable of handling a three-bottom plow and four-row planters and cultivators will likely prove more economical.

If the particular setup requires a standard type four-wheel or track-type tractor, then a choice must be made between sizes ranging from a machine rated at 10 hp. at the drawbar and capable of handling a two-bottom plow, to machines rated as high as 60 or 70 hp. at the drawbar and capable of pulling as many as 10 plow bottoms.

The trend in practically all kinds of farming at the present time is toward the use of smaller units. Only in a few specialized types of farming are the larger machines finding favor. For example, in large-scale wheat production, some growers prefer to use large tractors and perform the different operations of plowing, harrowing, and drilling all at one time. At harvest time these same tractors may pull a large combine, or two medium-size combines.

On the smaller wheat farms varying from a few hundred to perhaps 1,000 acres, the 15-30 to 20-40 size of tractor is preferred. It will handle a three- or four-base moldboard plow or a 6- to 8-ft. one-way-type disk plow, two 12- or 14-ft. grain drills, and a 12- to 20-ft. combine.

Certain other heavy tractive jobs, for which large tractors are frequently used, are subsoiling, land clearing and drainage, terracing, road construction and grading, and heavy hauling.

There is also a limited demand for large tractors for operating large-size threshing machines, sawmills, water pumps, rock crushers, and similar heavy belt-driven machines.

437. Make of Tractor.—The choice of a make of tractor is a factor of major importance. In this connection Fletcher and Kinsman[1] suggest that the following points be considered:

1. The cost of power during the life of the tractor, based upon initial cost, operating expense, repairs, and probable life.

[1] *Calif. Agr. Exp. Sta., Bull.* 415.

2. The design and construction of the tractor itself, including quality of materials, workmanship, accessibility, and dust exclusion.

3. General suitability of the tractor to the kind of work to be done.

4. The general satisfaction given by the tractor after working, for a considerable period, under similar conditions on other farms.

5. Nearness of a dependable dealer who keeps a good stock of repair parts.

6. Stability and business reputation of the manufacturer.

In selecting a tractor, construction and design should be observed closely. This is important from the standpoint of durability, service, accessibility, and adaptability to the kinds of work to be done. Parts requiring frequent adjustment should be accessible, and lubrication should be simplified but positive. Convenience and safety in operation are especially desirable. This includes ease of steering, control levers that are readily manipulated, good rear-wheel fenders, a well-located belt pulley to facilitate lining up, and a convenient and properly protected power take-off.

The past reputation and future stability of the manufacturer and the character and dependability of the local dealer are factors of prime importance in selecting the most suitable make of tractor. The tractor business is a highly competitive one and necessitates constant alertness and persistent effort on the part of the manufacturer to place upon the market a reliable machine at a reasonable price. He must ever be on the lookout for weaknesses in design and possibilities of improvement in construction and operation. He knows that his best salesman is the satisfied customer.

The success of a tractor often lies with the local dealer. He should be able to provide his customers with prompt, reliable, and competent service. The successful tractor dealer either must have a thorough knowledge of the merits, construction, and operation of the machine that he sells, or he should employ a man who is so trained. Only this type of dealer can secure the confidence of the trade and thereby carry on a successful business. In addition, an efficient tractor dealer maintains a reasonable supply of staple repair parts and always makes a special effort to obtain, without delay, those parts that he cannot afford to stock.

THE TRACTOR AND TYPES OF FARMING

Mechanical power is now being applied and utilized efficiently in practically all types of farming. This includes general grain and livestock farming, dairy farming, wheat growing, corn, cotton, rice, potato, and similar row-crop production, and fruit and vegetable growing.

Much comparatively up-to-date and, therefore, reliable information is available in the form of bulletins, circulars, and reports covering the accomplishments and operating costs of tractors in different types of farming. A limited amount of these data is presented below for the more common types of farming, namely, general grain and livestock farming, wheat production, corn production, and cotton production.

ANIMAL AND TRACTOR POWER COSTS

437a. Factors Affecting Power Costs.—According to Fletcher and Kinsman,[1] the cost of using any kind of power is made up of two major items, namely, fixed costs and direct operating costs. The fixed costs are those items which occur independently of the amount the power unit is used; the direct operating costs are those that vary with the amount of power that is developed.

In the case of tractor power, the fixed charges consist of interest on investment, housing, taxes, insurance, and depreciation, repairs, and replacements due directly to obsolescence or nonuse; and the direct operating costs include such items as fuel, lubricating oil, grease, wages of the operator, and depreciation, repairs, and replacements due directly to wear. If animal power is being considered, the fixed charges include interest on investment, housing, taxes, insurance, the value of feed consumed, and care required when not used for work, and the depreciation

TABLE XVIII.—HOURS OF USE AND COST OF KEEPING HORSES ON 35 FARMS IN CENTRAL ILLINOIS, 1931

	Average of 35 Farms Where *No Tractors* Were Used
Acres per farm	231
Crop acres per farm	189
Number of work horses per farm	8.2
Crop acres worked per horse	23.0
Amounts of feed fed:	
Corn, bushels	22.8
Oats, bushels	38.5
Hay, tons	0.9
Other roughage, tons	0.3
Pasture, days	198
Cost items per work horse:	
Feed	$39.45
Labor	10.60
Interest	4.82
Depreciation	4.32
Shelter	4.83
Harness	2.74
Miscellaneous[1]	.81
Total cost for year	67.57
Manure credit	5.23
Net cost for year	62.34
Average hours horses were worked per horse	652
Horse cost per hour	0.096

[1] The miscellaneous items consisted entirely of cash expenditures for veterinary services, medicines, and minor supplies.

[1] *Calif. Agr. Exp. Sta., Bull.* 115.

TABLE XIX.—AVERAGE COSTS, YEAR BASIS, OF OPERATING TWO- AND THREE-PLOW
STANDARD TRACTORS AND TWO-PLOW ALL-PURPOSE TRACTORS, 1931

	Two-plow standard[1]		Three-plow standard[1]		All-purpose	
	Total	Per hour	Total	Per hour	Total	Per hour
Number of tractors................	32	......	19	......	65	
Cost items:						
Fuel, oil, and grease............	$101.80	$0.265	$130.86	$0.326	$126.25	$0.246
Repairs.......................	35.30	0.092	34.39	0.086	19.60	0.038
Man labor (exclusive of driver)...	6.38	0.017	11.90	0.030	10.50	0.021
Miscellaneous..................	0.15	......	0.32	......		
Total operating costs..........	$143.63	$0.374	$177.47	$0.442	$156.35	$0.305
Shelter.......................	5.26	0.014	4.21	0.010	5.00	0.010
Depreciation..................	70.58	0.184	89.94	0.224	97.23	0.189
Interest......................	22.14	0.058	33.55	0.084	28.88	0.056
Total overhead...............	$ 97.98	$0.256	$127.70	$0.318	$131.11	$0.255
Total cost....................	$241.61	$0.630	$305.17	$0.760	$287.46	$0.560
Hours of drawbar work..........	342		341		488	
Hours of belt work..............	41		61		26	
Total hours...................	383		402		514	
Crop acres per farm.............	202		262		220	

[1] The standard-tractor figures are based on detailed cost records extending over a 5-year period (1927–1931), with fuel, oil, labor, and repair costs adjusted to 1931 prices.

TABLE XX.—INFLUENCE OF TOTAL HOURS OF USE ON HOUR-COST OF OPERATION OF
ALL-PURPOSE TRACTORS, 1931

Hours of use	Interest per hour	Interest, in percentage of total cost	Depreciation, per hour	Depreciation in percentage of total cost	Total cost per hour
200 to 349	$0.091	13.7	$0.0238	36.0	$0.662
350 to 499	0.066	11.2	0.0205	34.7	0.591
500 to 649	0.050	9.2	0.0186	34.1	0.546
650 to 799	0.042	8.3	0.0153	30.2	0.506
800 and over	0.032	7.0	0.0150	32.4	0.463

due directly to aging; while the direct operating costs consist of the extra feed and care required, operator's wages, and the extra depreciation that will take place while the animal is being worked.

Since the fixed costs must be prorated against the power actually developed in order to determine the cost per unit of power utilized, the number of hours the power equipment is used annually has a very decided bearing upon the total cost per horsepower-hour. If the power equipment is used only a few hours each year, the fixed costs represent the larger part of the cost of the power; while if the annual use is large, the fixed charges become small compared with the direct operating costs.

438. Horse and Tractor Costs in the Corn Belt.—The information given in Tables XVIII, XIX, and XX is taken from *Illinois Agricultural Experiment Station, Bulletin* 395, and covers detailed studies of a large number of Illinois farms.

439. Horse and Tractor Costs in the Cotton Belt.—The information given in Tables XXI and XXII is taken from *U. S. Department of Agriculture, Technical Bulletin* 497, and covers detailed studies made of a large number of Arkansas and Mississippi delta plantations.

TABLE XXI.—HOURS OF USE AND COST OF KEEPING MULES, 1929

Acres per plantation	941
Crop acres per farm	665
Number of mules per plantation	37
Crop acres per mule	18
Amounts of feed fed:	
Corn, bushels	40.6
Oats, bushels	18.7
Commercial feeds, pounds	166.6
Hay, tons	2.48
Other roughage, tons	0.08
Pasture, days	51.90
Cost items per mule:	
Feed	$112.30
Labor	9.31
Veterinary and medicine	1.18
Interest	6.47
Depreciation	9.63
Harness	3.63
Total cost for year[1]	$142.52
Manure credit	2.52
Net cost for year	$140.00
Average total hours of work per mule	1,121
Mule work, cost per hour	$ 0.125

[1] Does not include shelter, taxes, or insurance.

The reader should bear in mind that the all-purpose tractors referred to in Tables XIX, XX, and XXII are all of medium size. No specific information is available relative to the cost of operation of the small-size all-purpose tractor, but with the obviously smaller investment and lower fuel and oil consumption, the per-hour cost of operation should be from 25 to 35 per cent less than that of the medium size.

CAPACITY OF HORSES, TRACTORS, AND FIELD MACHINES

440. Capacity of Horses, Tractors, and Field Machines in the Corn Belt.—Tables XXIII and XXIV give the accomplishments of horses and

TABLE XXII.—AVERAGE ANNUAL COSTS OF OPERATING TRACTORS ON PLANTATIONS CLASSIFIED ACCORDING TO POWER TYPE, 1929

Item	All-purpose-tractor plantations			Mixed (ordinary and all-purpose) tractor plantations			Ordinary-tractor plantations		
	Cost		Proportion of all costs, %	Cost		Proportion of all costs, %	Cost		Proportion of all costs, %
	Per year (733 hr.)	Per hour used		Per year (733 hr.)	Per hour used		Per year (439 hr.)	Per hour used	
Fuel[1]	$186.49	$0.255	38.4	$190.51	$0.260	36.7	$148.06	$0.337	32.6
Oil[2]	33.32	0.046	6.9	34.27	0.047	6.6	26.98	0.062	5.9
Grease[3]	4.13	0.006	0.9	3.86	0.005	0.7	3.69	0.008	0.8
Chore labor[4]	10.82	0.015	2.2	9.62	0.013	1.9	7.60	0.017	1.7
Repairs and expert labor, in cash	42.22	0.057	8.7	66.21	0.090	12.8	81.97	0.187	18.1
Plantation repair labor[4]	0.32	......	0.1	0.31	0.001	0.1	0.28	0.001	0.1
Total variable cost	$277.30	$0.378	57.2	$304.78	$0.416	58.8	$268.58	$0.612	59.2
Depreciation[5]	179.63	0.245	37.0	186.61	0.255	36.0	158.97	0.362	35.0
Interest[6]	28.21	0.039	5.8	27.10	0.037	5.2	26.50	0.061	5.8
Total fixed cost	$207.84	$0.284	42.8	$213.71	$0.292	41.2	$185.47	$0.423	40.8
Total costs[7]	$485.14	$0.662	100.0	$518.49	$0.708	100.0	$454.05	$1.035	100.0

[1] Fuel at average prices per gallon at plantation—17.3 cents for gasoline and 12.2 cents for kerosene and fuel oil.

[2] Oil at average prices per gallon at plantation—61.6 cents for motor oil and 94.3 cents for transmission oil.

[3] Grease at average price of 12.3 cents per pound at plantation.

[4] Chore labor and plantation repair labor at average price of 20.6 cents per hour of labor.

[5] Annual charge per machine based on estimated life.

[6] Annual charge per machine based on 6 percent of one-half first cost.

[7] Not including shelter, taxes, or insurance for tractor or wages for operator.

tractors in performing various field operations in the corn belt as reported by *U. S. Department of Agriculture, Technical Bulletin* 384.

441. Capacity of Mules, Tractors, and Field Machines in the Cotton Belt.—Tables XXV and XXVI give the accomplishments of mules and tractors in performing various field operations in the delta sections of Arkansas and Mississippi as reported in *U. S. Department of Agriculture, Technical Bulletin* 497.

TRACTORS IN WHEAT PRODUCTION

No crop lends itself better to the application of mechanical power than wheat. In fact, the modern gas tractor has entirely replaced animal power for wheat production, particularly where this grain is raised to any extent as a cash crop.

TABLE XXIII.—ACCOMPLISHMENTS OF HORSES IN THE CORN BELT

Operation	Implement and size	Horses per team	Acres per team-hour
Plowing, fall	One bottom, 14 in	3	0.23
	One bottom, 16 in	3	0.27
	Two bottom, 12 in	4	0.38
	Two bottom, 14 in	4	0.41
	Two bottom, 12 in	5	0.51
	Two bottom, 14 in	5	0.53
	Four bottom, 14 in	12	1.11
Plowing, spring	One bottom, 14 in	3	0.23
	One bottom, 16 in	3	0.30
	Two bottom, 12 in	4	0.40
	Two bottom, 14 in	4	0.42
	Two bottom, 12 in	5	0.50
	Two bottom, 14 in	5	0.52
	Three bottom, 14 in	7	1.09
Disking, single	6 to 7 ft	4	1.48
	8 to 10 ft	4	1.75
	8 to 10 ft	6	2.32
	10 ft	8	2.90
Disking, tandem	7 ft	4	1.27
	7 ft	5	1.43
	8 ft	6	2.08
Harrowing, spike-tooth	8 to 10 ft	2	1.69
	8 to 10 ft	3	1.77
	12 to 15 ft	3	2.30
	16 to 18 ft	3	3.43
	12 to 15 ft	4	2.45
	16 to 18 ft	4	3.35
	20 to 22 ft	4	4.04
	24 to 26 ft	6	5.27
Drilling grain	7 to 8 ft	2	1.25
	10 ft	4	1.81
Corn planting	Two-row, 42-in. rows	2	1.52
Cultivating	One-row, 42-in. rows	2	0.67
	Two-row, 42-in. rows	3	1.35
	Two-row, 42-in. rows	4	1.42
Binding grain	6 ft	3	1.05
	7 ft	4	1.46
	8 ft	4	1.68
Binding corn	One-row, 42-in. rows	3	0.67
Mowing	5 ft	2	0.98
Raking	Dump rake—all sizes	2	1.98
	Side delivery rake	2	1.84

TABLE XXIV.—ACCOMPLISHMENTS OF TRACTORS IN THE CORN BELT

Operation	Size of implement	Two-plow standard tractor	Three-plow standard tractor	All-purpose tractor (med. size)
		Accomplishment per hour, acres		
Plowing, fall	Two-bottom, 14 in.[1]	0.78	1.05	0.79
Spring	Two-bottom, 14 in.	0.78	1.12	0.80
Disking, single	8 to 10 ft.	2.92	....	3.00
Disking, tandem	6 to 7 ft.	2.05	2.12	2.25
Disking, tandem	8 to 10 ft.	2.33	2.73	2.42
Harrowing, spike-tooth	20 to 22 ft.	5.12	....	6.00
Drilling grain	10 to 11 ft.	2.85	2.81	
Cultivating	Two-row, 42-in. rows	....	....	1.96
Binding grain	7 ft.	1.74	....	1.90
Binding grain	8 ft.	2.22	2.17	2.18
Binding grain	10 ft.	2.83	3.06	2.95

[1] Three-plow tractors used three-bottom 14-in. plow.

Wheat being a broadcast crop, all operations from land preparation to harvest involve the use of trailing implements. If the acreage is limited or the fields are comparatively small, it is advisable to use a lighter tractor and smaller implements. In large-scale operations, relatively large tractors can be used with large-capacity implements.

Another development that has done much to popularize the tractor for wheat production and greatly increase its efficiency is the recent introduction of the combined harvester-thresher into all the wheat-growing sections of the country. Although this machine has been in use for many years in certain wheat-growing sections, it often required the utilization of a large number of horses for pulling it. This meant more labor both in preparation and during actual operation, slower movement, and greater difficulty of manipulation.

Although wheat is one of the outstanding crops grown in the United States and Canada, there is some variation in the methods followed in its production in the different sections and states. These variations are largely in the methods of land preparation and tillage rather than in planting and harvesting, and are due primarily to the varying climatic and soil conditions. Owing to this diversity, it is somewhat difficult to establish definite figures concerning wheat production costs under the different conditions.

The following setups (Tables XXVII, XXVIII, XXIX, and XXX) have been prepared as more or less typical examples of the cost of wheat production in the principal wheat producing areas of the United States.

The rotations and land-preparation practices are fairly representative for the regions for which the costs are figured.

The typical wheat farm of the humid section of the hard-red-winter-wheat belt varies in size from 320 to 480 acres. The wheat acreage for

TABLE XXV.—ACCOMPLISHMENTS OF MULES IN THE COTTON BELT

Operation	Implement and size	Horses per team	Acres per team-hour
Stalk cutting......................	One row	2	0.84
Stalk cutting......................	Two row	3	1.63
Plowing...........................	One bottom—7 to 8-in. moldboard	2	0.14
Plowing...........................	One bottom—9 to 10-in. moldboard	2	0.17
Disking, single....................	5 to 6 ft.	4	1.04
Disking, single....................	7 to 8 ft.	4	1.23
Harrowing, spike-tooth.............	5 to 6 ft.	2	1.12
Harrowing, spike-tooth.............	7 to 8 ft.	2	1.42
Bedding, one furrow................	10- to 14-in. middle buster	2	0.67
Bedding, one furrow................	16-in. middle buster	4	0.61
Bedding, two furrows...............	7- to 10-in. plow	2	0.35
Harrowing beds, 36 to 42-in. rows.....	7- to 8-ft. spike-tooth harrow	2 or 3	1.22
Harrowing beds, 36 to 42-in. rows.....	9- to 10-ft. spike-tooth harrow	2 or 3	1.47
Harrowing beds, 36 to 42-in. rows.....	10- to 12-ft. spike-tooth harrow	4	2.02
Fertilizing—36 to 42-in. rows........	Two-row distributor	2	1.49
Planting, 36 to 42-in. rows..........	One-row planter	1	0.63
Planting, 36 to 42-in. rows..........	Two-row planter	2	1.29
Drilling grain.....................	6- to 8-ft. drill	2	0.93
Drilling grain.....................	6- to 8-ft. drill	4	1.11
Cultivating, 36 to 48-in. rows........	One-mule cultivator	1	0.63
Cultivating, 36 to 48-in. rows........	Two-mule cultivator	2	0.67
Cultivating, 36 to 48-in. rows........	Two-row cultivator	4	1.44
Mowing............................	4½-ft. mower	2	0.67
Mowing............................	5 to 5½-ft. mower	2	0.70
Mowing............................	6-ft. mower	2	0.89
Raking............................	8- to 9-ft. dump rake	2	1.40
Raking............................	10-ft. dump rake	2	1.59
Raking............................	Side delivery rake	2	1.27

such farms varies from 100 to 240 acres. The usual rotation consists of corn, oats, wheat, and then wheat or corn. Table XXVII shows the operations, labor requirements, and costs involved in producing wheat in this area.

TABLE XXVI.—ACCOMPLISHMENTS OF TRACTORS IN THE COTTON BELT

Operation	Size of implement	Accomplishment per hour, acres
Plowing...........................	Two-disk plow	0.55
Bedding...........................	Two-row middle buster	1.40
Disking, tandem....................	5 to 6 ft.	1.75
Disking, tandem....................	7 to 8 ft.	1.96
Harrowing, spike-tooth.............	11 to 14 ft	4.54
Planting..........................	Four-row	2.75
Drilling grain......................	6 to 8 ft.	1.75
Cultivating........................	Two-row	1.78
Cultivating........................	Four-row	3.08
Mowing............................	7 ft.	1.98

TABLE XXVII.—TRACTOR AND LABOR REQUIREMENTS AND COSTS INVOLVED IN PRODUCING 240 ACRES OF WHEAT WITH A 15-HP. (DRAW-BAR) TRACTOR

Operation	Tractor, hr.	Man-hr.[1]
Plowing with three-bottom 14-in. gang plow..............	200	215
Double disking with 8-ft. tandem disk harrow..............	100	105
Harrowing with 16-ft. spike-tooth harrow.................	50	52
Drilling with a 12-ft. grain drill.........................	80	90
Harvesting with a 12-ft. combine........................	80	170
Total...	510	632

[1] Man-hours include field servicing of equipment.

Costs Involved

Tractor at $0.80 per hour...........................	$ 408.00
Labor at $0.40 per hour.............................	252.80
Combine cost (less tractor and labor costs).............	300.00
Interest, taxes, depreciation, and repairs on balance of equipment..	75.00
Seed, 1¼ bu. per acre at $0.80 per bushel..............	240.00
Total operating costs.............................	$1,275.80
Operating costs per acre...........................	$ 5.31

The average yield for this area is 15 bu. per acre, one-third of which is delivered to the elevator for rent. This leaves a cost of 53 cents per bushel, plus the cost of hauling all the wheat to the elevator.

The typical wheat farm of the semiarid, hard-red-winter-wheat belt varies from 640 to 4,000 acres. A representative farm probably contains 1,280 acres, 640 acres of which are planted to wheat each year. The normal rotation is summerfallow, wheat, and wheat. The remainder will be in summerfallow or other kinds of crops. Table XXVIII shows the operations, labor requirements, and costs involved in producing wheat under these conditions.

TABLE XXVIII.—TRACTOR AND LABOR REQUIREMENTS AND COSTS INVOLVED IN
PRODUCING 640 ACRES OF WHEAT WITH A THREE- TO FOUR-PLOW 22-HP.
(DRAWBAR) TRACTOR

Operation	Tractor-hr.	Man-hr.[1]
Seedbed preparation on 320 acres of summerfallow:		
Weed killing with 8-ft. one-way disk plow..............	120	125
Deep plowing with 8-ft. one-way disk plow.............	160	175
Weeding with 12-ft. duck-foot weeder..................	80	85
Weeding with 12-ft. duck-foot weeder.................	80	85
Seedbed preparation on 320 acres where wheat follows wheat:		
Plowing with 8-ft. one-way disk plow.................	160	175
Weeding with 8-ft. one-way disk plow.................	120	125
Seeding 640 acres with a 14-ft. grain drill...............	160	180
Harvesting 640 acres with a 16-ft. combine..............	160	350
Total...	1,040	1,300

[1] Man-hours include field servicing of the equipment.

Costs Involved

Tractor at $0.80 per hour........................... $	832.00
Labor at $0.40 per hour............................	520.00
Combine cost (less tractor and labor cost).............	640.00
Interest, taxes, depreciation, and repairs on balance of equipment..	168.75
Seed, ¾ bu. per acre at $0.80 per bushel..............	384.00
Total operating costs.............................	$2,544.75
Operating cost per acre........................... $	3.98

The average yield for this area under the summerfallow rotation is 18 bu. per acre, one-third of which is delivered to the elevator for rent. This gives a cost of 33 cents per bushel, plus the cost of hauling all the wheat to the elevator.

A third setup (Table XXIX) involves the wheat farm of the semi-arid, hard-red-winter-wheat belt, having the farm size balanced for the use of a 30-hp. (drawbar) tractor. The rotation and cultural practices are the same as those of the farm operated by the three- to four-plow tractor. The farm size would probably be about 2,000 acres with 1,280 acres annually in wheat.

The spring wheat farm that uses a three- to four-plow tractor plow has a slightly lower cost per acre. This is due largely to the lower seedbed preparation requirement. The rotation that gives the best results is a corn or wheat crop followed by wheat. The land is prepared either by fall plowing or disking the old land. Table XXX shows the operations, labor requirements, and costs involved if wheat follows wheat and corn or a row crop.

TABLE XXIX.—TRACTOR AND LABOR REQUIREMENTS AND COSTS INVOLVED IN PRODUCING 1,280 ACRES OF WHEAT WITH A 30-HP. (DRAWBAR) TRACTOR

Operation	Tractor-hr.	Man-hr.[1]
Seedbed preparation on 640 acres of summerfallow:		
Weed killing with 2 8-ft. one-way disk plows............	130	140
Deep plowing with 2 8-ft. one-way disk plows..........	160	180
Weeding with 2 12-ft. duckfoot weeders................	75	80
Weeding with 2 12-ft. duckfoot weeders................	75	80
Seedbed preparation on 640 acres of wheat after wheat:		
Plowing with 2 8-ft. one-way disk plows................	160	180
Weeding with 2 8-ft. one-way disk plows................	130	140
Seeding 1,280 acres with 3 14-ft. grain drills.............	130	150
Harvesting 960 acres[2] with 20-ft. combine................	200	440
Total...	1,060	1,390

[1] Man-hours include field servicing of the equipment.

[2] Harvesting 1,280 acres with one combine requires too long harvest period. Hence 320 acres are custom-combined.

Costs Involved

Tractor at $1.25 per hour...........................	$1,325.00
Labor at $0.40 per hour............................	556.00
Combine cost (less tractor and labor costs)............	960.00
Combine hire on 320 acres at $1.50 per acre and 5 cents per bushel (18 bu. per acre)...........................	768.00
Interest, taxes, depreciation, and repairs on balance of equipment.......................................	300.00
Seed, ¾ bushel at $0.80 per bushel...................	768.00
Total operating cost.............................	$4,677.00
Cost per acre on production.......................	$ 3.65

The average yield for this area under the summer fallow rotation is 18 bu. per acre, one-third of which is delivered to the elevator for rent. This makes a cost of 30 cents per bushel, plus the cost of hauling all the wheat to the elevator.

TABLE XXX.—TRACTOR AND LABOR REQUIREMENTS AND COSTS INVOLVED IN PRODUCING 640 ACRES OF WHEAT WITH A THREE- TO FOUR-PLOW 22-HP. (DRAWBAR) TRACTOR

Operation	Tractor-hr.	Man-hr.[1]
Plowing 320 acres with a three-bottom 14-in. gang plow.....	150	157
Disking 320 acres with an 8-ft. tandem disk...............	110	115
Drilling 640 acres with a 14-in. grain drill................	160	180
Harvesting with 16-ft. combine.........................	160	350
Total...	580	802

[1] Man-hours include field servicing of the equipment.

Costs Involved:

Tractor at $0.80 per hour...........................	$ 464.00
Labor at $0.40 per hour............................	320.00
Combine cost (less tractor and labor cost).............	640.00
Interest, taxes, depreciation, and repairs on other equipment...	78.75
Seed wheat seeded at ¾ bu. per acre at $0.80 per bushel..	384.00
Total operating cost..............................	$1,886.75
Operating cost per acre...........................	$ 2.95

The average yield for this area under the summerfallow rotation is 12 bu. per acre, one-fourth of which is delivered to the elevator for rent. This makes a cost of 33 cents per bushel, plus the cost of hauling all the wheat to the elevator.

TRACTORS IN CORN PRODUCTION

Two significant developments affecting the economics of corn production are the introduction of a successful planting- and cultivating-type tractor and the development of a satisfactory corn-picking machine.

The corn grower has the choice of two- or four-row equipment for planting and cultivating the crop with mechanical power. The wide-tread type of tractor straddling two rows handles either two or four rows. The choice is determined largely by individual preference and certain local conditions.

Table XXXI[1] gives the labor and power requirements and per acre cost of growing and harvesting corn with the various types of power used in the corn belt. It will be observed that a certain amount of horse labor was required even when tractors were used. However, when all-purpose tractors were used, the amount of horse labor required was comparatively small.

Tables XXXII and XXXIII give some data concerning corn planting and cultivation studies made in Ohio by McCuen.[2]

[1] *U. S. Dept. Agr., Tech. Bull.* 384.
[2] *Agr. Eng.*, Vol. 8, No. 10.

TABLE XXXI.—LABOR AND POWER USED AND COST PER ACRE FOR GROWING AND
HARVESTING CORN

Husked by Hand

Power type	Work required per acre				Cost	
	Man- hr.	Horse- hr.	Tractor draw- bar, hr.	Belt (all) hr.	1929	1931– 1932
Two-plow tractor.............	12.6	24.6	1.7	[1]	$ 8.30	$4.73
Three-plow tractor...........	12.9	27.6	1.2	[1]	8.68	5.00
All-purpose tractor...........	11.2	17.6	2.8	[1]	7.23	4.23
Ordinary horse hitch..........	15.2	40.4	0.1	[1]	9.76	5.04
Big-team hitch...............	12.0	39.0	[1]	[1]	7.80	4.00

Husked by One-row Tractor Picker

Power type	Man- hr.	Horse- hr.	Tractor draw- bar, hr.	Belt (all) hr.	1929	1931– 1932
Two-plow tractor.............	10.4	20.1	3.0	[1]	$ 8.10	$5.02
Three-plow tractor...........	10.0	20.4	2.4	[1]	8.26	5.29
All-purpose tractor...........	9.0	9.2	5.0	...	6.77	4.46
Ordinary horse hitch..........	14.7	35.9	1.4	...	10.04	5.56
Big-team hitch...............						

Husked by Two-row Tractor Picker

Power type	Man- hr.	Horse- hr.	Tractor draw- bar, hr.	Belt (all) hr.	1929	1931– 1932
Two-plow tractor.............	9.2	15.5	2.2	...	$ 6.47	$3.93
Three-plow tractor...........	8.2	18.8	1.6	...	6.65	4.13
All-purpose tractor...........	7.3	5.4	4.5	...	5.47	3.68
Ordinary horse hitch..........	11.3	31.4	0.5	0.3	8.21	4.54
Big-team hitch...............						

Cut with Horse Binder for Silage

Power type	Man- hr.	Horse- hr.	Tractor draw- bar, hr.	Belt (all) hr.	1929	1931– 1932
Two-plow tractor.............	20.0	28.8	2.4	0.9	$12.47	$7.39
Three-plow tractor...........	18.4	30.3	1.5	0.9	11.89	7.12
All-purpose tractor...........	16.4	23.0	3.2	1.4	11.22	6.93
Ordinary horse hitch..........	21.3	45.8	0.2	0.6	12.90	6.89
Big-team hitch...............	20.2	48.9	0.3	0.6	12.08	6.47

Cut with Horse Binder and Shredded

Power type	Man- hr.	Horse- hr.	Tractor draw- bar, hr.	Belt (all) hr.	1929	1931– 1932
Two-plow tractor.............	26.0	29.6	2.2	1.3	$14.64	$8.63
Three-plow tractor...........	20.5	25.5	1.8	0.9	12.12	7.36
All-purpose tractor...........	18.4	19.1	2.9	1.2	10.78	6.52
Ordinary horse hitch..........	28.8	48.6	[1]	1.0	15.58	8.30
Big-team hitch...............						

[1] Less than 0.01 hr. per acre.

TABLE XXXII.—COMPARATIVE OPERATING DATA ON TWO-ROW AND FOUR-ROW TRACTOR CORN PLANTERS

Equipment	Acres per hour	Man-hours per acre	Fuel, oil, grease; cost per acre	Labor costs per acre	Total operating costs
Two-row planter........	1.50	0.666	$0.229	$0.266	$0.495
Four-row planter.......	3.26	0.612	0.112	0.224	0.354

TABLE XXXIII.—COMPARATIVE OPERATING DATA ON TWO-ROW AND FOUR-ROW TRACTOR CORN CULTIVATORS

Equipment	Acres per hour	Man-hours per acre	Fuel, oil, grease; cost per acre	Labor costs per acre	Total operating costs
Two-row Cultivator:					
First cultivation......	1.65	0.605	$0.180	$0.244	$0.424
Second cultivation....	1.87	0.534	0.151	0.214	0.365
Third cultivation.....	2.37	0.418	0.118	0.167	0.285
Four-row Cultivator:					
First cultivation......	2.82	0.354	0.088	0.141	0.229
Second cultivation....	2.42	0.413	0.083	0.165	0.248
Third cultivation.....	3.17	0.315	0.062	0.126	0.188

Table XXXIV[1] gives some information concerning the operation and cost of corn pickers in Iowa.

TABLE XXXIV.—TIME AND COST FIGURES FOR HARVESTING CORN WITH MECHANICAL PICKERS

	One-row picker		Two-row picker	
	57.20 acres	Per acre	85.64 acres	Per acre
Tractor-hours	97.00	1.70	115	1.35
Man-hours, net..................	139.50	2.45	115	1.35
Man-hours, total.................	228.62	4.09	255	2.98
Machine-hours...................	97.00	1.70	115	1.35
Fuel, gallons.....................	154.00	2.70	200	2.34
Tractor operating costs............	$ 29.87	$0.52	$ 31.00	$0.36
Labor costs......................	$ 93.70	$1.64	$102.40	$1.19
Total expenditures............	$123.57	$2.16	$133.40	$1.55
Yield, bushels...................	2,008	35.00	3,785	44.30
Cost per bushel..................	$0.062		$0.035	

[1] *Iowa State Coll., Agr. Eng. Dept.*, unpublished report.

TRACTORS IN COTTON PRODUCTION

The all-purpose tractor is playing a very important part in the economical production of cotton in certain sections of the cotton belt. Only the lack of a satisfactory and successful mechanical picker, which will do for the cotton grower what the combine does for the wheat farmer, prevents the large-scale cotton grower from securing the most economical production costs. Even though hand picking must be resorted to, large-scale cotton-production methods involving the use of mechanical power and two- or four-row equipment are proving successful.

A study made of large-scale cotton production in certain sections of Texas[1] brought out the following:

1. One tractor with two-row equipment will easily replace four horses.
2. One tractor with two-row equipment is capable of handling 100 acres of cotton up to picking, whereas most farmers were of the opinion that four horses could handle only about 80 acres.
3. One tractor with four-row planting and cultivating equipment is capable of handling 200 acres of cotton up to picking. A similar acreage requires 10 to 12 horses.

The author,[2] as a result of a field study made of the adaptability of the all-purpose tractor to cotton growing, secured the data given in Table XXXV. This covers the production of 68 acres using a tractor and two-row equipment.

TABLE XXXV.—PERFORMANCE DATA FOR A TWO-ROW TRACTOR AND EQUIPMENT IN PRODUCING 68 ACRES OF COTTON

Operation	Total hours required	Fuel used, gallons	Lubricating oil used, quarts	Acres per hour	Tractor-hours per acre	Man-hours per acre	Fuel used per acre, gallons
Disking........................	45.17	78.75	10	1.51	0.663	0.663	1.16
Bedding.......................	49.25	88.50	10	1.38	0.723	0.723	1.30
Planting, first time............	38.25	62.25	10	1.78	0.562	0.562	0.92
Planting, second time...........	38.75	56.00	9.5	1.50	0.666	0.666	0.82
Cultivating, first time..........	39.58	50.65	10	1.72	0.583	0.583	0.74
Cultivating, second time.........	41.08	53.00	3	1.66	0.603	0.603	0.78
Cultivating, third time..........	39.58	50.50	9	1.72	0.583	0.583	0.74
Cultivating, fourth time.........	37.00	51.25	2	1.84	0.543	0.543	0.75
Cultivating, fifth time...........	33.00	49.25	2	2.06	0.485	0.485	0.73
Cultivating, sixth time..........	31.00	42.00	7	2.20	0.456	0.456	0.62
Total.......................	392.66	582.15	72.5	....	5.870	5.870	8.56

Table XXXVI shows the actual cost of producing the crop and the items involved.

It is quite evident that picking and ginning costs make up an important part of the total expense of producing a crop of cotton.

[1] *Texas Agr. Expt. Sta., Bull.* 362.
[2] *Agr. Eng.*, Vol. 10, No. 6.

TABLE XXXVI.—SETUP SHOWING COST OF OPERATION OF ONE TRACTOR AND TWO-ROW EQUIPMENT IN PRODUCING 68 ACRES OF COTTON

Expenses:

582.15 gal. fuel at 17 cts. per gallon......................... $	98.97
18.13 gal. lubricating oil at 65 cts. per gallon....................	11.78
9 gal. transmission lubricant at $1.00 per gallon...............	9.00
8 lb. cup grease at 15 cts. per pound.......................	1.20
Repair parts for tractor and equipment.........................	8.00
Blacksmith labor, 27¾ hr. at 75 cts. per hour...................	20.81
Changing tools, 11 hr. at 20 cts. per hour......................	2.20
Tractor operator, 412.29 man-hours at 20 cts. per hour.............	82.46
Interest on investment in tractor and equipment at 6 per cent........	95.70
Depreciation of tractor and equipment, 20 per cent................	319.00
Seed for planting, 127.78 bu. at $3.00 per bushel.................	383.34
Chopping three times...	253.63
Poisoning and insect control, 15 acres at 60 cts. per acre (two applications)	18.00
Picking 48,457 lb. seed cotton at $1.00 per cwt....................	484.57
Ginning 17,306 lb. lint cotton at 90 cts. per cwt....................	155.75
Bagging and ties, 33 bales at $2.00 per bale......................	66.00
Taxes on land, 68 acres at $1.00 per acre........................	68.00
Total...	$2,078.41

Summary:

Fuel used per acre, gallons....................................	8.56
Cost of producing crop per acre................................ $	30.57
Cost of producing crop per acre (based on tractor cost items only—first 10 listed above).. $	9.55

Definite figures and actual costs are not available covering the capacity and operating costs of all-purpose tractors and four-row equipment in producing cotton. However, Tables XXXVII and XXXVIII present data that can be considered as conservative and reliable. The results are based upon a 200-acre setup involving the use of a 7-ft. disk harrow for cutting up stalks, weeds, and trash; a two-row middle buster for bedding; a four-row planter; and a four-row cultivator. Since practices vary in many sections, particularly with respect to land preparation, it will likely be necessary to revise some of the cost figures accordingly. However, this should not greatly affect the total or per acre cost.

TABLE XXXVII.—PERFORMANCE DATA FOR FOUR-ROW TRACTOR AND EQUIPMENT IN PRODUCING 200 ACRES OF COTTON

Operation	Total hours required	Fuel required, gallons	Lubricating oil required, gallons	Acres per hour	Tractor hours per acre	Man-hours per acre	Fuel required per acre, gallons
Disking...........	133	230	7	1.50	0.67	0.67	1.15
Bedding..........	143	260	7	1.40	0.72	0.72	1.30
Planting..........	67	120	4	3.00	0.34	0.68	0.60
Cultivating 4 times.	268	400	10	3.00	1.34	1.34	2.00
Total...........	611	1,010	28	....	3.07	3.41	5.05

TABLE XXXVIII.—SETUP SHOWING COST OF OPERATION OF ONE TRACTOR AND FOUR-ROW EQUIPMENT FOR 200 ACRES OF COTTON

Fuel—5.05 gal. per acre at 15 cts. per gallon	$151.50
Lubricating oil—28 gal. at 65 cts. per gallon	18.20
Transmission lubricant—10 gal. at $1.00 per gallon	10.00
Cup grease—15 lb. at 15 cts. per pound	2.25
Repairs, tractor repair labor, changing tools, sharpening shovels, etc.	100.00
Man labor—tractor operator and extra man in planting—740 hr. at 30 cts. per hour (includes 1 hr. per day for fueling and oiling tractor, etc.)	222.00
Interest on investment in tractor and equipment at 6 per cent	96.00
Depreciation on tractor and equipment at 20 per cent	320.00
Total	$919.95
Cost of tractor operation per acre	$ 4.60

Table XXXIX shows the results of a study of the use of tractors for cotton production in the delta sections of Arkansas and Mississippi.

TABLE XXXIX.—AVERAGE COST OF TRACTOR WORK ON 24 PLANTATIONS THAT HAD COTTON AND ON WHICH TRACTORS WERE THE PRINCIPAL KIND OF POWER, 1929

Operation	Tractor work per plantation		Operating cost	
	Acres[1]	Hours	Per year	Per acre
Flat breaking, two-disk plow	77	142	$ 91.93	$1.19
Disking, flat, and beds	573	318	191.91	0.33
Harrowing	216	72	40.66	0.19
Bedding	238	161	94.99	0.40
Fertilizing, four-row	80	30	17.55	0.22
Planting:				
Two-row	38	27	14.73	0.39
Four-row	161	54	31.53	0.20
Drilling	29	14	8.31	0.29
Cultivating:				
Two-row	423	232	120.54	0.28
Four-row	1,346	454	268.41	0.20

[1] Computed as acreage on which the operation was performed multiplied by the "times over."

References

An Economic Study of Tractors on New York Farms, *Cornell Univ., Agr. Exp. Sta., Bull.* 506.

A Study of the Cost of Horse and Tractor Power on Illinois Farms, *Univ. Ill., Agr. Exp. Sta., Bull.* 395.

Cost of Wheat Production by Power Methods of Farming, 1919–1929, *State Coll. Wash., Agr. Exp. Sta., Bull.* 255.

Diesel Power and Field Operating Costs, *Mont. State Coll., Agr. Exp. Sta., Bull.* 289.

Economic Aspects of Farm Mechanization, *Agr. Eng.*, Vol. 13, No. 12.

Farm Power in the Yazoo-Mississippi Delta, *Miss. State Coll., Agr. Exp. Sta., Bull.* 295.

Horses, Tractors, and Farm Equipment, *Iowa State Coll., Agr. Exp. Sta., Bull.* 264.

Large Scale Cotton Production, *A. & M. Coll. of Tex., Agr. Exp. Sta., Bull.* 362.

Latest Developments in the Motorization of Corn Production, *Trans. Amer. Soc. Agr. Eng.,* Vol. 21.

Life, Service, and Cost of Service of Farm Machinery, *Iowa State Coll., Eng. Exp. Sta., Bull.* 92.

Making Cotton Cheaper, *Miss. A. & M. Coll., Agr. Exp. Sta., Bull.* 290.

Mechanical Equipment in Corn Cultivation, *Trans. Amer. Soc. Agr. Eng.,* Vol. 21.

Motorizing the Corn Crop in Ohio, *Trans. Amer. Soc. Agr. Eng.,* Vol. 18.

Potato Growing with Tractor Power, *Pa. State Coll., Agr. Exp. Sta., Bull.* 306.

Reducing Cotton Production Costs by the Utilization of Improved Machinery, *Agr. Eng.,* Vol. 10, No. 6.

Relation of Farm Power and Farm Organization in Central Indiana, *Purdue Univ., Agr. Exp. Sta., Bull.* 332.

Some Experiences in Industrialized Farming, *Agr. Eng., Vol.* 11, No. 2.

Sources of Power on Minnesota Farms, *Univ. Minn., Agr. Exp. Sta., Bull.* 262.

The Cost of Agricultural Production, *Trans. Amer. Soc. Agr. Eng.,* Vol. 21.

The Economic Relation of Tractors to Farm Organization in the Grain Farming Areas of Eastern Washington, *Wash. State Coll., Agr. Exp. Sta., Bull.* 310.

The Efficient Use of Animal Power, *Trans. Amer. Soc. Agr. Eng.,* Vol. 15.

The Tractor on California Farms, *Univ. Calif., Agr. Exp. Sta., Bull.* 415.

Tractor and Horse Power in the Wheat Area of South Dakota, *S. D. State Coll., Agr. Exp. Sta., Circ.* 6.

Tractors and Trucks on Louisiana Rice Farms, 1929, *La. State Univ., Agr. Exp. Sta., Bull.* 218.

Trends in Large Scale Wheat Farming, *Agr. Eng.,* Vol. 11, No. 2.

Use of Machinery in Cotton Production, *Agr. Eng.,* Vol. 13, No. 4.

Utilization and Cost of Power on Corn Belt Farms, *U. S. Dept. Agr., Tech. Bull.* 384.

Utilization and Cost of Power on Mississippi and Arkansas Delta Plantations, *U. S. Dept. Agr., Tech. Bull.* 497.

CHAPTER XXXI

MATERIALS OF CONSTRUCTION—POWER TRANSMISSION

A large part of the material used in the construction of internal-combustion engines consists of iron in some form or another. Therefore, some knowledge of its source, manufacture, the methods of treatment, and the characteristics and properties of the different forms is of fundamental importance. The source of all iron and steel is iron ore, which is dug from the earth's surface or from underground mines. In this ore, the iron is in the form of oxides, that is, compounds of iron and oxygen.

442. Pig Iron.—The first step in the manufacture of iron and steel is the removal of the oxygen from the iron ore by melting it in a blast furnace with a fuel, usually coke. Limestone is mixed with the ore and fuel to take up the oxygen. The process and resulting products are as follows:

Iron oxide + limestone + coke + hot-air blast = iron + slag + gases (carbon monoxide, carbon dioxide, nitrogen)

The iron thus obtained is cast into bars and is commonly known as *pig iron*. Pig iron is not pure iron by any means for it contains about 3.5 per cent of carbon and small amounts of silicon, manganese, phosphorus, and sulphur. From pig iron are manufactured the various kinds of iron and steel, such as gray, white, and chilled cast iron, malleable iron, wrought iron, and steel.

CAST IRON

443. Process of Manufacture.—The manufacture of cast iron from pig iron is very similar to the manufacture of pig iron from ore, except that it is more of a refining process for the purpose of securing castings of a definite composition. The process involves the melting of the pig iron with coke and a small amount of limestone in a cupola or air furnace. Sometimes scrap iron is added to secure a special grade of cast iron. The molten iron is drawn off into special molds which form the castings into their final shape for a given machine. Cast iron contains from 2.5 to 4.5 per cent of carbon and very small amounts of silicon, manganese, phosphorus, and sulphur.

444. Kinds of Cast Iron.—Gray-iron castings are made by allowing the metal in the molds to cool slowly so that the carbon is retained in the form of graphite. Gray iron contains from 2.5 to 4.5 per cent carbon and has the following characteristics:

1. Very brittle, due to high carbon content
2. Will not bend or twist.
3. Very soft and machines easily.
4. Easily cast into any shape.
5. Cannot be welded in a forge but only by means of an electric arc or an oxygen-acetylene flame.
6. Low tensile strength—20,000 lb. per square inch.

Gray-iron castings are used for gas-engine cylinders, pistons, and frames.

White cast iron is made by cooling the casting rapidly so that the iron and carbon remain combined chemically as cementite (Fe_3C) and other compounds. White cast iron is harder and more brittle than gray cast iron.

Chilled cast-iron castings are produced by lining the sand molds with iron at those points where the chill is desired. The molten iron, coming in contact with this cold iron lining, cools rapidly so that those portions of the casting become very hard and resistant to wear, due largely to the fact that the carbon remains in chemical combination and the crystals so form themselves that they are long and narrow and arranged perpendicular to this chilled surface. Some characteristics of this material are:

1. Chilled part very hard.
2. Very brittle and will not bend.
3. Cannot be welded except by electric arc or oxygen-acetylene flame.
4. Cannot be machined but must be ground.

The chilling process is used for moldboards, shares and other parts of chilled plows, some bearings, cams, car-wheel treads, and brake shoes.

MALLEABLE IRON

445. Process of Manufacture.—Pig iron is first melted and cast into ordinary white-iron castings of the desired size and shape but containing only about 2.5 per cent carbon. The castings are then annealed by placing them in iron boxes, packing them with iron oxide and sand, placing the boxes in a large oven or furnace and heating to about 1400°F. for 3 to 5 days. They are then cooled slowly. This reduces the carbon content of the outside or skin of the casting to about 0.2 per cent but the inside or center remains the same. The usual characteristics of malleable iron are:

1. Will bend, twist, and resist shock.
2. Is soft and easily filed or machined.
3. Cannot be welded except with electric arc or oxygen-acetylene flame.
4. Higher tensile and compressive strength than plain cast iron.

Malleable castings are used extensively in farm machinery, and for small pinions, pipe fittings, and various kinds of hardware.

WROUGHT IRON

Pig iron likewise forms the raw material in the manufacture of wrought iron, and the process is essentially a purifying one. The difference between this process and the manufacture of steel, as described later, is that the metal is never completely melted. The furnace used is called a reverberatory furnace. The fire is in a compartment at one end, and the flames pass over the hearth of the furnace which lies beyond. The arched roof over the hearth reflects the heat down upon the charge of pig iron and iron ore, and the products of combustion pass to the chimney.

Under the influence of the heat the oxygen from the ore oxidizes the impurities in the pig iron and the ore itself is reduced. The impurities form a slag. By constant stirring with iron rods through openings in the side of the furnace the entire charge is exposed to the heat and the impurities in the pig iron finally reduced to a very small amount.

Since the metal is heated only enough to reach a pasty condition and not enough to melt completely, the slag does not rise in a distinct layer to the top but permeates the entire mass. Near the end of the process the workmen gather the pasty mass into balls. This process is called puddling. The balls of iron are removed from the furnace and, while still hot, most of the slag is squeezed out by hydraulic presses or steam hammers.

As already stated, some of the slag remains in the mass, and the resulting network of slag in the iron, after it has been hammered and rolled, gives wrought iron its characteristic fibrous structure. This slag, which consists largely of silicates and phosphates of iron and manganese, makes up about 2.8 per cent of the material. Aside from the presence of the slag, the percentage composition of wrought iron is essentially that of low-carbon steel. Wrought iron is perhaps the purest form of iron. It is tough, malleable, and ductile and is used largely for nails, stay bolts, rivets, water and steam pipes, boiler tubes, horseshoes, and for general forging purposes. It is easy to weld in a common forge.

STEEL

Steel is iron containing a small amount of carbon—less than 1.5 per cent. This carbon is in a combined state called iron carbide, whereas in ordinary cast iron, it is in the form of graphite and is mixed with the iron to form a physical mixture.

The general process of manufacture consists of heating the pig iron by one of four processes in order to remove the carbon and other impurities, and then adding to or inoculating the mass with a certain amount of an iron and carbon mixture to give the steel the correct percentage of carbon. There are four methods of manufacturing steel: the Bessemer process, the open-hearth process, the crucible process, and the electric process. The Bessemer process is not extensively used in the United States; by far the largest amount of steel is made by the open-hearth process.

STEEL MANUFACTURE

446. Open-hearth Process.—The impurities of the pig iron are oxidized by the addition of iron oxide and diluted by the addition of scrap steel. The charge, consisting of pieces of pig iron, iron ore, and steel scrap, is melted by the flame of a blast of mixed air and gas passing over the hearth of a saucer-shaped furnace. The process takes a much longer time than the Bessemer process, and samples of the metal are taken from time to time to determine its composition. Open-hearth steel is used for bridges, armor-plate, and the better class of structural work, as well as for conversion into high-grade tool steel.

447. Crucible Steel.—For uses demanding the greatest uniformity and freedom from undesirable impurities, further refining than is obtained in the process just described is necessary. The crucible and electric processes are the two most important ones in this respect.

In the crucible process wrought iron or steel is remelted in a graphite crucible. When wrought iron is used, the proper percentage of carbon is secured by the addition of charcoal to the iron before melting. As very pure wrought iron can be obtained, a high degree of purity can be secured in the steel. When steel is used as the crucible charge, it has usually been made by heating wrought-iron bars for a long time in contact with carbon. This results in the formation of iron carbide, particularly near the surface, and the object in remelting in the crucible, in this case, is to secure greater uniformity in the product. The crucible process is used chiefly to manufacture the various alloy steels but is being replaced by the electric-furnace process.

448. Electric Process.—In the electric-furnace process, the furnace proper is similar to the open-hearth furnace. The metal is placed in the furnace in a molten condition and the two graphite electrodes lowered into it. The resistance offered to the flow of the current develops an intense heat, and the sulphur and phosphorus are oxidized by the iron oxide and lime thrown in. A slag is formed on the top as a result of the oxidation of these impurities and the carbon is likewise almost entirely burned out. The proper amount of carbon is then introduced by the use of a recarburizing material similar to that used in the open-hearth process. By the electric-furnace process, Bessemer and open-hearth steels can be quickly and cheaply converted into high-grade carbon and alloy steels of any desired composition.

449. Kinds of Steel and Their Uses.—Steel varies greatly in its character and properties, depending upon the carbon content, heat-treatment given, and alloying metals added to it.   According to carbon content, steel may be classified as follows:

Steel	Per Cent
Low carbon, mild or soft	0.05 to 0.30
Medium carbon or half	0.30 to 0.80
High carbon or hard	0.80 to 1.50

As shown, the higher the carbon content, the harder and more brittle the steel.   The following table gives the carbon content of steel as used for different purposes:

Per Cent	Steel Products
0.05 to 0.10	Wire, nails, boiler plate, rivets, bolts
0.10 to 0.20	Rivets, screws, machine parts to be casehardened
0.20 to 0.25	Ordinary forgings, structural steel, cold-rolled shafting
0.25 to 0.40	Axles, gears, crankpins, shafts, connecting rods
0.40 to 0.70	Railroad rails, steel castings
0.70 to 0.80	Anvil facings, band saws, cold chisels, wrenches
0.80 to 0.90	Punches and dies, circular saws, rock drills
0.90 to 1.00	Springs, machinists' hammers, punches, and dies
1.00 to 1.10	Springs, lathe parts, taps
1.10 to 1.20	Ball-bearing races, wood chisels, wood-working tools, thread dies, spring steel
1.20 to 1.50	Files, hack saws, ball and roller bearings

HEAT-TREATMENT OF STEEL

Not only does the carbon content of steel influence its character and properties but steel parts and tools, when subjected to certain heat-treatments, become possessed of specific properties. Some of the heat-treatments and their effects and the methods employed are described below.

450. Annealing.—The metal is heated to 1450 to 1700°F. (depending upon the carbon content and the size of the casting) and allowed to remain at maximum temperature for one to several hours according to the size of the piece.   It is then allowed to cool slowly.   This treatment refines the coarse structure of the steel casting or part, removes the existing coarseness of grain, removes strains due to uneven cooling, and increases the tensile strength and resistance to shock.   Wire and similar cold-rolled or cold-drawn objects are annealed to remove brittleness caused by mechanical treatment.

451. Hardening.—The metal is heated to a high temperature (2000°F.) and cooled suddenly by quenching in water or oil. This increases the hardness, tensile strength, and brittleness. Its purpose is to adapt high-carbon steel for cutting-tool purposes.

452. Tempering.—The hardened or quenched steel is first reheated to a temperature below the former hardening temperature for the purpose of partly restoring its ductility and softness. It is then cooled. The rate of cooling is immaterial. The piece is quenched in water when the proper heat is reached as indicated by the color.

453. Casehardening.—This is a process of introducing additional carbon into the outer shell of the steel piece. It produces a fine-grained core and a very hard and close-grained outer shell. Stock containing 0.1 to 0.2 per cent carbon is used. The process is as follows:

1. Heat and apply the carbonizing material—usually charcoal, charred leather, crushed bone, horn, and so on.
2. Allow the piece to cool to black in daylight.
3. Reheat to critical temperature of core and quench in oil or water.
4. Reheat to critical temperature of shell and quench.

For quick, superficial casehardening, the steel piece is heated to about 1700°F., and powdered potassium cyanide and potassium ferrocyanide are applied.

Gears, bearings, and parts requiring very hard external surfaces, because of exposure to high pressure, are usually casehardened.

STEEL ALLOYS

A steel alloy is a mixture containing a high percentage of steel and a small amount of one or more of the common alloy metals, namely, manganese, nickel, chromium, vanadium, and tungsten. The common steel alloys, their characteristics, and uses are as follows:

1. *Manganese Steel.*—Commercial contains 12 to 13 per cent manganese. High tensile strength, hard and resistant to wear, and ductile. Uses: rock crushers, railroad frogs and crossings, railroad curve rails, burglarproof safes.
2. *Nickel Steel.*—Two to four per cent nickel. Increases hardness, toughness, and tensile strength. Decreases ductility. Used largely with chromium. Uses: structural bridge work, railroad-curve rails, steel castings, shafting, frame and engine parts for automobiles, axles.
3. *Chrome Steel.*—One to two per cent chromium. Gives a very hard steel. Used largely for armor-piercing projectiles. Also used for armor-plate and burglarproof safes.
4. *Vanadium Steel.*—Three-twentieths to one-quarter per cent vanadium. It improves the general physical properties of steel but is somewhat expensive.
5. *Chrome-vanadium Steel—Nickel-vanadium Steel—Chrome-nickel Steel.*—Used for ball and roller bearings, axles, crankpins, gun barrels, crankshafts.
6. *Tungsten Steel.*—Increases hardness. Used for high-speed cutting tools.

BEARING METALS—MISCELLANEOUS ALLOYS

Table XL gives the percentage composition and uses of a number of common bearing metals, solder, and similar alloys.

POWER TRANSMISSION

There are a number of methods of transmitting power from the engine or other source to the driven machine. Some of the more common are

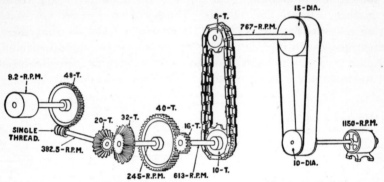

Fig. 443.—Sketch showing different methods of power transmission.

(1) direct drive, (2) pulleys and belts, (3) sprocket wheels and chains, and (4) gears. Figure 443 shows all of these methods in combination.

TABLE XL.—Composition of Bearing Metals and Other Alloys

Material or alloy	Copper, per cent	Tin, per cent	Lead, per cent	Zinc, per cent	Antimony, per cent	Phosphorus, per cent	Remarks
Babbitt (high grade)	4.00 to 5.00	90.00	0.35	......	4.00 to 5.00	......	For bearings
Babbitt (low grade)	5.00 to 6.50	86.00	0.35	......	6.00 to 7.50	......	For bearings, subject to heavy pressure
Babbitt (hard)......	2.25 to 3.75	60.00	26.00	......	9.50 to 11.50	......	For light pressures
Brass (red).........	83.00 to 86.00	4.50 to 5.50	4.50 to 5.50	4.50 to 5.50			
Brass (yellow).......	65.00 to 75.00	1.00	2.00	30.00			
Bronze (hard cast)	90.00	6.50	1.50	2.00	....	......	General utility
Bronze (phosphor)	79.00 to 82.00	9.00 to 11.00	9.00 to 11.00	......	....	0.10 to 0.25	For heavy loads and severe use
Solder..............	....	45.00 to 50.00	50.00 to 55.00	......	0.15		

454. Direct Drive.—A direct power connection, although impractical for some kinds of work, offers certain advantages, such as (1) little power loss, (2) less trouble, and (3) compactness. Figure 444 illustrates such an application.

455. Pulleys and Belts.—The use of pulleys and belts for transmitting power is well adapted to many farm and other power operations. The equipment involved is not complicated, and the power can be transmitted short or long distances, that is, from a few feet or less to 100 ft. or more, with practically the same general layout. The principal objection to this method is that it is not entirely positive and there may be some loss of power due to slippage.

456. Kinds of Belting.—The three most common belting materials are leather, rubber, and canvas. Leather is considered an excellent belting material for general use. It is strong, very durable, does not stretch, and is not injured by oils and greases. On the other hand, its high first cost and inability to withstand exposure to excessive moisture and steam do not warrant its use under all conditions.

Fig. 444.—Centrifugal pump driven direct by a gas engine.

Leather belting can be secured in both single and double strength, the latter consisting of two layers of leather glued together. In using leather belts always run them with the hair side next to the pulley. This is the smoother side of the belt. The reason for this is that the hair side is harder and firmer and will not stretch so much as the flesh side. Therefore, the hair side would likely crack in a short time if it were not placed next to the pulley. To keep leather belts flexible and in good condition apply neat's-foot oil. Never use mineral oils.

457. Rubber Belting.—Rubber belting is now used extensively for farm power transmission. It costs less than leather, withstands exposure to moisture and temperature variation, and is flexible. On the other hand, rubber belts will not stand exposure to oils and greases of any kind and should not be allowed to rub on or against any stationary object. The seam side should be run away from the pulley. Rubber belting really consists of from two to eight layers or plies of canvas, impregnated and held together with vulcanized rubber. It can be secured in widths varying from 1 in. to 1 ft. or more.

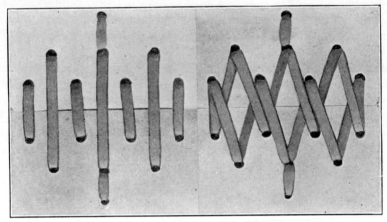

Fig. 445.—The double straight method of lacing a belt.

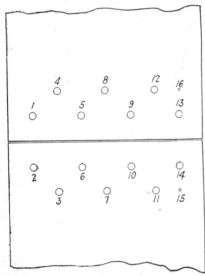

Fig. 446.—Hole arrangement for double straight and double hinge methods of belt lacing.

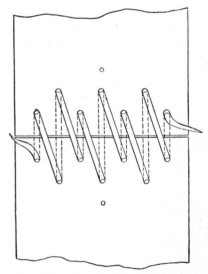

Fig. 447.—The double straight lacing half completed.

458. Canvas Belting.—Canvas belting is used largely for transmitting power from tractors to threshers, silo fillers, and similar mobile machines. It is made of from four to eight plies of heavy canvas duck, stitched together with strong twine and treated with a vegetable oil and a special paint.

Canvas belts are made endless for the special power jobs mentioned above. The material in various widths can be obtained in rolls and pieces cut to the desired length for any purpose. A canvas belt is strong and durable, and withstands exposure to oil, moisture, and steam. On the other hand, it may stretch and shrink under certain conditions, it often becomes stiff, and it frays on the edges and ends, unless used with care or watched closely.

459. Belt Lacing.—Endless belts of either leather, rubber, or canvas can be secured in almost any length or width and are preferable because a laced joint is seldom perfectly smooth and flexible. A poor joint produces a bumping or flapping noise as it passes over the pulley. All belting materials, however, are subject to stretching and contraction as a

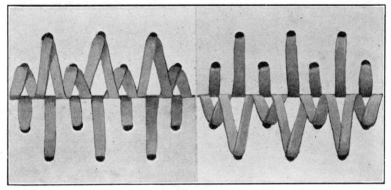

Fig. 448.—The double hinge method of lacing a belt.

result of continued use, exposure to moisture, and temperature variation. Therefore, if the pulleys cannot be adjusted but are fixed, the proper tension can be maintained only by shortening the belt.

The method of lacing a belt depends largely upon the kind and size of the belting material. For a long, wide belt, such as a thresher belt, running over medium to large pulleys, a good leather lacing such as the double straight (Fig. 445) or the double hinge (Fig. 448) works very well. The latter is especially recommended for rubber or canvas belts which fray out at the ends.

460. Double Straight Lacing.—The double straight lacing (Fig. 445) is a good, strong lacing for wide belts but is not so flexible as the double hinge. To make this lacing, arrange the holes the same as for the double hinge (Fig. 446) except that the finishing holes 15 and 16 are at the center instead of at one side. Begin the lace at the center, and work to the outside, going one way with one end and the opposite way with the other until the lace has passed through all holes once and the two ends are up as shown in Fig. 447. It will be noted that the lacing is arranged diagonally on the top side and straight on the pulley side. Now, work back to the center with each end by doubling back through each hole. The completed lacing will appear as in Fig. 445. The lacing should be entirely straight on the pulley side as well as doubled between every pair of holes.

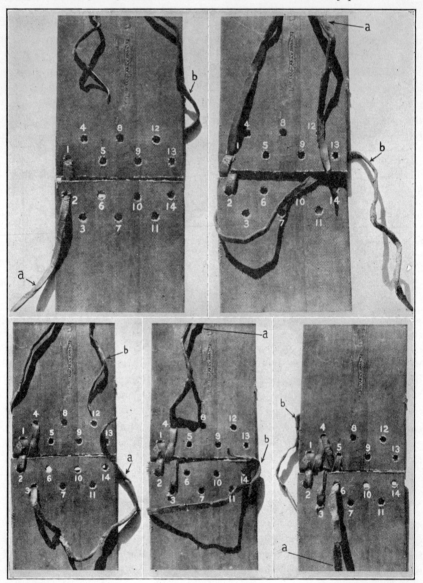

Fig. 449.—Showing how the double hinge lacing is started and carried out.

461. Double Hinge Lacing.—The double hinge lacing (Fig. 448) is somewhat difficult to master but if correctly made is very strong, neat, and flexible. Two precautions must be observed, namely, (1) always start at the left-hand side of the belt (Fig. 449) and work towards the right, and (2) avoid letting the lace get twisted and keep it flat.

To make the lacing, proceed as follows: Arrange holes as shown in Figs. 446 and 449. Designate one end of lace as *a* and the other end as *b*. With pulley side of belt down, bring end *a* of lace up through hole 1, then down between ends of belt and up through hole 2. Draw lace through, until ends are even. Now bring end *b* up between ends of belt and run end *a* down between the ends lapping end *b* on top of *a* and putting it through hole 2. Then bring end *a* up through hole 1; bring *b* up between ends of belt and run *a* down between the ends, placing *a* under *b*. Run *b* down through hole 4 and *a* up through hole 3. Then run *a* down between the belt ends and *b* up between them, lapping *b* over *a*. Run *b* down through hole 3 and *a* up through hole 4. Then run *a* down between the belt ends and *b* up between

FIG. 450.—A hand-operated machine used for metal belt lacing.

them, lapping *b* over *a*. Put *b* down through hole 5 and *a* up through hole 6. Then run *a* down between the ends of the belt and *b* up between them, lapping *b* over *a* as before. Put *b* down through hole 6 and *a* up through hole 5. Continue in this manner until the full width of the belt is laced. Holes 15 and 16 are extra ones into which the last inch or two of the ends are inserted to keep them from flapping. These finishing holes should be very small so that they will pinch the ends and hold them tight. It should

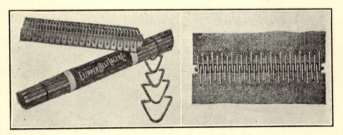

FIG. 451.—One type of metal belt lacing.

be noted that the lace passes through all holes twice, with the exception of the two finishing holes. Also remember that the lace must always pass between the ends of the belt before it goes through a hole, that is, the lace never goes straight across to an opposite hole on the same side of the belt.

462. Metal Belt Fasteners.—Metal belt fasteners and lacings (Figs. 450 and 451) are coming into extensive use, especially for short and medium-sized belts. Metal-laced joints are strong and flexible and require less time.

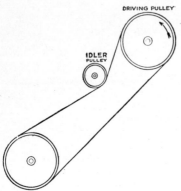

FIG. 452.—Proper use of idler pulley.

GENERAL PRECAUTIONS FOR USING BELTS

1. Excessively tight belts cause injurious strains on the belts and machinery and produce hot bearings.

2. Excessively loose belts slip readily and have a flapping, unsteady motion.

3. Belts should be run as nearly horizontal as possible to secure good pulley contact.

4. The lower side of a belt should be the driving side.

5. Place idler pulleys on the slack side of a belt nearer the driven pulley (see Fig. 452).

USEFUL BELTING RULES

1. To calculate the speed or size of pulleys remember

Revolutions per minute of driven pulley ×
its diameter = r.p.m. of driving pulley × its diameter

Example 1.—A tractor has a 16-in. pulley running at 600 r.p.m. What size pulley is required for a thresher having a cylinder speed of 1,200 r.p.m.?

Let d = diameter of thresher pulley. Then

$$1,200 \times d = 600 \times 16$$

or

$$d = \frac{600 \times 16}{1,200}$$
$$= 8 \text{ in.}$$

Example 2.—Referring to Fig. 443, the speed of the final drive shaft, if the motor speed is 1,150 r.p.m., is computed as follows:

Speed of final drive shaft = $1,150 \times {}^{10}\!/_{15} \times {}^{8}\!/_{10} \times {}^{16}\!/_{40} \times {}^{32}\!/_{20} \times {}^{1}\!/_{48}$ = 8.2 r.p.m.

For a worm and worm wheel the speed ratio is the ratio of the number of separate threads on the worm to the number of teeth on the wheel. The number of separate threads on the worm may be determined by following one thread around 360 deg. The number of threads lying between the starting and finishing points indicates the number of threads on the worm.

2. To calculate the horsepower a belt will transmit or the width of belt required to transmit a given horsepower, use the following formula

$$\text{Hp.} = \frac{WPDN}{12,000}$$

where W = width of belt in inches.

P = number of plies of belt.

D = diameter of pulley in inches.

N = r.p.m. of pulley.

For single leather belt $P = 4$.

For double leather belt $P = 8$.

3. To calculate the speed or rate of travel in feet per minute of a belt over a given pulley, use the following formula:

$$\text{Belt speed (ft. per min.)} = \frac{D \times N \times 3.1416}{12}$$

where D = diameter of pulley in inches.

N = r.p.m. of pulley.

4. To calculate the length of belt required for a given purpose, proceed as follows:

$$\text{Required length in feet} = \frac{(D + d) \times 3.1416}{24} + 2L$$

where D = diameter of driving pulley in inches.

d = diameter of driven pulley in inches.

L = distance between pulley centers in feet.

PULLEYS

463. Kinds of Pulleys.—Pulleys are made of wood, cast iron, steel, and fiber or paper. Wood pulleys of any size are built up in sections and split, usually in two parts which are held together by long bolts. They are suitable for line shafts, especially where it is desired to remove or replace the pulley without taking down the shafting. Wood pulleys are usually held in place by means of wood bushings.

Cast-iron pulleys are often faced (lagged) with leather or heavy canvas to reduce belt slippage. Solid cast-iron pulleys are used extensively on engines and such machines as threshers, and corn huskers. They are always made to fit a certain size of shafting and are held in place by a key or setscrew or both.

Steel pulleys are either one piece like cast-iron pulleys, or split similar to wood pulleys. If split, they are held in place by metal bushings.

Fiber or composition pulleys are now coming into extensive use on agricultural machines and for many other power purposes, especially where a small pulley is required with good belt-gripping qualities. Like cast-iron pulleys, they must fit the shaft on which they are to be used and are keyed rigidly to it.

Pulley sizes are designated by the diameter and face width. For example, a 12 by 6-in. pulley means a pulley 12 in. in diameter with a 6-in. face.

A tight- and loose-pulley arrangement (Fig. 453) is often used where it is desirable to stop a machine without removing the belt or stopping the engine or motor. One pulley is keyed to the shaft and the other is free to turn on it. To stop the driven machine, the belt, while still running, is merely thrown over onto the loose pulley by means of a shifting device.

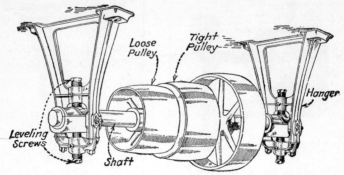

Fig. 453.—Line shaft and hangers with tight- and loose-pulley arrangement.

464. Line Shafting.—A line shaft is often convenient and necessary where it is desired to operate several stationary machines from one source of power or to secure the proper speed ratio.

TABLE XLI.—Horsepower Transmitted by Cold-rolled Steel Shafting at Different Speeds

Diameter, inches	100 r.p.m.	200 r.p.m.	300 r.p.m.	400 r.p.m.	500 r.p.m.
½	0.13	0.25	0.38	0.50	0.63
⅝	0.24	0.49	0.73	0.98	1.22
¾	0.42	0.84	1.27	1.68	2.10
⅞	0.67	1.33	2.00	2.67	3.24
1	1.00	2.00	3.00	4.00	5.00
1⅛	1.42	2.85	4.27	5.69	7.11
1¼	1.95	3.90	5.85	7.80	9.75
1⅜	2.59	5.18	7.77	10.36	12.95
1½	3.40	6.80	10.20	13.60	17.00
1⅝	4.30	8.60	12.80	17.10	21.00
1¾	5.40	10.80	16.10	21.00	27.00
1⅞	6.60	13.10	19.70	26.00	33.00
2	8.00	16.00	24.00	32.00	40.00
2⅛	9.60	19.20	29.00	38.00	48.00
2¼	11.40	23.00	34.00	45.00	57.00
2⅜	13.40	27.00	40.00	54.00	67.00
2½	15.60	31.00	47.00	62.00	78.00
2⅝	18.10	36.00	54.00	72.00	90.00
2¾	21.00	41.00	62.00	83.00	104.00
2⅞	24.00	48.00	72.00	95.00	119.00
3	27.00	54.00	81.00	108.00	135.00

Shafting is made of cold-rolled steel and comes in various sizes and lengths up to about 20 ft. The size depends upon the power to be transmitted as shown by Table XLI.

Fig. 454.—Wall hanger for a line shaft.

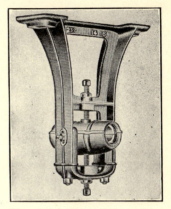

Fig. 455.—Ceiling hanger for a line shaft.

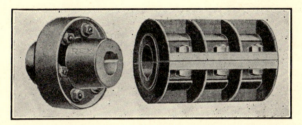

Fig. 456.—Flange and compression couplings for line shaft.

465. Line-shaft Supports and Accessories.—A line shaft may be supported either by plain bearings, better known as journal boxes or pillow blocks resting on wooden brackets, or by specially constructed iron brackets or hangers. The latter are made in the wall type (Fig. 454), or the ceiling type (Fig. 455). They are equipped with adjusting screws by means of which the bearings may be accurately lined up. The type and size of bearing hanger to use depend upon the location of the shaft and diameter of the pulleys to be used. Large hangers are necessary to give sufficient wall clearance for large pulleys. The hanger bearings are always provided with some means of lubrication. Line-shaft bearings should be

Fig. 457.—Collar for line shaft.

spaced not more than 10 ft. apart. A support every 8 ft. will give satisfactory results with almost any size and length of shafting. When a very long shaft, requiring two or more lengths, must be installed, the ends are joined by means of special shaft couplings (Fig. 456). To prevent side movement of a line shaft, two or more collars (Fig. 457), held in place by setscrews, should be located next to one or both of the end bearings.

GEARS AND GEARING

A gear is a wheel, made usually of iron or steel or of certain other materials, and having tooth-like projections that engage directly with

Fig. 458.—Spur gear and pinion.

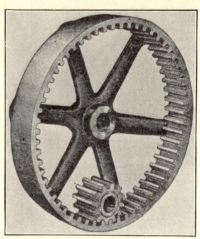

Fig. 459.—Internal spur gear and pinion.

similar projections on another gear and thereby drive it. A pinion is any small gear. Gears are said to transmit power positively, because there

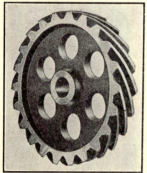

Fig. 460.—A spiral or helical spur gear.

is no slippage as with belts. The use of gears and gearing is confined largely to the transmission of motion and power from one part to another of the same machine, rather than from one machine to another. Gears also offer a simple means of speed reduction or increase with respect to the different parts of a machine.

466. Spur Gear.—A spur gear is one on which the teeth are so arranged that its axis of rotation is parallel to the axis of rotation of the gear with which it meshes. The plain external spur gear (Fig. 458) has its teeth on the outer face. They are straight teeth arranged parallel to the wheel axis. An internal spur gear (Fig. 459), as the name implies, has its teeth arranged on an inner face of the offset rim. A spiral or helical spur gear (Fig. 460) has angular or twisted teeth. This arrangement permits more teeth to mesh at one time, thus producing smoother and quieter action.

467. Bevel Gear.—A bevel gear has its teeth so arranged at an angle with its rotating axis that they mesh with another similar gear

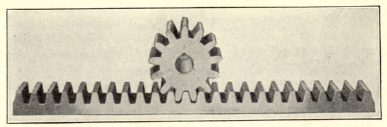

Fig. 461.—Rack and pinion.

and drive two shafts that form an angle with each other. Plain bevel gears (Fig. 462) have straight teeth, while spiral bevel gears (Fig. 463) have twisted teeth. The latter are used extensively for final-drive gears in automobiles to secure smoothness and quiet action. Bevel gears always have a tendency to slide out of mesh and thereby produce end thrust on the shafts involved.

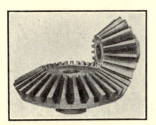

Fig. 462.—Plain bevel gear and pinion.

Fig. 463.—Spiral bevel gear and pinion.

Therefore, some sort of a well-supported thrust bearing (Fig. 464) is always necessary behind each gear.

468. Worm Gear.—A worm gear (Fig. 465) consists essentially of two parts, the worm and the worm wheel. The worm is nothing more or less than a coarsely threaded screw, while the worm wheel, usually larger in diameter, is very much like a spiral spur gear. The use of a worm gear permits a large speed variation between the driving and driven members without the need of large parts that require considerable space or clearance. Worm gearing has certain disadvantages that make its use limited. (1) It will not usually operate in a reverse manner; that is, the worm must always drive the worm wheel rather than the worm wheel drive the worm; (2) its efficiency is considerably lower than that of other kinds of gearing; and (3) it should have continuous and thorough lubrication while in operation.

469. Gear Materials.—Plain spur gears, operating at low speeds and used where weight is not an important factor, are usually made of cast iron. This material, however, is not recommended for gears exposed to sand, grit, or other abrasives that cause rapid wear.

Fig. 464.—Bevel gearing showing tapered roller thrust bearings.

Gears operating at medium to high speeds and requiring protection and good lubrication are made of steel or steel alloys and are hardened or heat-treated.

Other materials used for small gears in certain machines where the strain is light are bronze, fiber, Bakelite, and rawhide. The principal advantage of these is that they reduce noise.

SPROCKET WHEELS AND CHAINS

Sprocket wheels and chains, like gears, are a positive means of power transmission and are used largely for driving different parts of a machine, particularly when these parts are several feet apart or considerable speed

Fig. 465.—Worm and worm-wheel drive.

reduction is desired. A chain drive possesses certain distinct advantages as follows:

1. It is positive and there is no slippage.
2. It can be coupled and uncoupled readily.
3. Chains absorb shock.
4. A small variation in the distance between sprocket centers is not harmful.

Figure 466 illustrates an ordinary sprocket wheel. The sprockets are the projections on the wheel rim. Their size, shape, and spacing depend upon the kind of chain, speed, amount of power transmitted, and other factors.

470. Kinds of Chain.—There are a large number of different kinds of chains used for the many applications of this method of power trans-

FIG. 466.—Sprocket wheel.

FIG. 467.—Open or detachable–link chain with malleable-links.

mission. Likewise, the same type of chain may be made in several distinct sizes and variations. The more common kinds being used for agricultural purposes are (1) the detachable-link chain, which may be malleable (Fig. 467) or steel (Fig. 468); (2) the pintle chain (Fig. 469); (3) the finished steel roller chain (Fig. 470); and (4) the silent chain (Fig. 471).

FIG. 468.—Open or detachable-link chain with steel links.

The detachable-link chain is used largely for light work and medium speeds where the chains are exposed, as on grain binders, threshers, and combines. The steel link is used more extensively because it is cheaper to manufacture and stronger.

The pintle chain is made of malleable links held together by pins. It is used largely for heavy-duty slow-speed work where the chain is exposed.

Steel roller chain is made of special steel, finished and polished, and equipped with hardened rollers. It is more efficient, stronger, and wears longer, even under high speeds. It is used largely for heavy-transmission installations such as the final drives of trucks and tractors. It should be well protected from dirt and grit or enclosed and run in oil.

Silent chain is a special type having built-up toothlike links running over toothed wheels similar to gear wheels. This chain is a sort of

FIG. 469.—Pintle chain.

FIG. 470.—Steel roller chain.

FIG. 471.—Silent chain drive.

flexible, metal, corrugated belt. It is used for light, high-speed work such as driving the cam and accessory shafts of automobiles and trucks.

BEARINGS

In general, all bearings may be classed, according to construction, as either plain or antifriction. For plain bearings, the most common bearing materials are wood, babbitt, and bronze.

471. Plain Bearings—Wood.—Wood as a bearing material is used to a limited extent where a bearing is exposed to dirt and grit, and therefore is subject to rapid wear and frequent replacement. Figure 472 shows the wood bushings used in a disk harrow. A specially treated hard wood is recommended.

472. Babbitt.—Babbitt is an alloy containing tin, copper, lead, and antimony and is used extensively for line-shaft and general-transmission bearings, engine-crankshaft and con-necting-rod bearings, generator bear-ings, and bearings in many agricultural machines. Babbitt is comparatively cheap and soft enough so that it does not wear the journal and at the

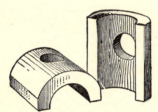

FIG. 472.—Wood bushings for bearings.

FIG. 473.—Babbitt-lined split bearing.

same time hard enough to withstand excessive pressure. When a babbitt bearing wears out, it is inexpensive to replace either by means of an entirely new babbitt shell, or by melting and repouring the bearing. Babbitt also possesses certain other advantages. It retains

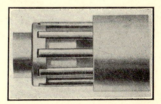

FIG. 474.—Babbitt-lined solid bearing.

FIG. 475.—A plain roller bearing.

the lubricant well, even when smooth and well run in, and it melts at a temperature varying from 350 to 450°F.; consequently, danger from fires, due to excessive overheating of a bearing, is avoided.

A plain babbitt bearing such as a line-shaft boxing (Fig. 473) consists of a cast-iron framework lined with the bearing metal. A solid bearing

(Fig. 474) is made of one piece. It is not so satisfactory as the split bearing (Fig. 473), which permits of easy disassembly, and adjustment by means of shims.

473. Bronze.—Bronze is used extensively for small bearings subject to high speeds or excessive pressures, as piston-pin bearings, certain bearings in farm machines, transmission bearings, and the like. The bronze liner is usually in the form of a solid or a split bushing pressed in place or anchored by a setscrew, pin, or key. The bushing should not be loose or allowed any movement. A bronze bearing wears very slowly, especially if kept well lubricated.

ANTIFRICTION BEARINGS

Antifriction bearings are bearings that are specially constructed to give less power loss and greater efficiency than plain bearings. They are classed as either (1) roller or (2) ball bearings.

474. Roller Bearings.—There are three kinds of roller bearings: (1) the plain roller (Fig. 475), (2) the spiral roller (Fig. 476), and (3) the tapered roller (Fig. 477). A plain roller bearing consists of a

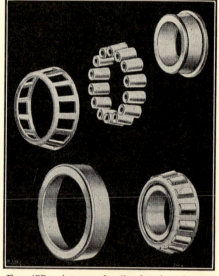

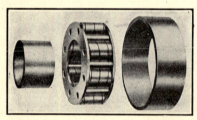

FIG. 476.—A spiral-roller, or Hyatt, bearing.

FIG. 477.—A tapered roller bearing showing parts and construction.

number of solid cylindrical steel rollers assembled as shown. This type is used primarily in agricultural machinery.

The spiral-roller bearing, better known as the Hyatt bearing, consists of a number of hollow alloy-steel rollers assembled as illustrated. Each roller is constructed in the form of a helix. The complete bearing always includes an outer race and may or may not include an inner race.

Hyatt roller bearings are designed for both heavy-duty and light high-speed work. They are best adapted to installations producing radial loads only. If side thrust is present, thrust washers or an additional thrust ball bearing is necessary. These bearings are used in tractor transmissions, threshing machines, combines, and similar heavy agricultural machines.

The tapered roller bearing, as shown by Fig. 477, is made up of a number of conical rollers assembled in a cage, and revolving about the inner and outer races whose bearing surfaces are likewise conical. The outer race is sometimes termed the cup, and the inner race the cone. The rollers and races are made of high-carbon alloy steel.

A tapered roller-bearing installation (Fig. 464) requires at least two sets of rollers assembled in such a way that they act against each other. Either one or both bearings are usually adjustable, permitting the taking up of any wear. This type of bearing will carry a radial load as well as one producing a side or angular thrust. Tapered roller bearings are used extensively in tractors, trucks, automobiles, and heavy agricultural machinery.

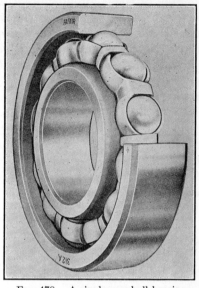

Fig. 478.—A single-row ball bearing.

475. Ball Bearings.—A ball bearing (Fig. 478) consists of one or more rows of highly polished, hardened steel balls held in a cage and rolling

Fig. 479.—Two-row ball bearing.

between an inner and an outer race. It may carry an ordinary radial load, a combined radial and thrust load, or a straight thrust load.

The double-row radial bearing (Fig. 479) is used for heavy-duty purposes or where the outside diameter of the bearing must be limited owing to interference of other parts.

Ball bearings are now being used very extensively for nearly every known purpose.   In the agricultural field they are being used for tractor crankshaft mountings, transmissions, steering mechanisms and wheels, thresher cylinder bearings, ensilage cutters, and combines.

476. Care of Roller and Ball Bearings.—Good antifriction bearings are made of high-quality materials and require very precise workmanship.   It is, therefore, more expensive for the manufacturer of a certain machine to equip it with these bearings, but they usually assure more satisfactory service.   They practically eliminate overheating of bearings; they wear very slowly and seldom produce any noise; they greatly reduce friction and power losses; and their lubrication is easier to care for.

Any ball or roller bearing will give excellent service if the following precautions are observed:

1. It must be properly installed so that it is not subjected to any abnormal strains.
2. It must be well enclosed and completely protected from dirt, grit, or other foreign matter.
3. If adjustable, any wear or looseness should be taken up or removed.
4. It should be properly lubricated as often as necessary.

References

Cornell Univ., N. Y. State Coll. Agr., Ext. Bull. 72.
DAVIDSON: "Agricultural Machinery."
FAVARY: "Motor Vehicle Engineering—Engines."
KRANICH: "Farm Equipment."
LEIGHOU: "Chemistry of Materials."
PULVER: "Materials of Construction."
RADEBAUGH: "Repairing Farm Machinery and Equipment."
SMITH: "Farm Machinery and Equipment."

INDEX

A

Afterfire, 419
Agricultural workers, 2
  acres handled by, 5
  average power utilized per year by, 5
  relative number required, 1
  released from agriculture, 2
Air cleaners, 119
  need for, 119
  requirements of, 119, 126
  types of, 120
    combination, 125
    inertia, 120
    oil, 123
    water, 122
Air cooling, 128
  advantages of, 129
Air heating, 115
Air injection, 159
Air-fuel mixtures, 94
Alcohol, 86
  cooling with, 139
Alloys, 31, 457
  aluminum piston, 31
  steel, 457
All-purpose tractor, 403, 407
  characteristics of, 407, 408
  corn production with, 445
  cotton production with, 448
  transmissions in, 319
Ampere, definition of, 171
Ampere-hour, 171
  rating for storage cell, 187, 189
Animal power, 7, 8
  advantages and disadvantages of, 14
Antifreezing mixtures, 139
  requirements of, 139
Antiknock fuels, 90
Armature, testing of, 255
Auxiliary air valve, 104

B

Babbitt metal, 458
Backfire, 419

A

Ball bearings, 475
  for crankshaft, 61
Battery, 169
  definition of, 169
  storage, 178
  troubles in, 193
Bearings, 472
  antifriction, 474
  babbitt, 473
  ball, 475
  bronze, 474
  metals for, 458
  plain, 473
  roller, 474
  tractor transmission, 330
  wood, 473
Belt lacing, 461
  double hinge, 462
  double straight, 461
  metal, 463
Belt speeds, 327, 465
Belting, care of, 464
  kinds of, 459
  rules for, 464
Bosch magneto, 240
Brake, for clutch, 303
  transmission, 325
Brake mean effective pressure, 366
Breaker points, 210, 212, 237
  adjustment of, 237, 252
  for high-tension magneto, 237, 252
  multiple types of, 212
  timing of, 250
Breather cleaner, 125
British thermal unit, 93, 365

C

Cams, 67
Camshaft, 62, 69
  speed of, 69
  for tractor engine, 62
Carbon deposits, 291
Carburetion, 98
  principles of, 98
  systems of, 98

477